程序员典藏

Android
开发入门百战经典

张亚运 ◎ 著

清华大学出版社
北京

内容简介

本书内容系统全面，采用层层递进的方式进行讲解，让读者理解起来更为容易。全书分为10章，主要包括Android Studio的常用操作和技巧、Android的属性和布局、Android的基础控件、Android的系统组件、Android几种常用的数据存储方式、Android动画、Android网络、Android手机的基本功能及多媒体操作等。

另外，本书还创新地引入了扫描二维码查看动态图的功能，让纸质图书也能和读者交互起来，提升阅读的乐趣。本书适用于广大初、中级Android开发者。对于初级开发者，本书对常用核心的基础知识通过实例的形式进行了系统的讲解，保证初学者学习后可迅速上手进行Android应用开发；对于中级开发者，本书有助于查缺补漏、夯实基础。另外，本书还可以作为高等学校电子信息类专业和计算机类专业本科生的教材以及Android应用开发技术人员的参考书。

本书封面贴有清华大学出版社防伪标签，无标签者不得销售。
版权所有，侵权必究。侵权举报电话：010-62782989 13701121933

图书在版编目(CIP)数据

Android开发入门百战经典 / 张亚运著. —北京：清华大学出版社，2017
（程序员典藏）
ISBN 978-7-302-47345-9

Ⅰ. ①A… Ⅱ. ①张… Ⅲ. ①移动终端－应用程序－程序设计 Ⅳ. ①TN929.53

中国版本图书馆CIP数据核字（2017）第109645号

责任编辑：张　敏
封面设计：杨玉兰
责任校对：徐俊伟
责任印制：沈　露

出版发行：清华大学出版社
网　　址：http://www.tup.com.cn，http://www.wqbook.com
地　　址：北京清华大学学研大厦A座　　邮　编：100084
社 总 机：010-62770175　　邮　购：010-62786544
投稿与读者服务：010-62776969，c-service@tup.tsinghua.edu.cn
质量反馈：010-62772015，zhiliang@tup.tsinghua.edu.cn

印 装 者：三河市中晟雅豪印务有限公司
经　　销：全国新华书店
开　　本：185mm×260mm　　印　张：22.75　　字　数：554千字
版　　次：2017年8月第1版　　印　次：2017年8月第1次印刷
印　　数：1～3000
定　　价：69.80元

产品编号：072883-01

前　言

不知不觉，Android 伴随我已经走过了四五个年头，它是忠实的伙伴、可靠的朋友。当初学习 Android 起源于对移动互联网事业的憧憬和向往，一旦进入 Android 的世界就变得一发不可收拾。在移动互联网事业如火如荼的今天，富有聪明才智、充沛精力的年轻人不在此开辟一片属于自己的天地，也许会成为日后一大憾事。

研究生期间，在研究学习 Android 之余喜欢将自己的学习总结以博客的形式发布到 CSDN，本意为自己记录学习点滴之用，无意中竟获得了一些关注和支持，也因此结交了很多志同道合的朋友。当然，期间也有一些出版社和培训机构的朋友找过我，也就有了后来出版了两本电子书《Android 百战经典》和《Android 控件操作之二十四章经》。这两本书的出版激发了我持续写作的兴趣，写作的过程是总结的过程、创作的过程，也是提高的过程。脑中的知识在笔尖流淌，智慧的火花在指尖碰撞。写作并不一定是专业作家才可以做的事，只要坚持写作、总结，相信你也可以做到。

"兴趣是最好的老师"，培养兴趣是做事前的第一步，当然想让无趣的事情强制变得有趣也是不可能的一件事。我认为开发本身是一件非常有趣的事情，记得第一次为一个 Button 添加了单击事件监听，当其成功响应的时候是多么令人欣喜，我第一次真正操控了机器！最能让人感到愉悦的应该就是操控感了，在现实的世界里想要操控别人已经变得不可能，在代码的世界里，你就是"King of the World"！

本书告别枯燥繁冗的理论讲解，能用代码说话的坚决不用文字，同样，能用图表表达的地方尽量避免文字。我相信，密密麻麻的文字往往是吓退读者的罪魁祸首。本书的样例都是笔者多年总结、积累的非常实用而又有趣的实例，这些实例都是围绕 Android 最核心、最常用的知识点展开，让读者在感到有趣的同时接收新鲜知识的灌溉。

本书适用对象

本书适用于初中级 Android 开发者。对于初级开发者。本书对常用核心的基础知识通过实例的形式进行了系统的讲解，保证一本书即可上手进行简单 Android 应用的开发；对于中级开发者，本书有助于查缺补漏、夯实基础。我也相信，阅读有趣的实例可以为开发者带来新的灵感。本书还适用于在开发道路上犹豫不觉得小白们，相信你搭上了这辆车，一定不会后悔。

本书特色

- 本书和市面上绝大多数理论堆砌的书不同，以有趣的实例结合通俗易懂的讲解带领读者在感受到开发乐趣的同时学习到核心有用的知识。
- 本书创新地引入了扫描二维码查看动态图的功能，让纸质图书也能和读者交互起来，提升阅读的乐趣。我相信一张动态图的表达效果胜过一百个字，相信读者到时也会"一目了然"。
- 本书系统而全面，从 Android 开发工具的安装、实用技巧到 Android 的布局、控件、组件、存储、网络等，涵盖 Android 开发的方方面面，一本书即可带领你充分领略 Android 开发的魅力。
- 本书基于最新的 Android 7.0 和最新的 Android Studio 2.2.3 进行开发和讲解。

本书内容

本书内容系统全面，采用层层递进的方式进行讲解，让读者理解起来更为容易。本书一共分成 10 章，同时每章的内容也都是按照难度的递增进行讲解，让读者有个容易的开始同时也拥有一个充实的结尾。

第 1、2 章主要对 Android 和 Android Studio 进行介绍，着重对 Android Studio 的常用操作和技巧进行了详细讲解，开发者熟练使用 IDE 可以有效提升开发效率、避免低级错误的发生。

第 3 章主要对 Android 的属性和布局进行讲解。属性和布局是 Android 开发中最基本的部分，这也是检验一名 Android 开发者是否合格的最低标准。这部分主要讲解几个核心属性和核心布局方式的使用，读者可以认真学习、总结、理解。

第 4、5 章主要对 Android 基础控件进行讲解。控件运用相当于武术修炼中的"外功"，控件的方法也可以认为是武林秘笈中的各个招式，对于核心控件的常用方法要予以熟练理解并掌握，这两章主要结合有趣实用的例子进行讲解，相信读者不会感到枯燥无味。

第 6 章对 Android 系统组件进行详细的讲解。系统组件是 Android 的根基，所有的应用都围绕基本系统组件展开，对系统组件的深入学习和理解是修炼"内功"的过程，也是初级开发者和中高级开发者拉开距离的部分，读者要充分重视这部分内容。

第 7 章主要讲解 Android 几种常用的存储数据方式，通过经典实例的方式向读者讲解常用存储方法的使用。

第 8 章对 Android 动画进行了讲解。没有动画过渡的应用是僵硬、死板的。如今的 Android 应用无一例外地在交互上添加了动画。尝试着在你的应用中添加动画，它会让交互过渡更平滑，用户体验更棒。

第 9 章对 Android 网络进行讲解。没有网络的 Android 手机就好像鱼儿离开了水，因此，Android 开发者在开发过程中都会不可避免地涉及到网络操作。

第 10 章主要对 Android 手机的基本功能及多媒体进行实战操作。与功能手机相比，智能手机最鲜明的特点即是其人性化的基本功能和丰富的媒体功能。本章对常用 API 进行了系统的讲解。

本书的知识比较系统，建议读者按照章节的顺序进行阅读，循序渐进地掌握 Android 核心知识。打开本书，你已经迈开了成功的一小步。

另外，全书在描述中有中英混用的描述，凡是中英混用的都是些特定术语，无需统一。

致谢

感谢清华大学出版社的编辑，没有她的积极指导和帮助，就没有这本书的诞生；感谢在编写过程中给予指导的各位志同道合的朋友，是你们让这本书更具活力；最后还要感谢我的爸爸妈妈，感谢他们不遗余力的付出和无微不至的关怀。

张亚运

推 荐

亚运的这本《Android 开发入门百战经典》是一本偏应用层面的书籍，本书集中火力于应用开发中最基础、最核心的 UI 控件使用、动画绘制、四大组件以及网络操作这几个部分，所述内容覆盖到位，讲解思路新颖。另外，通过本书引入的扫描二维码查看案例动态图这一贴心功能，可知作者在如何更好传授知识方面用心良苦。

——邓凡平 资深 Android 开发专家、《深入理解 Android》作者

如何实现 Android 开发从入门到精通呢？我建议大家仔细读一下亚运同学的《Android 开发入门百战经典》这本书。本书图文并茂，并结合有趣的实例帮助读者开启了 Android 开发之旅。读者读完之后可以系统掌握 Android 开发的核心知识，同时结合书中的例子快速上手，逐步深入，并最终精通 Android 开发。

——王天青 DaoCloud 首席架构师

《Android 开发入门百战经典》内容丰富连贯，从 Android 职业路线出发，基础与实例相结合。真正贯彻了"从开发中来，到开发中去"的高质量应用型学习思想。摒弃传统图书的长篇理论，深入浅出地讲解了大量生动有趣的案例。可谓"凡技术点必出案例，凡案例必配实图"。不仅如此，还创新引入了扫描对应二维码查看案例动态图的功能。如果说技术的提升是由质变到量变的过程，那么相信跟随着张老师的脚步，更多的开发者可以从这本书中得到升华。

——姚尚朗 极客学院

目 录

第 1 章 认识 Android ············ 001

1.1 Android 系统 ················ 001
- 1.1.1 Android 的系统架构 ······ 001
- 1.1.2 Android 的历史 ·········· 002
- 1.1.3 Android 系统的优势 ······ 002

1.2 Android Studio 安装 ·········· 004
- 1.2.1 Android Studio 安装 ······ 004
- 1.2.2 SDK 更新 ················ 005

1.3 第一个 Android 项目 ·········· 005
- 1.3.1 创建一个新项目 ·········· 005
- 1.3.2 创建 Android 模拟器 ····· 007

第 2 章 Android Studio 使用技巧 ··· 010

2.1 Android Studio 基本配置 ······ 010
- 2.1.1 改变主题 ················ 010
- 2.1.2 改变字体大小和样式 ······ 011
- 2.1.3 改变 Logcat 窗口字体、主题 ···················· 013
- 2.1.4 显示行号 ················ 015
- 2.1.5 自动导包 ················ 016

2.2 Android Studio 常用快捷键 ···· 016
- 2.2.1 Ctrl 组合快捷键 ·········· 016
- 2.2.2 Ctrl+Alt 组合快捷键 ······ 020
- 2.2.3 Ctrl+Shift 组合快捷键 ···· 022
- 2.2.4 其他组合快捷键 ·········· 024

2.3 Android Studio 调试 ·········· 026
- 2.3.1 Logcat 调试 ·············· 026
- 2.3.2 断点调试 ················ 027
- 2.3.3 高级调试 ················ 029

第 3 章 Android 属性和布局 ······ 032

3.1 Android 项目文件结构 ········ 032
- 3.1.1 布局属性 ················ 032
- 3.1.2 配置属性 ················ 034
- 3.1.3 其他文件 ················ 036

3.2 Android 布局属性值 ·········· 037
- 3.2.1 Android padding 属性用法 ················ 038
- 3.2.2 Android margin 属性用法 ·· 038

3.3 Android 布局之线性布局——LinearLayout ·············· 041
- 3.3.1 LinearLayout 基础用法 ···· 041
- 3.3.2 LinearLayout 嵌套 ········ 043

3.4 Android 线性布局的重要属性 ·· 045
- 3.4.1 gravity 属性 ·············· 045
- 3.4.2 layout_weight 属性 ········ 048
- 3.4.3 weightSum 属性 ·········· 052

3.5 Android 布局之相对布局——RelativeLayout ·········· 053

3.6 Android 布局之帧布局——FrameLayout ················ 056

3.7 Android 布局优化 ············ 059
- 3.7.1 过度绘制 ················ 059
- 3.7.2 布局优化之 include 标签 ·· 061

第 4 章 Android 基础控件操作实战 …… 064

- 4.1 炫酷之星——TextView 控件 …… 064
  - 4.1.1 常用属性介绍 …… 064
  - 4.1.2 TextView 实战演练 …… 064
- 4.2 用户之窗——EditText 控件 …… 069
  - 4.2.1 常用属性介绍 …… 069
  - 4.2.2 EditText 实战演练 …… 070
  - 4.2.3 EditText 实战进阶 …… 074
- 4.3 交互之王——Button 控件 …… 078
  - 4.3.1 Button 单击事件响应 …… 078
  - 4.3.2 clickable 属性设置无效分析 …… 084
  - 4.3.3 Button 实战进阶 …… 086
- 4.4 执行中的指示器——ProgressBar · 088
  - 4.4.1 ProgressBar 样例 …… 088
  - 4.4.2 ProgressBar 基础用法 …… 089
  - 4.4.3 ProgressBar 模拟下载 …… 093
- 4.5 对话框之父——Dialog …… 095
  - 4.5.1 AlertDialog …… 096
  - 4.5.2 单选和多选对话框 …… 099
  - 4.5.3 ProgressDialog 进度对话框 …… 101
  - 4.5.4 定制对话框 …… 104

第 5 章 Android 控件进阶操作实战 …… 108

- 5.1 控之经典——ListView …… 108
  - 5.1.1 ArrayAdapter 适配器 …… 109
  - 5.1.2 SimpleAdapter 适配器 …… 110
  - 5.1.3 BaseAdapter 适配器 …… 112
- 5.2 控之经典——ListView 进阶 …… 116
- 5.3 控之经典——GridView …… 121
- 5.4 控之经典——GridView 进阶 …… 126
  - 5.4.1 GridView 动态图删除子项 …… 126
  - 5.4.2 GridView 动态图增加子项 …… 130
- 5.5 新控件——RecyclerView 控件 …… 132
  - 5.5.1 RecyclerView 线性布局 …… 134
  - 5.5.2 RecyclerView 网格布局 …… 137
  - 5.5.3 RecyclerView 瀑布流布局 …… 137
- 5.6 多页面切换器——ViewPager 控件 …… 139
  - 5.6.1 ViewPager 的基本用法 …… 140
  - 5.6.2 ViewPager 导航条 …… 143

第 6 章 Android 系统组件操作实战 …… 148

- 6.1 Activity 生命周期 …… 148
- 6.2 指向器——Intent …… 152
- 6.3 指向器——Intent 隐式启动方式 ·· 156
- 6.4 Mini 型 Activity——Fragment …… 160
  - 6.4.1 静态方式 …… 160
  - 6.4.2 动态方式 …… 162
- 6.5 Mini 型 Activity——Fragment 生命周期 …… 166
- 6.6 FragmentPagerAdapter&FragmentStatePagerAdapter …… 172
  - 6.6.1 FragmentPagerAdapter 实现页面切换 …… 173
  - 6.6.2 FragmentStatePagerAdapter 实现页面切换 …… 176
- 6.7 Android 广播接收器之 BroadcastReceiver …… 179
  - 6.7.1 静态注册 BroadcastReceiver …… 179
  - 6.7.2 动态注册 BroadcastReceiver …… 180
  - 6.7.3 广播接收器 BroadcastReceiver 实用实例 …… 182
- 6.8 Android 自定义广播 Broadcast …… 186
  - 6.8.1 普通广播发送和接收实例 …… 187
  - 6.8.2 有序广播发送和接收实例 …… 188

6.9 Android Service——startService 和 bindService ……………………… 191
    6.9.1 startService 启动服务 ……… 192
    6.9.2 bindService 启动服务 ……… 195

第 7 章 Android 存储操作实战 ……… 200

7.1 轻型存储器——SharedPreferences ……………… 200
    7.1.1 SharedPreferences 基本用法 ……………………… 200
    7.1.2 SharedPreferences 实现自动登录功能 ……… 204
7.2 Android 数据库 SQLite …………… 209
    7.2.1 SQLiteOpenHelper 类 ……… 210
    7.2.2 SQLiteDatabase 类 ………… 212
7.3 数据中心——ContentProvider …… 218

第 8 章 Android 动画操作实战 ……… 223

8.1 Android 传统动画——Tween（补间动画）………………… 223
    8.1.1 AlphaAnimation——渐变动画 ……………………… 224
    8.1.2 RotateAnimation——旋转动画 ……………………… 229
    8.1.3 ScaleAnimation——尺寸动画 ……………………… 232
    8.1.4 TranslateAnimation——位移动画 ……………………… 235
8.2 Android 传统动画进阶 …………… 238
    8.2.1 动画插值器 Interpolator …… 238
    8.2.2 动画监听器 AnimationListener ……………… 241
    8.2.3 动画集 AnimationSet ……… 243
    8.2.4 LayoutAnimationController 组件动画 ……………………… 246
8.3 Android 传统动画——Frame Animation（帧动画）…………… 248

8.4 Android 属性动画——ObjectAnimator …………………… 252
    8.4.1 属性动画与传统动画的区别 ……………………… 252
    8.4.2 旋转动画 ………………… 254
    8.4.3 尺寸动画 ………………… 254
    8.4.4 渐变动画 ………………… 254
    8.4.5 XML 方式实现属性动画 … 257
8.5 Android 属性动画——ValueAnimator …………………… 259
8.6 Android 属性动画集 ……………… 262
    8.6.1 简单的组合方式 ………… 262
    8.6.2 PropertyValuesHolder 方式 ……………………… 264
    8.6.3 AnimatorSet 方式 ………… 265
8.7 Android 属性动画实现浮动菜单 … 266

第 9 章 Android 网络操作实战 ……… 271

9.1 Android 网络核心控件 WebView · 271
    9.1.1 简单的 WebView ………… 271
    9.1.2 丰富 WebView 功能 ……… 273
9.2 WebView 滚动事件 ……………… 276
    9.2.1 WebView 滚动监听的实现 ……………………… 276
    9.2.2 WebView 一键回到顶部功能实现 ……………… 278
    9.2.3 WebView 退出记忆功能实现 ……………………… 280
    9.2.4 WebView 联合滚动实现 … 281
9.3 网络连接类——HttpURLConnection ……………… 283
    9.3.1 HttpURLConnection 打印网页 ……………………… 284
    9.3.2 HttpURLConnection 下载图片 ……………………… 287
    9.3.3 HttpURLConnection 保存图片 ……………………… 290
9.4 Android Handler 消息处理机制 …… 294

9.4.1 消息类 Message ………… 295
9.4.2 消息处理类 Handler ……… 295
9.4.3 Handler 实现倒计时功能 ·· 298
9.4.4 Handler 延迟操作 ………… 301
9.4.5 Handler postDelay
实现循环调用 ………… 303
9.4.6 Looper 用法 ………… 304
9.5 Android 异步操作类 AsyncTask … 307
9.5.1 AsyncTask 基本用法 ……… 308
9.5.2 AsyncTask 实用实例 ……… 310

第 10 章 Android 手机基本功能及多媒体操作实战 ………… 315

10.1 Android 拨打电话功能实例 …… 315
10.2 Android 发送短信功能实例 …… 319
    10.2.1 直接发送短信 ………… 320
    10.2.2 跳转到短信发送界面 …… 322
10.3 Android 播放音乐功能实例 …… 323
10.4 Android 播放视频功能实例 …… 329
10.5 Android 录制音频功能实例 …… 335
10.6 Android 拍照功能实例 ………… 341
    10.6.1 Intent 方式 ………… 342
    10.6.2 借助 Camera 类 ………… 344

第 1 章 认识 Android

1.1 Android 系统

Android 是谷歌推出的基于 Linux 的手机平台，作为开源的移动操作系统，不存在任何阻碍移动产业创新和发展的专利权障碍，因此 Android 一经面世就获得了空前的发展，在移动操作系统市场份额一度超过 80%，处于绝对的垄断地位。

现如今 Android 已经不局限于手机系统，越来越多的车载、穿戴、电视设备也集成了 Android 系统。我们相信随着物联网的不断深入和发展，Android 系统将会以更多样的形式融入到我们的生活、学习和工作之中。因此，学习 Android 不会过时，正当其时！

1.1.1 Android 的系统架构

Android 的系统架构，如图 1.1 所示。

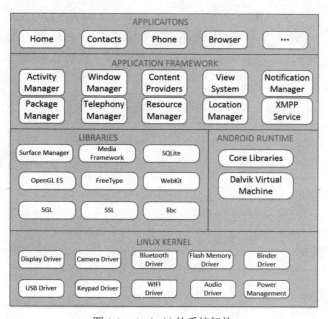

图 1.1 Android 的系统架构

和其他操作系统一样，Android 的系统架构也采用了分层的结构。从架构图来看，Android 分为四个层次，下面分别总结这几个层。

1. 应用程序（Applications）

Android 应用程序的源程序除了包含 Java 代码之外，还包含各种资源文件（放置于 res 目录中）。将源程序进行编译可以得到一个 APK 安装包，这个安装包可以安装到 Android 手机上，将对应一个 Android 应用程序。Android 软件开发者可以使用应用程序框架层提供的 API 快速开发 Android 应用，这也是 Android 的巨大潜力所在。

2. 应用程序框架（Application Framework）

Android 的应用程序框架层有供 Android 开发者所调用的丰富 API，该层实际上就是一个应用程序的框架。框架中不仅包含各种 API，同时还内置了方便开发者开发的各种控件，例如 Views（视图组件），其中又包含了 List（列表）、Grid（栅格）、Button（按钮）、TextView（文本框）等，甚至还内置了一个浏览器。有了这些基本控件，开发者可以更快速地构建应用程序，提高开发效率。

3. 各种库（Libraries）和 Android 运行环境（Android Runtime）

对应一般的嵌入式操作系统，本层相当于中间件层次。Android 中的本层分成两个部分，一个是各种库（Libraries），另一个是 Android 运行环境。本层的大多数代码是由 C 和 C++ 语言实现。Android 运行环境指的是 Android 虚拟机技术 Dalvik。

4. 操作系统层（OS）

Android 基于 Linux，使用的是 Linux 2.6 操作系统，以它作为底层。Android 对操作系统的使用包括了核心和驱动程序两个部分，其中驱动程序有显示驱动、蓝牙驱动、相机驱动、网络驱动和各种传感器设备驱动等。

1.1.2 Android 的历史

自 2008 年 9 月发布 Android 第一版，时至今日，Android 已经发展到了 7.0 时代。从 2009 年 5 月开始，Android 版本开始使用甜点作为版本代号：1.5（纸杯蛋糕）、1.6（甜甜圈）、2.0（泡芙）、2.2（冻酸奶）、2.3（姜饼）、3.0/3.2（蜂巢）、4.0（冰激凌三明治）、4.1/4.2（果冻豆）、4.4（奇巧）、5.0（棒棒糖）、6.0（棉花糖）、7.0（牛轧糖），历代发布的时间请参考表 1.1。

表 1.1 历代 Android 版本时间表

版 本 号	API Level	发 布 时 间
1.1	1	2008 年 9 月
1.5（纸杯蛋糕）	3	2009 年 4 月
1.6（甜甜圈）	4	2009 年 9 月
2.0（泡芙）	5	2009 年 10 月
2.2（冻酸奶）	8	2010 年 5 月
2.3（姜饼）	10	2010 年 12 月
3.0（蜂巢）	11	2011 年 2 月
3.2（蜂巢）	13	2011 年 7 月
4.0（冰激凌三明治）	14	2011 年 10 月
4.1（果冻豆）	16	2012 年 6 月
4.2（果冻豆）	17	2012 年 10 月
4.4（奇巧）	19	2013 年 9 月
5.0（棒棒糖）	21	2014 年 11 月
6.0（棉花糖）	23	2015 年 5 月
7.0（牛轧糖）	24	2016 年 8 月

1.1.3 Android 系统的优势

对于想要从事移动开发的读者来讲，开始都会万分纠结的问题就是到底该学习 iOS 还是Android系统，对于一些初学者来说，这一定是一个单选题。本书主要对Android知识进行讲解，当然要自卖自夸一下。比较上述两系统而言，学习Android系统可以有如下优势：

1. 更容易上手

对于初学者来说，最缺乏的就是基础知识，最渴望的就是快速上手，最苦恼的莫过于一头雾水。Android系统使用Java语言进行开发，对计算机语言稍有基础的同学而言，Java语言都不会陌生，Java语言也是常年霸占计算机语言流行榜No.1的位置，国内Java语言学习风气浓厚，随便百度一下即可获得海量Java语言学习资源。同时Java语言也是以其简单、易用而闻名，所以对于初学者来说，这第一个骨头并不难啃。而iOS系统采用Object-C进行开发（2014年推出了Swift作为新的开发语言），相对Java语言来说，其学习难度要大不少，此外志同道合者较少，因此，学习资料也就相对缺乏，一起讨论交流的朋友也比较少。因此，就上手难易程度来说，Android系统确实优于iOS系统。

2. 更宽广的就业方向

即使不了解开发的人都会知道，iOS是闭源的系统，开发者除了能开发iOS应用什么也做不了，而Android是开放的系统，源代码公开，从上层的应用开发、到Framework层再到底层驱动都可以进行研究和学习，任何一个环节、任何一个模块都可以作为今后从业的方向。此外，学习好Android的开发语言Java，就算以后不从事移动端开发，还可以转向Web开发等，而学习iOS开发语言就只能从事iOS相关专业开发了。因此学习Android开发将拥有更宽广的就业渠道，更丰富的研究方向。

3. 更多的学习资源

在百度搜索中输入"Android学习资料"关键字并搜索，你可以获得8 640 000个相关结果，而输入"iOS学习资料"，仅获得两百多万个相关结果。对于初学者最好的老师——搜索引擎来说，显然，它知道Android的知识更多一些。此外，根据2016年TIOBE世界编程语言排行榜，Java语言以20.5%占有率的绝对优势占据榜首，而iOS的开发语言Swift和Object-C则排在了第14和15位，两者之和还不到3%的占有率。因此，学习Android系统你将拥有更多志同道合的朋友，从他们那里你可以获得更多帮助和指导。最后，由于Android的开放性，较iOS来讲，Android拥有绝对数量优势的优秀开源项目，有一定基础的开发者可以登录github浏览这些项目，提升自己的开发能力。

4. 学习成本

学习iOS系统，至少得配备一台Mac作为开发工具，配备一台iPhone作为调试工具，这两种开发工具都价值不菲，对于一穷二白的初学者来说，经济上的拮据是不可避免的问题。对于缺乏定力和恒心而半途而废的初学者来说，损失就更大了。而学习Android系统只需一台具有Windows操作系统的电脑就好了（基本每个人都有），对于调试工具可以选择模拟器，也可以花几百元买一台入门级Android手机，所以前期投入很少，不存在任何风险。因此，想学习Android开发马上就可以开始，不需要太大的经济投入，没有经济压力和风险。

综上，对于踌躇不前、犹豫不决的初学者来说，何不先选择Android系统学习一下呢？因为它简单、易上手且无须任何前期投入。我也相信，鉴于Android系统的开放性、流行性，只要尝试，你肯定会爱上它，因为作者本人就是这么掉进"陷阱"里来的。

1.2 Android Studio 安装

俗话说："工欲善其事，必先利其器。"要想获得快速的开发效率和学习速度，选择一样合适的开发工具是首先要做的事情。很长一段时间，开发者都习惯了使用 Eclipse 并结合 ADT 插件来开发 Android 应用，但这一习惯将随着 Android Studio 的不断强大而必须改变了。自从 2013 年 5 月 16 日，在 I/O 大会上推出的 Android Studio 雏形，到现在更新到了最新的 2.2.2.0 版本，Android Studio 越来越稳定，功能也越来越强大，是时候该享受全新的开发工具了。

1.2.1 Android Studio 安装

安装 Android Studio，首先需要下载安装包，这里推荐一个下载地址（Android Studio 中文社区）http://www.android-studio.org/。可以看出，当前最新的 Android Studio 已经更新到了 2.2.2.0 版本，单击如图 1.2 所示的"下载"按钮，下载安装包到本地。

图 1.2 Android Studio 下载

双击下载完成的安装包安装 Android Studio，如图 1.3 所示，单击 Next 按钮跳转到下一界面，如图 1.4 所示。

图 1.3 Android Studio 安装一

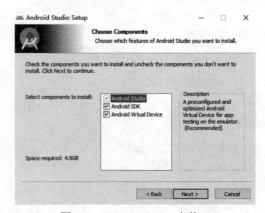

图 1.4 Android Studio 安装二

将默认一起安装 Android SDK 和 Android Virtual Device。不断单击 Next 按钮到安装成功页，如图 1.5 所示。

图 1.5 Android Studio 安装成功页

1.2.2 SDK 更新

单击 Finish 按钮进入 Android Studio，刚进去可能需要更新 SDK，如图 1.6 所示。

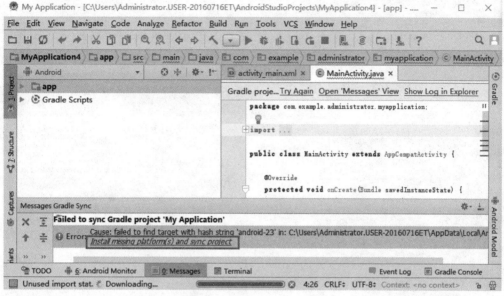

图 1.6　Android Studio 页

单击 Install missing platform(s) and sync project 按钮更新 SDK，如图 1.7 所示。

单击 Finish 按钮完成 SDK 更新，届时 Android Studio 安装完毕。

1.3　第一个 Android 项目

万事开头难，凡事都有套路，勇敢迈出第一步就成功一大半了。本节将带领初学者迈出属于自己的一小步。本书的开发工具采用 Android Studio，因此，首先通过图文讲解如何使用 Android Studio 新建我们的第一个 Android 项目。

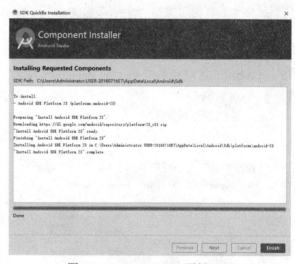

图 1.7　Android Studio 更新 SDK

1.3.1　创建一个新项目

安装成功 Android Studio 会默认生成一个 Android 项目，这里新建一个属于我们自己的项目。

Step 01　选择 File → New → New Project，如图 1.8 所示。

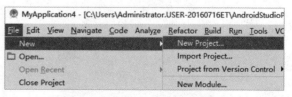

图 1.8　Android Studio 新建项目

Step 02 选择 New Project 后会弹出新建项目页,在 Application name 中输入项目名称,在 Package name 中输入包名(需要单击右边的 Edit 按钮),单击右下角的 Next 按钮进入下一步骤,如图 1.9 所示。

Step 03 本书开发的是手机应用,因此选择最上方的 Phone and Tablet、Minimum SDK,即最小支持的 SDK,其余选择默认即可,单击右下角的 Next 按钮,如图 1.10 所示。

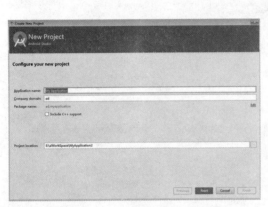
图 1.9 Android Studio 新建项目页一

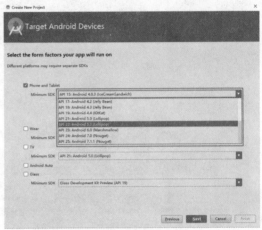
图 1.10 Android Studio 新建项目页二

Step 04 这个界面用来选择生成项目时默认 Activity 的样式,Android Studio 提供了丰富的 Activity 模板供我们选择,有 Basic Activity(基本 Activity)、Empty Activity(空 Activity)、Google Maps Activity(谷歌地图 Activity)、Login Activity(登录 Activity)等,这里选择 Empty Activity,继续单击 Next 按钮,如图 1.11 所示。

Step 05 这里有两个文本框,Activity Name 文本框用来输入默认的 Activity 名,Layout Name 文本框用来输入默认 Activity 的默认布局名称,这里都选择默认的即可,单击右下角的 Finish 按钮,等待 Android Studio 生成项目即可,如图 1.12 所示。

图 1.11 Android Studio 新建项目页三

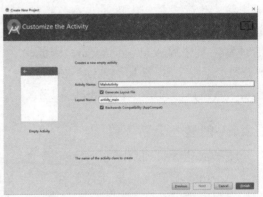
图 1.12 Android Studio 新建项目页四

Step 06 等待 Gradle 编译完成,如图 1.13 所示。生成的 Android Studio 页面如图 1.14 所示。

图 1.13　Gradle Build 编译

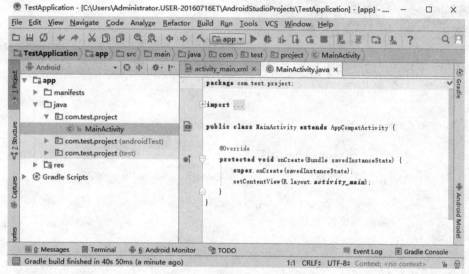

图 1.14　Android Studio 页面

1.3.2　创建 Android 模拟器

单击工具栏中的 Run 按钮，如图 1.15 所示。

弹出 Android 模拟器选择框，此时看到提示：No USB devices or running emulators detected，也就是没有检测到 Android 模拟器，因此需要单击左下角的 Create New Virtual Device 按钮创建一个新的模拟器，如图 1.16 所示。

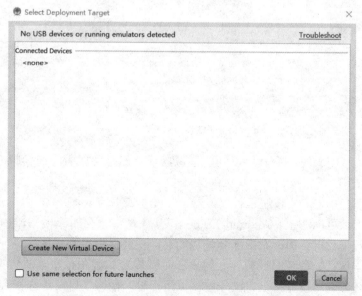

图 1.15　Android Run 按钮　　　　图 1.16　Android 模拟器选择框

单击 Create Virtual Device 按钮，创建一个 Android 模拟器，如图 1.17 所示，这里需要选择手机模拟器，Android 提供了众多型号的模拟器供开发者选用，选择一款你喜欢的手机作为模拟器，单击 Next 按钮，如图 1.18 所示。

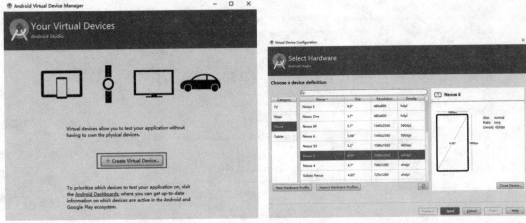

图 1.17　Android 模拟器创建一　　　　图 1.18　Android 模拟器创建二

在左下角会看到提示信息：A system image must be selected to continue，也就是说必须先安装一个系统镜像，单击 Download 按钮后如图 1.19 所示。单击 Next 按钮安装系统镜像，如图 1.20 所示。

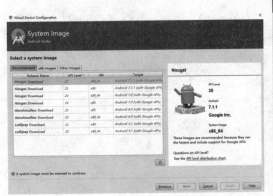 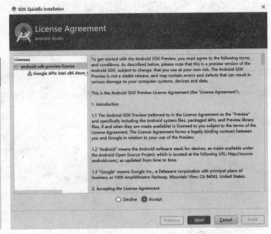

图 1.19　Android 模拟器创建三　　　　图 1.20　Android 模拟器创建四

安装完成后就可以创建模拟器了，创建完成后，在模拟器列表就出现了新的模拟器，如图 1.21 所示。

单击 OK 按钮运行模拟器，如图 1.22 所示。

可以看出，Android 7.0 的新 UI 还是很漂亮的，模拟器由两部分组成：左半部分是模拟器手机界面；右半部分是功能栏，功能栏由上到下依次是：屏幕开关、声音上键、声音下键、屏幕逆时针旋转、屏幕顺时针旋转、截图按钮、放大按钮、返回键、Home 键、多任务键和设置按钮。在开发模拟运行时，根据项目需要选择合适的操作。

再稍等片刻，TestApplication 项目将运行起来，如图 1.23 所示，程序员们熟悉的

"Hello World!"在模拟器中显示出来了。

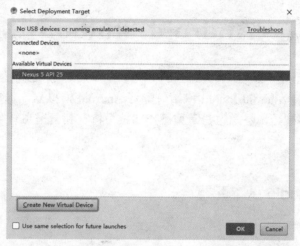

图 1.21　Android 模拟器选择

图 1.22　Android 模拟器

图 1.23　Android 模拟器项目运行

第 2 章　Android Studio 使用技巧

正所谓"磨刀不误砍柴工"，Android 开发中最重要的利器就是 Android Studio。上一章介绍了如何安装 Android Studio 和如何配置 Android 模拟器。本章主要讲解 Android Studio 的常用操作和技巧，熟悉这些常见操作和技巧将有利于提高开发效率，减少开发时低级错误的发生。

2.1　Android Studio 基本配置

2.1.1　改变主题

安装成功时，Android Studio 默认的主题名为 IntelliJ，其效果如图 2.1 所示。

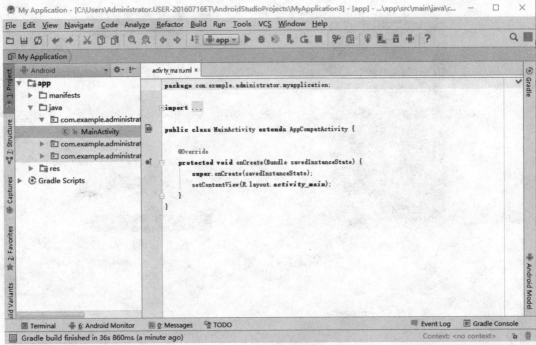

图 2.1　Android Studio 默认主题

这种主题比较亮，对于长时间盯着屏幕的程序员来说，最辛苦的莫过于眼睛了。考虑到这种情况，Android Studio 提供了"护眼模式"Darcule 主题，使用该主题，应按如下几个步骤操作：

Step 01　选择 File → Settings，如图 2.2 所示。

Step 02　此时将跳转到设置页面，选择 Appearance 标签中的 Theme 值，如图 2.3 所示。

Step 03　选择 Darcula 并单击 Apply 按钮即可使用这个主题，如图 2.4 所示。

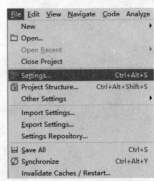

图 2.2　File 菜单栏

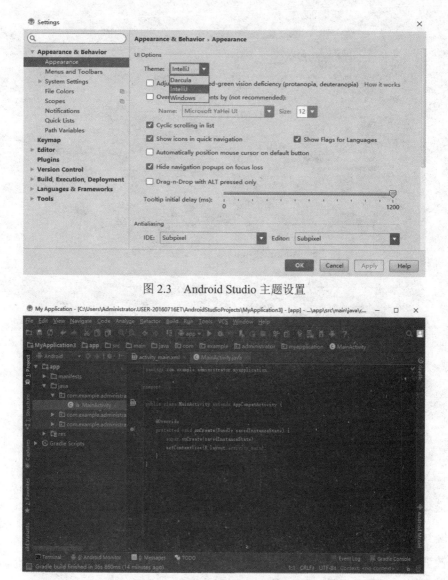

图 2.3　Android Studio 主题设置

图 2.4　Android Studio Darcula 主题

此时的主题变成了"黑色风格",对于视力都不太好的程序员来说,这简直是巨大的福利。

2.1.2　改变字体大小和样式

上面更改了 Android Studio 的主题样式,不过,可以看出默认的字体也是比较小的,长时间浏览小字体也会造成视觉疲劳,下面通过图文讲解如何改变字体样式和大小。

改变字体大小和样式包括改变菜单字体和样式与改变编辑器字体和样式两部分。

1. 改变菜单字体大小和样式

选中如图 2.5 所示的 Override default fonts by(not recommended)复选框,这时就可以修改菜单字体的大小和样式了。左边的 Name 下拉列表框用来选择字体样式,右边的 Size 下拉列表框用来选择字体大小,如图 2.6 所示。

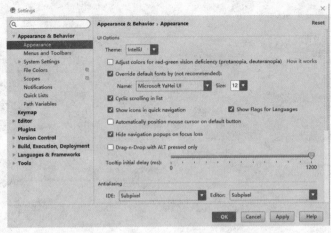

图 2.5　Android Studio 设置字体大小和样式一

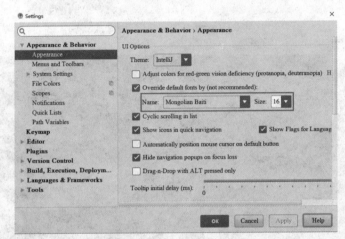

图 2.6　Android Studio 设置字体大小和样式二

可以看出，此时菜单字体的大小和样式都改变了。

2. 改变编辑器字体大小和样式

选择 Editor → Colors&Fonts → Font，如图 2.7 所示。

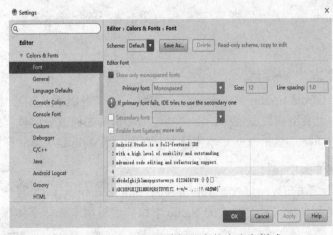

图 2.7　Android Studio 编辑器字体大小和样式一

默认情况下，字体大小和样式是不允许修改的，先单击 Save As 按钮将 Scheme 另存，如图 2.8 所示。

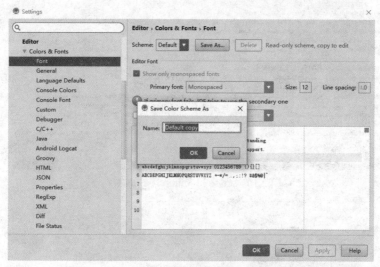

图 2.8　Android Studio 编辑器字体大小和样式二

单击 OK 按钮，这时字体样式和大小就可编辑了，选择心仪的字体大小和样式，如图 2.9 所示。

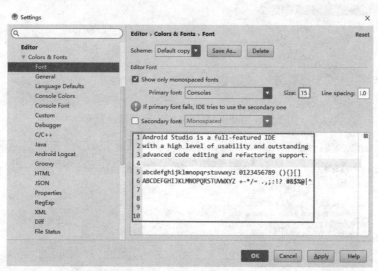

图 2.9　Android Studio 编辑器字体大小和样式三

单击 Apply 按钮，即可应用该字体样式和字体大小。

2.1.3　改变 Logcat 窗口字体、主题

除了编辑窗之外，开发者用的最多的应该就是 Logcat 窗口了，默认该窗口如图 2.10 所示。

可以看出，默认的字体很小，不便于调试阅读。和编辑区一样，这里的字体和主题也是可以自定义的。通过 File → Settings → Editor → Colors&Fonts → Console Font，可以打开 Android Studio Logcat 字体设置页面，如图 2.11 所示。

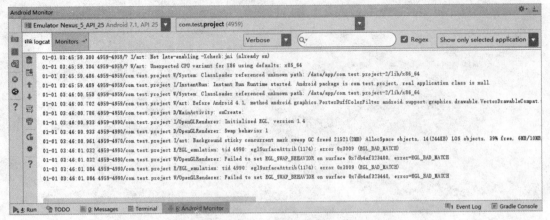

图 2.10　Android Studio Logcat 窗口

图 2.11　Android Studio Logcat 字体设置一

同样，无法改变已存在主题的字体大小和样式（置灰的），首先要自定义一下自己的主题，单击 Save As 按钮，另存当前主题，这时字体样式和大小就可以自定义了，如图 2.12 所示。

图 2.12　Android Studio Logcat 字体设置二

通过下面的即时效果页可以看到改变字体样式和大小后的效果，同时还可以改变 Line spacing 中的值来改变行距。单击 Apply 和 OK 按钮后再次查看 Logcat 窗口，如图 2.13 所示。

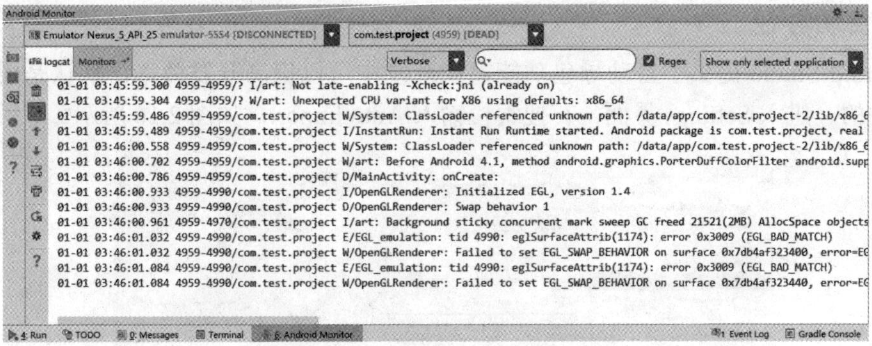

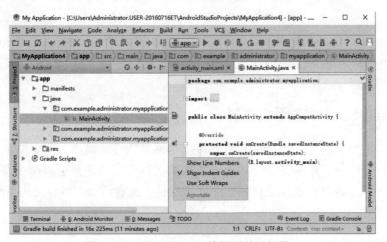

图 2.13　Android Studio Logcat 窗口

可以看出此时的字体大小、样式都改变了。

2.1.4　显示行号

默认在编辑页中没有显示代码行号，右击编辑页的右边栏，可弹出如图 2.14 所示的快捷菜单。

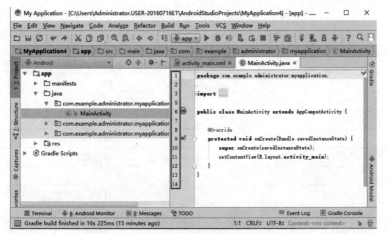

图 2.14　Android Studio 编辑页显示行号一

选择 Show Line Numbers 即可显示行号，如图 2.15 所示。

图 2.15　Android Studio 编辑页显示行号二

2.1.5 自动导包

Android Studio 提供了自动导包功能，在开发中可以提高开发效率，默认是没有打开这个开关的，进入 Settings 界面，在搜索栏输入 Auto Import，如图 2.16 所示。

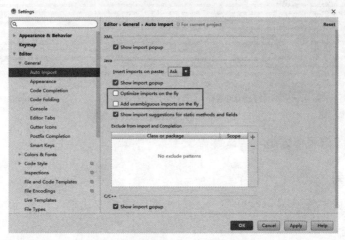

图 2.16　Android Studio 自动导包设置页

选中 Optimize imports on the fly（移除无用的包）和 Add unambiguous imports on the fly（自动导包）复选框，这时再复制代码时就可以自动导入需要的包和删除无用的包了。

2.2　Android Studio 常用快捷键

快捷键又称"热键"，多个按键的组合可以实现某些快速操作，例如 Windows 中最常用的 Ctrl+C 和 Ctrl+V。熟练使用快捷键可以大大提高开发效率并可以减少某些错误的发生。Android Studio 也默认提供了众多快捷键方式供开发者调用，推荐使用 Android Studio 默认风格的快捷键。

2.2.1　Ctrl 组合快捷键

这里将快捷键进行分类，方便学习和记忆，Ctrl 组合快捷键如表 2.1 所示。

表 2.1　Ctrl 组合快捷键

快　捷　键	说　　明
Ctrl+C	复制
Ctrl+V	粘贴
Ctrl+X	剪切
Ctrl+D	在当前行下方复制一行
Ctrl+Y	删除当前行
Ctrl+G	快捷行数定位
Ctrl+Z	撤销
Ctrl+E	查看最近打开的文件
Ctrl+/	注释一行，再按一次反注释
Ctrl+N	查找类名、文件名

快 捷 键	说　　明
Ctrl+O	显示父类中可覆写的方法
Ctrl+F	类内搜索
Ctrl+R	查找替换
Ctrl+J	自动代码
Ctrl+H	显示类继承结构图
Ctrl+W	选中代码，类似双击效果
Ctrl+F12	快速查找类内方法

对于初学者来说，一下子记住这么多快捷键简直就是噩梦，其实也没必要一次就背下来，只需要在开发中尽量使用快捷键并打印一份快捷键表放在电脑旁，经常使用和查阅，一段时间后就会使用了。

下面挑选一些常用的快捷键进行图文讲解。

1. Ctrl+G

同时按下 Ctrl+G 快捷键弹出快速定位框，在框中输入行数，单击 OK 按钮即可快速切换到对应的行数，如图 2.17 所示。

2. Ctrl+E

同时按下 Ctrl+E 快捷键，弹出最近打开文件的列表，可以快速选择最近曾经打开的文件，如图 2.18 所示。

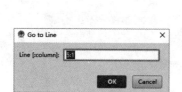
图 2.17　Go to Line

图 2.18　Recent Files

3. Ctrl+/

选中某一行，同时按下 Ctrl+/ 快捷键可以注释这一行，如图 2.19 所示。

```
//    setContentView(R.layout.activity_main);
```
图 2.19　注释代码

4. Ctrl+F

同时按下 Ctrl+F 快捷键，将在编辑页的顶部弹出类内快速搜索栏，如图 2.20 所示。

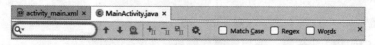
图 2.20　类内快速搜索栏

可以快速定位类内的某个单词，支持联想查找。如图 2.21 所示，输入 prote，将会高亮显示 protected，同时注意到搜索栏中有三个复选框，选中第一个 Match Case 复选框将会对大小写敏感。

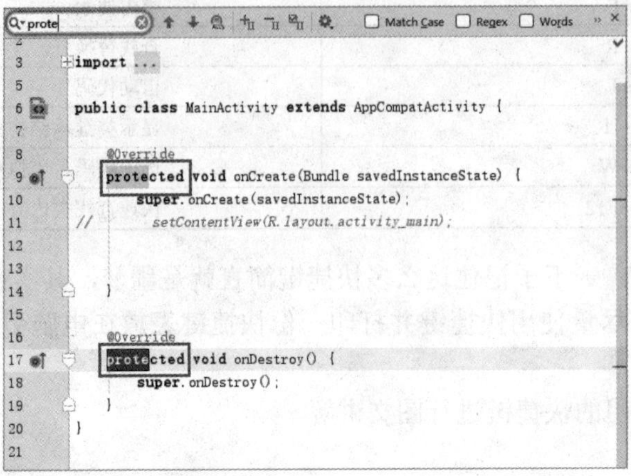

图 2.21　类内快速搜索示意图

5. Ctrl+R

Ctrl+F 快捷键常和 Ctrl+R 快捷键配合使用，用来快速查找并全部替换。如图 2.22、图 2.23 所示，先使用快捷键 Ctrl+F 搜索出所有 protected，然后使用快捷键 Ctrl+R 弹出替换栏，在替换栏文本框中输入替换后的单词并单击 Replace all 按钮，即可将类中所有的 protected 替换成 public，十分快捷。不过，在实际开发中要谨慎使用，避免引入不容易察觉的问题。

图 2.22　类内快速搜查找替换工具栏

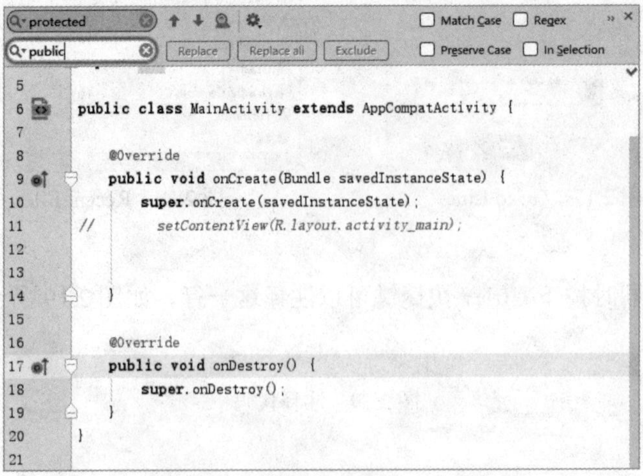

图 2.23　类内快速搜查找替换示意图

6. Ctrl+J

同时按下 Ctrl 和 J 快捷键，弹出快捷代码框，如图 2.24 所示。

对于一些常用的代码，Android Studio 中进行了封装，直接选中即可快速生成，在开发中十分实用，这里以打印 log 和弹出 Toast 为例，首先按下 Ctrl+J 快捷键，弹出如图 2.24 所示的快捷代码框，然后直接输入 logd 这一快捷代码的命令，如图 2.25 所示。

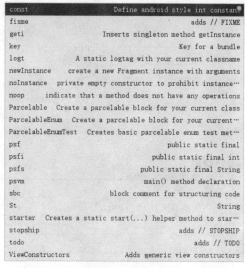

图 2.24　快捷代码框

图 2.25　快捷代码 logd

这时按下 Enter 键，即可快速生成一行 Log 代码，如图 2.26 所示。

打印 Log 需要 TAG，在类的最上方输入快捷代码 logt，即可快速生成一个 TAG，如图 2.27 所示。

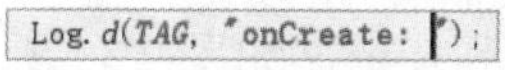

图 2.26　快捷代码 logd 示意图　　　图 2.27　快捷代码 logt

按下 Enter 键，如图 2.28 所示。

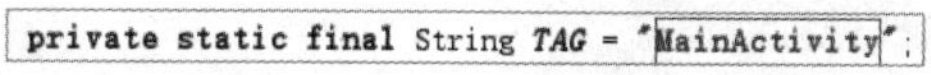

图 2.28　快捷代码 logt 示意图

同样，先按下 Ctrl+J 键，弹出快捷代码框，然后直接输入 toast，如图 2.29 所示。

图 2.29　快捷代码 toast

按下 Enter 键，如图 2.30 所示。

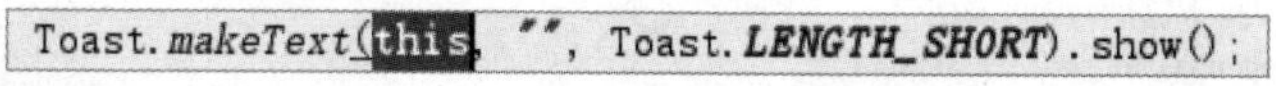

图 2.30　快捷代码 toast 示意图

此时快速生成了一行 Toast 语句，在引号中输入要 Toast 显示的信息即可，十分快捷方便。

7. Ctrl+F12

在类中方法比较多的情况下，同时按下 Ctrl 和 F12 键可以快速查看类中所有的方法，如图 2.31 所示。

弹出这个框的同时可以直接输入想要搜索的方法，进行快速匹配，如图 2.32 所示。

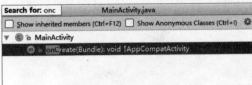

图 2.31 类中方法查看　　　　　　　　图 2.32 类中方法搜索匹配

2.2.2 Ctrl+Alt 组合快捷键

Ctrl+Alt 组合快捷键如表 2.2 所示。

表 2.2 Ctrl+Alt 组合快捷键

快　捷　键	说　　明
Ctrl+Alt+T	选中代码块，按下此快捷键可快速添加 if、try-catch 等语句
Ctrl+Alt+L	格式化代码
Ctrl+Alt+Space	弹出提示
Ctrl+Alt+V	快速声明一个变量
Ctrl+Alt+S	快速打开设置界面
Ctrl+Alt+H	查看此方法的引用
Ctrl+Alt+O	优化导入的包

下面通过图文详细讲解这些常用快捷键的用法。

1. Ctrl+Alt+T

选中一块代码，同时按下 Ctrl、Alt 和 T 键，弹出"包裹"列表框，如图 2.33 所示。

选择需要包裹的类型即可包裹选中的代码，这里以 try-catch 为例，单击选中即可，如图 2.34 所示。

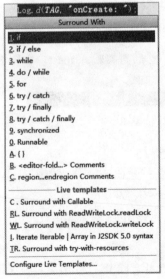

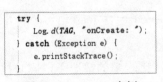

图 2.33 Ctrl+Alt+T　　　　　　　　图 2.34 try-catch 实例

可以看出，自动为选中的那行代码添加了 try-catch 语句进行包裹。

2. Ctrl+Alt+L

可以用此快捷键对当前类的所有代码进行格式化。代码格式化前如图 2.35 所示。

```
@Override
public void onCreate(Bundle savedInstanceState) {
    super.onCreate(savedInstanceState);
setContentView(R.layout.activity_main);

    Log.d(TAG, "onCreate: ");
        Toast.makeText(this, "", Toast.LENGTH_SHORT).show();
}
```

图 2.35　代码格式化前

编写代码时可能不会太注意格式问题，导致代码排版比较乱，不便于阅读。编写完毕可以通过此快捷键进行快速格式化，使用快捷键对代码格式化后如图 2.36 所示。

```
@Override
public void onCreate(Bundle savedInstanceState) {
    super.onCreate(savedInstanceState);
    setContentView(R.layout.activity_main);

    Log.d(TAG, "onCreate: ");
    Toast.makeText(this, "", Toast.LENGTH_SHORT).show();
}
```

图 2.36　代码格式化后

此时的代码就十分整齐了，阅读起来也十分方便。

3. Ctrl+Alt+V

此快捷键可以快速声明一个变量，例如在代码中输入一个字符串，并按下这个快捷键即可快速声明一个字符串变量，如图 2.37 所示。

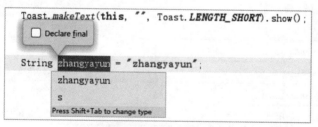

图 2.37　快速生成字符串变量

查看动态图，请扫描图 2.38 中的二维码。

图 2.38　快速生成变量二维码

4. Ctrl+Alt+H

选中某一个方法按下这个快捷键,在左边栏上弹出此方法的调用关系,如图 2.39 所示。此快捷键在开发中十分常用。

图 2.39　代码调用关系框

5. Ctrl+Alt+O

这个快捷键可以自动导包或删除无用的包,例如图 2.40 所示的代码中有一些不用的包。这时按下此快捷键即可自动删除这些无用的包,如图 2.41 所示。

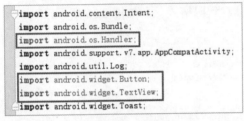

图 2.40　代码引入包

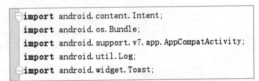

图 2.41　删除无用包

从图 2.41 可以看出三个无用的包被移除了。

2.2.3　Ctrl+Shift 组合快捷键

Ctrl+Shift 组合快捷键如表 2.3 所示。

表 2.3　Ctrl+Shift 组合快捷键

快　捷　键	说　　明
Ctrl+Shift+N	查找文件
Ctrl+Shift+Space	自动补全代码
Ctrl+Shift+/	注释代码块或反注释代码块
Ctrl+Shift+Insert	选择并插入剪贴板中的内容
Ctrl+Shift+Backspace	回到上次编辑的地方
Ctrl+Shift+ 上键	代码块整体上移
Ctrl+Shift+F	全局搜索
Ctrl+Shift+ 加号 / 减号	收起或展开方法
Ctrl+Shift+F12	关闭所有窗口

下面通过图文来讲解常用快捷键的用法。

1. Ctrl+Shift+/

和 Ctrl+/ 类似,都是实现注释代码的功能,Ctrl+Shift+/ 实现代码块的注释,如图 2.42 所示。

再次按下这个快捷键将反注释掉这部分代码，如图 2.43 所示。

图 2.42　注释代码块　　　　　　　　图 2.43　反注释代码块

2. Ctrl+Shift+F

按下这个快捷键将弹出全局搜索框，如图 2.44 所示。

这个快捷键在开发中经常使用，可以通过关键字快速搜索需要的信息，选中第一个复选框，代码的大小写敏感。单击右边的标签即可查看关键字的预览，如图 2.45 所示。

图 2.44　全局搜索框　　　　　　　　图 2.45　全局搜索预览框

3. Ctrl+Shift+ 加号 / 减号

若方法是收起的，同时按下 Ctrl+Shift++ 快捷键会将方法展开，如图 2.46 所示。

相反，若方法是展开的，同时按下 Ctrl+Shift+－快捷键则会收起方法，如图 2.47 所示。

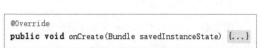

图 2.46　方法展开　　　　　　　　　图 2.47　方法收起

查看动态图,请扫描图 2.48 中的二维码。

图 2.48　方法收起展开二维码

2.2.4　其他组合快捷键

其他组合快捷键如表 2.4 所示。

表 2.4　其他组合快捷键

快　捷　键	说　　　明
Shift+Click	选中标签页关闭页面
Alt+Insert	生成构造方法获 getter、setter 方法等
双击 Shift	全局搜索类或文件
Shift+F6	重构 - 重命名
Alt+1	打开隐藏工程面板
Alt+ ↑ / ↓	方法间上、下快速定位移动
Shift+F2	高亮错误或快速定位错误位置
Alt+ 鼠标左键拖动	多行编辑
Ctrl+ 鼠标左键	进入到该方法或类

下面通过图文讲解这些快捷键的用法。

1．Alt+Insert

同时按下 Alt 和 Shift 键,弹出快速代码生成框,有 Constructor 构造方法、getter/setter 方法、toString（）方法等。

这里以生成构造方法为例,选择 Constructor 选项,如图 2.49 所示。

选中两个属性并单击 OK 按钮,如图 2.50 所示。

图 2.49　Android Studio 快速代码生成框

图 2.50　Android Studio 快速生成 Generate 方法

单击 OK 按钮后如图 2.51 所示。

可以看出，自动生成了包含两个属性的构造方法，很是方便快捷。生成 Getter/Setter 方法和生成 Generate 方法比较类似，同样选中这两个属性并按下快捷键，选中 Getter and Setter，如图 2.52 所示。

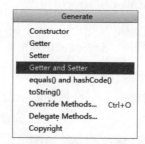

图 2.51　Android Studio 快速生成构造方法示意图　　图 2.52　Android Studio 快速生成 Getter/Setter 方法

按下 Enter 键，如图 2.53 所示。

单击 OK 按钮即生成这两个属性的 Getter 和 Setter 方法，如图 2.54 所示。

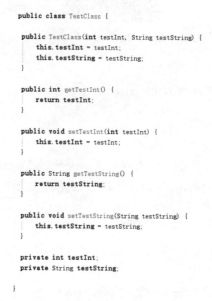

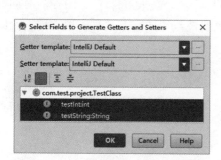

图 2.53　Android Studio 快速生成 Getter/Setter 方法选择　　图 2.54　Android Studio 快速生成 Getter/Setter 方法示意图

可以看出，Android Studio 为我们自动生成了 Generator 方法、Getter 和 Setter 方法，此快捷键在创建 JavaBean 时经常会用，可大大提高编码效率，减少编码错误。

查看动态图，请扫描图 2.55 中的二维码。

2. Alt+ 鼠标

按下 Alt 键并结合鼠标可以同时选中多行，如图 2.56 所示。

图 2.56 中一次选中了多行，此时可以进行多行编辑，如图 2.57 所示。

查看动态图，请扫描图 2.58 中的二维码。

图 2.55　Android Studio 快速生成 Getter/Setter 方法二维码

```
private static final String TAG = "MainActivity";
private static final int TEST = 1;
private static final int TEST1 = 1;
private static final int TEST2 = 1;
private static final int TEST3 = 1;
private static final int TEST4 = 1;
private static final int TEST5 = 1;
```

图 2.56　Android Studio 选中多行

```
public static final String TAG = "MainActivity";
public static final int TEST = 1;
public static final int TEST1 = 1;
public static final int TEST2 = 1;
public static final int TEST3 = 1;
public static final int TEST4 = 1;
public static final int TEST5 = 1;
```

图 2.57　Android Studio 编辑多行

3. Ctrl+ 鼠标左键

此快捷键可以查看鼠标选中的类或方法，扫描图 2.59 中的二维码，查看这个快捷键的使用方法。

图 2.58　Android Studio 编辑多行二维码

图 2.59　Android Studio Ctrl+ 鼠标左键二维码

2.3　Android Studio 调试

编写代码很多时候都是"差强人意"，很难一次获得想要的结果，出现错误的时候需要查找错误的原因，这种查找的过程称为"程序调试"。一般来讲程序员 10% 的时间写代码，90% 的时间都在调试，因此要认识到调试的重要性。调试的方式有多种，这里介绍最常用的两种：Logcat 调试方式和断点调试方式。

2.3.1　Logcat 调试

Logcat 调试方式很简单，在可能出现错误的地方将变量的值打印出来，方便分析总结错误原因。

这里编写一个简单的 Java 程序，代码如下：

```java
private int calculateMultiply(int i) {
    return i * i;
}

int[] testInts = new int[10];

public void test(View view) {
    int i = 0;
    while (i <10) {
        i++;
        int i2 = calculateMultiply(i);
        testInts[i-1] = i2;
        Log.d(TAG, "onCreate: " +testInts[i-1]);
    }
}
```

使用 while 循环不断计算 i*i 的值并通过 Log 打印出来，这时查看 Logcat 窗口中的

Log 信息如图 2.60 所示。

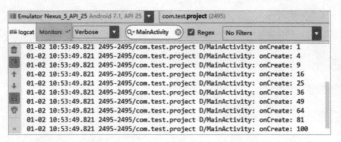

图 2.60　Logcat 窗口中的 Log 信息

这里通过 TAG（MainActivity）来过滤日志信息，可以看出所有的 i*i 都被打印出来了，根据打印的值即可初步判断是否发生错误。

2.3.2　断点调试

断点调试相对于 Logcat 调试要复杂一些，与 Logcat 显示运行后的结果相比，断点调试可以暂停程序的运行而获得运行中的结果。断点调试可以分成几个步骤，下面一一介绍。

1. 添加断点

在想要调试的代码行的左边栏单击即可添加一个断点，如图 2.61 所示。

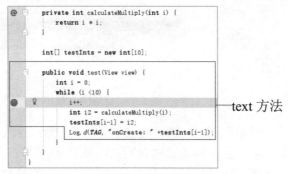

图 2.61　添加断点

从图 2.61 可以看出，在 i++ 处添加了一个断点，下面就可以单击工具栏中的调试按钮开始调试。

2. 开始调试

单击工具栏中的调试按钮，如图 2.62 所示。

此时进入调试准备阶段，查看底部的监视窗，如图 2.63 所示。

图 2.62　调试按钮　　　　　　图 2.63　调试监视窗一

图 2.61 中所示的 test 方法是按钮的单击事件监听，因此单击模拟器中的按钮触发进入调试阶段，如图 2.64 所示。

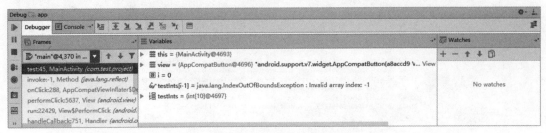

图 2.64　调试监视窗二

由图 2.64 可以看出，此时由 Console 切换到了 Debugger 标签，我们可以根据窗口内的工具栏按钮或快捷键来控制程序的运行。常用的调试方式有三种（Step Over、Step Into、Step Out），下面一一介绍。

Step Over，可以控制程序向下运行一步。有两种方式可以操作，其中一种是单击工具栏中的按钮，如图 2.65 所示。

当然也可以使用快捷键 F8 来控制，这时查看代码编辑区，如图 2.66 所示。

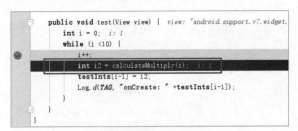

图 2.65　Step Over 按钮　　　　　图 2.66　使用快捷键 F8 代码调试下一步

从图 2.66 可以看出，此时代码已由断点处向下运行了一行（蓝色背景行）。

Step Into，这个操作可以进入到调试中遇到的方法体中。例如上面的操作中遇到了 calculateMultiply 方法，想要进入到这个方法体中就可以单击工具栏中的 Step Into 按钮，如图 2.67 所示。

同样也可以使用快捷键 F7 进行操作，再次查看代码编辑区，如图 2.68 所示。

图 2.67　Step Into 按钮　　　　　图 2.68　代码调试进入方法中

从图 2.68 可以看出，此时代码调试运行到方法 calculateMultiply 中了，在程序的右边显示出了这时方法的参数 i 的值。

若上面的步骤中进入了一个比较繁复的方法，而我们没有耐心一步步执行到最后，可以使用 Step Out 按钮跳出来，如图 2.69 所示。

同样可以使用 Shift+F8 快捷键进行操作，再次查看代码编辑区，如图 2.70 所示。

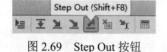

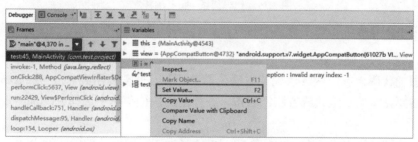

图 2.69　Step Out 按钮　　　　　　　　图 2.70　代码调试 - 跳出方法

这时又跳到了刚才进入的地方，这时再按 F8 快捷键又可以向下运行了。

2.3.3　高级调试

1. 变量值设置

对于有些 for 循环或 while 循环，一步步执行可能需要耗费很多时间，例如上面的 while 循环，我们想查看 i 为 9 时的值，若一步步执行就需要执行 9 遍，会比较烦，有没有比较好的方式呢？当然，我们可以设置变量的值，在监视面板中选择要改变数值的变量，右击，如图 2.71 所示。

图 2.71　代码调试 - 变量值设置一

在弹出的快捷菜单中选择 Set Value，输入数值，如图 2.72 所示。

按 Enter 键即可设置成功，这时再次到代码中查看，如图 2.73 所示。

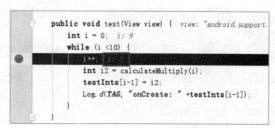

图 2.72　代码调试 - 变量值设置二　　　　图 2.73　代码调试 - 变量值设置三

可以看出，此时 i 变量的值已经变成了 9。

2. 断点跳转

一般来讲，一次调试过程可能涉及多个断点，这时就可能需要断点间跳转的功能，例如在图 2.74 中的程序中添加了两个断点。

假设运行到第一个断点查看变量值之后，想迅速跳转到第二个断点，这时就可以单击调试框中的 Run to Cursor 按钮快速跳转，如图 2.75 所示。

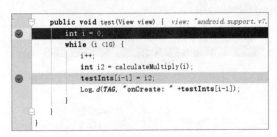

图 2.74　代码调试 - 断点跳转一

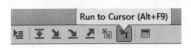

图 2.75　代码调试 - 断点跳转二

当然也可以通过快捷键 Alt+F9 来实现，如图 2.76 所示。

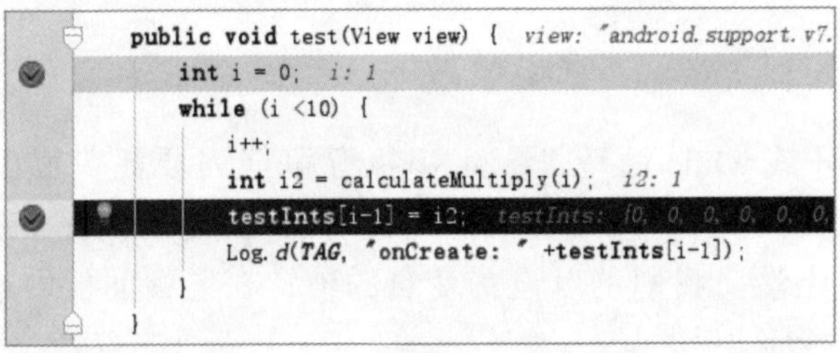

图 2.76　代码调试 - 断点跳转三

可以看出，直接跳转到了第二个断点。

3. 表达式 / 方法值计算

若调试的代码中有一些表达式或方法值需要计算，就需要用到 Evaluate Expression 功能。选中需要计算的表达式或方法，右击，在弹出的快捷菜单中选择 Evaluate Expression，如图 2.77 所示。

在弹出的对话框中单击 Evaluate 按钮即可计算表达式的值，如图 2.78 所示。

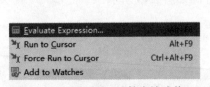

图 2.77　代码调试 - 计算表达式值一

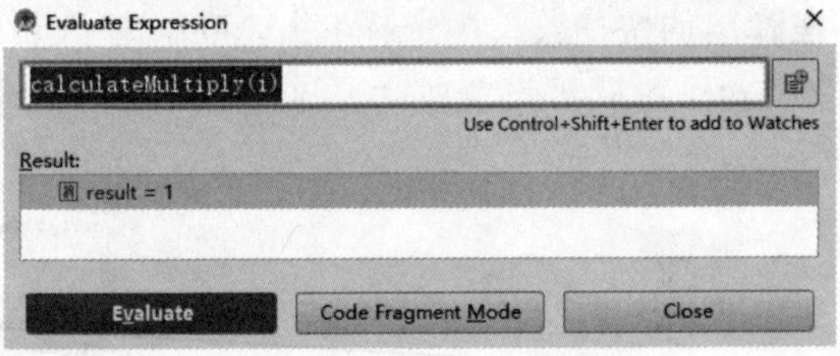

图 2.78　代码调试 - 计算表达式值二

在 Result 栏中显示出计算值。

4. 查看所有断点

单击调试监视框左边栏的 View Breakpoints 即可查看所有断点，如图 2.79 所示。

当然也可以使用快捷键 Ctrl+Shift+F8 来打开。此外，还有一个可以查看所有断点的入口，选择 Run → View Breakpoints，打开查看界面，如图 2.80 所示。

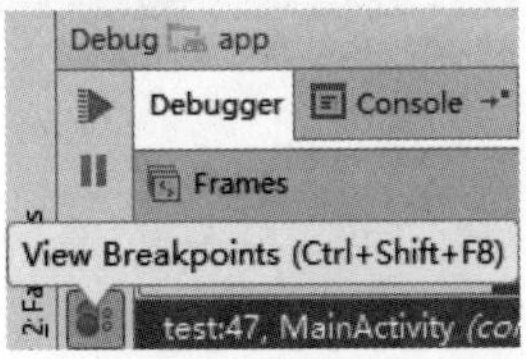

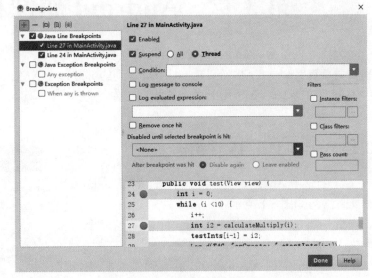

图 2.79　代码调试 - 查看所有断点

图 2.80　代码调试 - 查看所有断点界面

这时在左边栏中就可以看到所有的断点，在右下角还可以看到断点在代码中的位置，单击左上角的"＋"和"－"可以添加或删除断点。

5. 停止调试

不想继续调试时，单击左边栏的 Stop 按钮停止调试，如图 2.81 所示。

停止后的监视窗如图 2.82 所示。

图 2.81　代码调试 - 停止调试

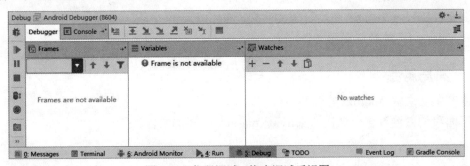

图 2.82　代码调试 - 停止调试后视图

第 3 章　Android 属性和布局

大致可以认为，Android APP 由两部分组成：属性、布局部分和逻辑代码部分。属性和布局负责 Android APP 的 UI，即用户看到的部分，由 XML 语言编写；逻辑代码部分则由 Java 语言编写，负责 APP 的逻辑控制工作。

3.1　Android 项目文件结构

新建一个 Android 项目，查看左侧的 Android 项目文件结构如图 3.1 所示。

最外层的根目录为 app，app 目录中有三个子文件夹。

manifests 文件夹：Android 系统配置文件夹，包含一个 AndroidManifest.xml 文件。

java 文件夹：存放 Java 代码的文件夹。新建项目时默认生成了三个文件夹，com.first.project 文件夹用来存放 Java 文件，这里包含一个名为 MainActivity 的 Java 文件，是新建项目时默认生成的。第二个和第三个文件夹为测试代码文件夹，不是十分常用。

res 文件夹：存放 Android 项目的资源文件。它包含四个文件夹：drawable（图片资源文件夹）、layout（布局资源文件夹）、mipmap（图片资源文件夹，存放项目图标）、values（存放数值资源文件）。

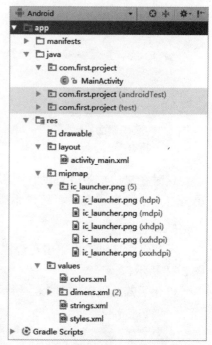

图 3.1　Android 项目文件结构

3.1.1　布局属性

本节主要讲解 Android 的属性和布局知识。

下面我们着重看这个默认的 activity_ma-in.xml 布局文件：

```
<?xml version="1.0" encoding="utf-8"?>
<RelativeLayout xmlns:android="http://schemas.android.com/apk/res/android"
    xmlns:tools="http://schemas.android.com/tools"
    android:layout_width="match_parent"
    android:layout_height="match_parent"
    android:paddingBottom="@dimen/activity_vertical_margin"
    android:paddingLeft="@dimen/activity_horizontal_margin"
    android:paddingRight="@dimen/activity_horizontal_margin"
    android:paddingTop="@dimen/activity_vertical_margin"
    tools:context="com.first.project.MainActivity">

    <TextView
```

```
            android:layout_width="wrap_content"
            android:layout_height="wrap_content"
            android:text="Hello World!" />
</RelativeLayout>
```

默认的布局文件采用 RelativeLayout 相对布局，在相对布局中仅添加了一个 TextView 文本控件，布局文件中默认生成的一些属性不太常用，可以手动去除。

去除后代码如下：

```
<?xml version="1.0" encoding="utf-8"?>
<RelativeLayout xmlns:android="http://schemas.android.com/apk/res/android"
    android:layout_width="match_parent"
    android:layout_height="match_parent">

    <TextView
        android:layout_width="wrap_content"
        android:layout_height="wrap_content"
        android:text="Hello World!" />
</RelativeLayout>
```

最上方的 <?xml version="1.0" encoding="utf-8"?> 是 xml 文件的 title，里面的 version 表示 xml 版本号；encoding 表示文本编码类型，这里默认设置为 utf-8。

RelativeLayout 是父布局的标签，表示相对布局。对于相对布局后面的章节还会详细介绍；xmlns 属性的全称是 xmlnamespace，添加了这个属性才可以使用 Android 的属性（即 Android: 开头的属性）；layout_width 表示父布局的宽属性，属性值为 match_parent 表示占据整个界面的宽；layout_height 表示父布局的高属性，属性值为 match_parent 表示占据整个界面的高。

TextView 是文本控件，同样也设置了两个属性 layout_width 和 layout_height，即宽和高。注意：这两个属性是所有布局控件的必备属性，这里设置了这两个属性的值都为 wrap_content，即包裹文本内容；text 属性表示文本的值，这里设置了 Hello World，这时 TextView 中会显示 Hello World。Android Studio 提供了即时布局渲染显示的功能，即在 Android Studio 的最右边，如图 3.2 所示。

图 3.2 Android Studio 预览窗口

单击最右侧的 Preview 按钮即可弹出这个布局如预览窗口，可以看出；在这个预览窗口有一个手机界面，这个手机界面的左上方有一个 TextView 文本控件，文本控件会显示我们设置的 Hello World 文本，但是这个文本字体太小了，可以添加属性 textSize 设置 TextView 控件中文本的字体大小，代码如下：

```
<?xml version="1.0" encoding="utf-8"?>
<RelativeLayout xmlns:android="http://schemas.android.com/apk/res/android"
```

```
        android:layout_width="match_parent"
        android:layout_height="match_parent">

    <TextView
        android:layout_width="wrap_content"
        android:layout_height="wrap_content"
        android:text="Hello World!"
        android:textSize="40sp" />
</RelativeLayout>
```

上述代码为这个 TextView 添加了 textSize 属性并设置其值为 40sp，添加后再次查看预览窗口如图 3.3 所示。

此时字体就变得很大了，当然除了这些属性之外，还有较多的属性供开发者选用，常用的 TextView 属性会在后面的 TextView 章节详细介绍。

3.1.2 配置属性

AndroidManiifest.xml 文件是每个 Android 项目必须要包含的文件（项目唯一），创建项目时默认就会生成这个文件，它配置了 Android 运行的基本属性，具有很重要的作用。灵活配置文件中的属性可以处理复杂的页面逻辑操作、简化代码复杂度、提高灵活度等。

图 3.3　Android Studio 预览窗口查看

下面看创建项目时默认生成的配置文件：

```
<?xml version="1.0" encoding="utf-8"?>
<manifest xmlns:android="http://schemas.android.com/apk/res/android"
    package="com.first.project">

    <application
        android:allowBackup="true"
        android:icon="@mipmap/ic_launcher"
        android:label="@string/app_name"
        android:supportsRtl="true"
        android:theme="@style/AppTheme">
        <activity android:name=".MainActivity">
            <intent-filter>
                <action android:name="android.intent.action.MAIN" />

                <category android:name="android.intent.category.
                LAUNCHER" />
            </intent-filter>
        </activity>
    </application>
</manifest>
```

此文件也是用 XML 语言编写的，因此在这个文件的最上方也添加了一个 xml 标签并设置了其 version 和 encoding 属性，这里还添加了几个标签，含义如下：

manifest 标签：整个文件的父标签。这个标签中添加了两个属性：xmlns 是命名空间属性，同样需设置这个属性的属性值为 http://schemas.android.com/apk/res/android 才可以使用"android:"为首的属性；package 属性表示包名，这个属性的属性值为新建项目设置的。

Application 标签：表示整个 Android 应用，在项目中是唯一的。在这个标签中添加了几个属性：allowBackup 属性设置为 true 表示允许备份应用的数据；icon 属性设置了这个 APP 在桌面上显示的 icon 图标；label 属性设置 APP 在桌面上显示的名称，可以查看图 3.4。

这两个属性是可以自定义的，例如修改这两个属性如下：

```
<application
    android:allowBackup="true"
    android:icon="@drawable/QQ"
    android:label="QQ"
    android:supportsRtl="true"
    android:theme="@style/AppTheme">
```

这时再次运行项目观察桌面图标和名称，如图 3.5 所示。可以看出整个项目的 icon 和项目名称都改变成了自定义的结果；supportsRtl 属性设置为 true 表示支持从右向左布局（阿拉伯语言会用到），这个属性不太常用；theme 属性为主题，整个项目的主题。

图 3.4　Android Studio Preview 查看

图 3.5　Android 替换 icon 图标

activity 标签：一个 Activity 可以理解成一个 APP 的界面，这里只添加了一个 name 属性，设置属性值为 .MainActivity，这里的这个"."表示当前 Activity 的包名，即 name 属性值为"包 .Activity 名"。注意，项目中的每一个 Activity 都需要在这个布局文件中添加配置。

intent-filter 标签：顾名思义，此标签可以添加过滤，一般在隐式启动时用来过滤和匹配 Activity，隐式启动在后面的章节会介绍。

action 标签：添加这个标签并设置其 name 属性为 android.intent.action.MAIN，表示这个 Activity 为项目的主 Activity，项目启动时会首先启动这个 Activity。

category 标签：添加这个标签并设置其 name 属性为 android.intent.category.LAUNCHER，项目将在程序列表中显示，即图 3.5 显示在桌面上的图标，若去除这个标签该项目将不会出现在程序列表中。

3.1.3 其他文件

1. 颜色资源文件

Android 的颜色属性也是通过 xml 资源来管理的，默认生成的文件（colors.xml）如下：

```xml
<?xml version="1.0" encoding="utf-8"?>
<resources>
    <color name="colorPrimary">#3F51B5</color>
    <color name="colorPrimaryDark">#303F9F</color>
    <color name="colorAccent">#FF4081</color>
</resources>
```

同样，最上面一行代码为 xml 文件的 title，color 文件的父标签是 resource 资源标签，每一个 color 标签中添加了一个 name 属性，通过这个 name 属性即可以找到这个颜色值（@color/name 值），name 值在项目中应该是唯一的。这个 color 标签之间的十六进制数值为颜色值，这种数值不便于理解和记忆。Android Studio 还提供了颜色选择器供开发者调用，单击编辑器左边栏的小颜色色块即可成功打开颜色编辑器，颜色色块如图 3.6 所示。

颜色编辑器界面如图 3.7 所示。

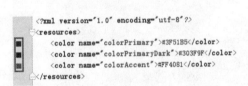

图 3.6　Android Studio 颜色色块

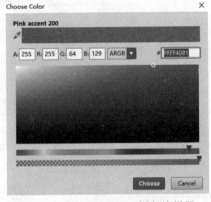

图 3.7　Android Studio 颜色编辑器

既可以在如图 3.7 所示界面右上角直接输入颜色色值，也可以分别输入 A、R、G、B 的值合成颜色数值，同时拖动下面的 Due 和 Opacity 来调整色调和透明度。输入或调整完毕后，单击 Choose 按钮即可设置成功，设置后观察左边的小色块，如图 3.8 所示。

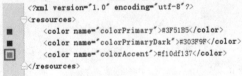

图 3.8　Android Studio 色块

可以看出，颜色值和小色块都对应改变了。

2. 尺寸资源文件

这个资源文件（dimens.xml）定义了布局文件中的各种尺寸的大小，默认生成的代码如下：

```xml
<resources>
    <!-- Default screen margins, per the Android Design guidelines. -->
    <dimen name="activity_horizontal_margin">16dp</dimen>
    <dimen name="activity_vertical_margin">16dp</dimen>
</resources>
```

和颜色资源文件相似，尺寸资源 dimen 标签同样也是由 resource 资源标签包裹，同样 dimen 尺寸标签中定义了 name 属性，这个 name 属性也是唯一的，两个 dimen 标签内部的数值即为这个尺寸的大小，考虑到 Android 手机屏幕分辨率的多样性，这里采用了 dp 作为单位。字符串大小一般采用 sp 作为单位。

3. 字符资源文件

字符资源文件（strings.xml）代码如下：

```
<resources>
    <string name="app_name">FirstProject</string>
</resources>
```

resource 标签中添加了一个 <string> 标签，同样为了找到这个字符串也为其添加了 name 属性并设置了唯一的属性值，其属性值为 string 型。

4. 样式资源文件（styles.xml）

样式资源文件可以方便地定义 Android 项目的样式和外观，默认生成的代码如下：

```
<resources>
    <!-- Base application theme. -->
    <style name="AppTheme" parent="Theme.AppCompat.Light.DarkActionBar">
        <!-- Customize your theme here. -->
        <item name="colorPrimary">@color/colorPrimary</item>
        <item name="colorPrimaryDark">@color/colorPrimaryDark</item>
        <item name="colorAccent">@color/colorAccent</item>
    </style>
</resources>
```

同样 resource 标签作为父标签，<style> 标签中添加了一个 name 属性方便引用，parent 属性表示继承关系。在这个标签中添加了多个 <item> 标签，这些标签共同构成整个样式表。

3.2 Android 布局属性值

padding 和 margin 属性在开发中十分常用，padding 意为"填充"，一般用来在控件内部填充布局，而 margin 意为"边缘"，一般指的是控件外部距父控件的距离，可以结合下面的图片来理解，如图 3.9 所示。

图中序号如表 3.1 所示。

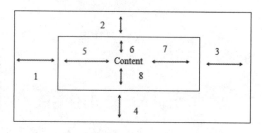

图 3.9　Android 布局示意图

表 3.1　Android 布局示意图含义表

图中序号	属　　性	说　　明
1	layout_marginLeft	左外边距
2	layout_marginTop	外顶部边距
3	layout_marginRight	右外边距
4	layout_marginBottom	外底部边距

续表

图中序号	属　　性	说　　明
5	paddingLeft	内左边距
6	paddingTop	内顶部边距
7	paddingRight	内右边距
8	paddBottom	内底部边距

3.2.1　Android padding 属性用法

下面通过一个实例来看一下这些属性的用法，首先看一下 padding 属性的用法：

```xml
<?xml version="1.0" encoding="utf-8"?>
<LinearLayout xmlns:android="http://schemas.android.com/apk/res/android"
    android:layout_width="match_parent"
    android:layout_height="match_parent"
    android:orientation="vertical">

    <TextView
        android:layout_width="wrap_content"
        android:layout_height="wrap_content"
        android:background="#96e25f"
        android:paddingBottom="80dp"
        android:paddingLeft="20dp"
        android:paddingRight="60dp"
        android:paddingTop="40dp"
        android:text="Hello World!" />
</LinearLayout>
```

上述代码为 TextView 控件添加了四个相关的 padding 属性，并设置了不同的属性值，为了方便观察还为用来演示的 TextView 控件添加了背景色（设置了 background 属性）。查看 Android Studio 的预览窗口即可实时查看效果图，如图 3.10 所示。

可以看出，和设置属性值一致，左上右下四个方向的 padding 值依次变大。

3.2.2　Android margin 属性用法

下面看一下 margin 属性的用法：

图 3.10　Android padding 属性示意图

```xml
<?xml version="1.0" encoding="utf-8"?>
<LinearLayout xmlns:android="http://schemas.android.com/apk/res/android"
    android:layout_width="match_parent"
    android:layout_height="match_parent"
    android:orientation="vertical">

    <TextView
        android:layout_width="wrap_content"
```

```
        android:layout_height="wrap_content"
        android:layout_marginLeft="10dp"
        android:layout_marginTop="30dp"
        android:background="#96e25f"
        android:paddingBottom="80dp"
        android:paddingLeft="20dp"
        android:paddingRight="60dp"
        android:paddingTop="40dp"
        android:text="Hello World!" />
</LinearLayout>
```

由于 LinearLayout 中控件默认在左上角显示，因此这里添加了两个 margin 属性，分别是 layout_marginLeft（距左边界的距离）和 layout_marginTop（距上边界的距离），效果如图 3.11 所示。

修改代码如下：

```
<?xml version="1.0" encoding="utf-8"?>
<LinearLayout xmlns:android="http://schemas.android.com/apk/res/android"
    android:layout_width="match_parent"
    android:layout_height="match_parent"
    android:gravity="bottom|right"
    android:orientation="vertical">

    <TextView
        android:layout_width="wrap_content"
        android:layout_height="wrap_content"
        android:layout_marginBottom="30dp"
        android:layout_marginRight="10dp"
        android:background="#96e25f"
        android:paddingBottom="80dp"
        android:paddingLeft="20dp"
        android:paddingRight="60dp"
        android:paddingTop="40dp"
        android:text="Hello World!" />
</LinearLayout>
```

为了查看 layout_marginBottom（距离底部边界）和 layout_marginRight（距离右部边界）的效果，这里为 LinearLayout 添加了 gravity 属性并设置其值为 bottom|right（控件将置于右下角），再次查看预览窗口，如图 3.12 所示。

可以看出，TextView 位于右下角，距离其父布局边界底部边界 30dp，距离父布局右边边界 10dp。

当然除了上面指定具体"上下左右"边界的值，还提供了 padding 和 layout_margin 属性，这时"上下左右"都是相同的值了，下面通过一个实例看这两个属性的效果：

```
<?xml version="1.0" encoding="utf-8"?>
<LinearLayout xmlns:android="http://schemas.android.com/apk/res/android"
    android:layout_width="match_parent"
    android:layout_height="match_parent"
```

```
    android:orientation="vertical">

    <TextView
        android:layout_width="wrap_content"
        android:layout_height="wrap_content"
        android:layout_margin="40dp"
        android:background="#96e25f"
        android:padding="40dp"
        android:text="TextView1" />
</LinearLayout>
```

图 3.11　Android Margin 属性示意图一　　　图 3.12　Android Margin 属性示意图二

上述代码为 TextView 添加了 padding 属性为 40dp，这时候 TextView 的"上下左右"内间距相同，都为 40dp；为 TextView 添加了 layout_margin 属性并设置了其值为 40dp，这时距左边距 40dp，距上边距 40dp，效果如图 3.13 所示。

可以看出，此时 TextView 控件距离模拟器的上边界和左边界的距离都相同了，且 TextView 里的文字位于 TextView 正中。

再次修改下代码为：

```
<?xml version="1.0" encoding="utf-8"?>
<LinearLayout xmlns:android="http://schemas.android.com/apk/res/android"
    android:layout_width="match_parent"
    android:layout_height="match_parent"
    android:gravity="bottom|right"
    android:orientation="vertical">

    <TextView
        android:layout_width="wrap_content"
        android:layout_height="wrap_content"
        android:layout_margin="40dp"
        android:padding="40dp"
        android:text="TextView1" />
</LinearLayout>
```

上述代码为 LinearLayout 添加了 gravity 属性其值为 bottom|right（右下），这时显示效果如图 3.14 所示。

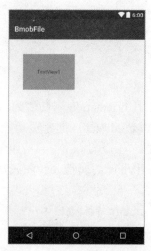 

图 3.13　Android Margin 属性示意图三　　　图 3.14　Android Margin 属性示意图四

可以看出，此时 TextView 距离模拟器底边界和右边界距离相同。

3.3　Android 布局之线性布局——LinearLayout

第一节介绍了新建 Android 项目时默认生成的布局文件，默认的布局文件采用的相对布局 RelativeLayout，在这个布局中也默认添加了一个 TextView 控件。本节讲解的 LinearLayout 线性布局和 RelativeLayout 相对布局类似，同属于布局容器，也可以包裹[1]普通 UI 控件（TextView、Button 等）。

线性布局是按照水平或垂直的方式将布局元素（控件或布局）按照顺序依次排列，后面的元素将位于前面元素的下方或右方。至于是下方或右方，和布局设置的方向有关，可以通过设置属性 orientation 的值为 horizontal（水平）或 vertical（垂直）来实现。

3.3.1　LinearLayout 基础用法

下面通过一个实例来看 orientation 属性值为 vertical 时的效果：

```
<?xml version="1.0" encoding="utf-8"?>
<LinearLayout xmlns:android="http://schemas.android.com/apk/res/android"
    android:layout_width="match_parent"
    android:layout_height="match_parent"
    android:orientation="vertical">

    <TextView
        android:layout_width="match_parent"
        android:layout_height="wrap_content"
        android:text="Hello World!"
        android:textSize="30sp" />

    <TextView
```

---

[1]　"包裹"是指一部代码被另一部分代码包裹起来，形成一个整体，实现某些功能。

```
        android:layout_width="match_parent"
        android:layout_height="wrap_content"
        android:text="Hello World!"
        android:textSize="30sp" />
</LinearLayout>
```

上述代码添加了一个 LinearLayout 布局容器并设置其 orientation 属性值为 vertical，按照上面的解释，所有的子控件都应该垂直排列，通过 Android Studio 的预览功能来看效果，如图 3.15 所示。

可以看出，两个 TextView 控件垂直排列，通过 TextView 的属性 text 的属性值可以看出，前一个 TextView 在后一个 TextView 的上方。

下面修改这个布局文件的 orientation 属性值为 horizontal，代码如下：

```
<?xml version="1.0" encoding="utf-8"?>
<LinearLayout xmlns:android="http://schemas.android.com/apk/res/android"
    android:layout_width="match_parent"
    android:layout_height="match_parent"
    android:orientation="horizontal">

    <TextView
        android:layout_width="wrap_content"
        android:layout_height="wrap_content"
        android:text="Hello01!"
        android:textSize="30sp" />

    <TextView
        android:layout_width="wrap_content"
        android:layout_height="wrap_content"
        android:text="Hello02!"
        android:textSize="30sp" />
</LinearLayout>
```

通过代码可以看出，将 orientation 属性值设置成了 horizontal 水平布局，由于是水平布局，因此再设置 TextView 的宽为 match_parent（占据整个宽）就不合适了，因为这样会造成控件重叠，因此同时修改两个 TextView 的 layout_width 为 wrap_content（包裹内容），再次查看预览如图 3.16 所示。

图 3.15　Android 布局 LinearLayout 之 vertical　　图 3.16　Android 布局 LinearLayout 之 horizontal

可以看出，两个 TextView 水平显示，下面的 TextView 在上面 TextView 的右边。注意 orientation 属性为 Linearlayout 的必设属性。

3.3.2 LinearLayout 嵌套

除了单个 LinearLayout 的使用，也支持 LinearLayout 的嵌套，下面通过一个实例来看如何通过 LinearLayout 的嵌套来实现模拟微信的底部 Tab：

```xml
<?xml version="1.0" encoding="utf-8"?>
<LinearLayout xmlns:android="http://schemas.android.com/apk/res/android"
    android:layout_width="match_parent"
    android:layout_height="70dp"
    android:background="#ffffffff"
    android:orientation="horizontal">

    <LinearLayout
        android:id="@+id/ll_chat"
        android:layout_width="0dp"
        android:layout_height="fill_parent"
        android:layout_weight="1"
        android:gravity="center"
        android:orientation="vertical">

        <ImageView
            android:id="@+id/img_chat"
            android:layout_width="40dp"
            android:layout_height="40dp"
            android:background="#0000"
            android:src="@drawable/chat_yes" />

        <TextView
            android:layout_width="wrap_content"
            android:layout_height="wrap_content"
            android:text=" 微信 "
            android:textColor="#b6b3b3" />
    </LinearLayout>

    <LinearLayout
        android:id="@+id/ll_frd"
        android:layout_width="0dp"
        android:layout_height="fill_parent"
        android:layout_weight="1"
        android:gravity="center"
        android:orientation="vertical">

        <ImageView
            android:id="@+id/img_frd"
            android:layout_width="40dp"
            android:layout_height="40dp"
            android:background="#0000"
            android:src="@drawable/frd_no" />
```

```xml
        <TextView
            android:layout_width="wrap_content"
            android:layout_height="wrap_content"
            android:text="通讯录"
            android:textColor="#b6b3b3" />
    </LinearLayout>

    <LinearLayout
        android:id="@+id/ll_find"
        android:layout_width="0dp"
        android:layout_height="fill_parent"
        android:layout_weight="1"
        android:gravity="center"
        android:orientation="vertical">

        <ImageView
            android:id="@+id/img_find"
            android:layout_width="40dp"
            android:layout_height="40dp"
            android:background="#0000"
            android:src="@drawable/find_no" />

        <TextView
            android:layout_width="wrap_content"
            android:layout_height="wrap_content"
            android:text="发现"
            android:textColor="#b6b3b3" />
    </LinearLayout>

    <LinearLayout
        android:id="@+id/ll_me"
        android:layout_width="0dp"
        android:layout_height="fill_parent"
        android:layout_weight="1"
        android:gravity="center"
        android:orientation="vertical">

        <ImageView
            android:id="@+id/img_me"
            android:layout_width="40dp"
            android:layout_height="40dp"
            android:background="#0000"
            android:src="@drawable/me_no" />

        <TextView
            android:layout_width="wrap_content"
            android:layout_height="wrap_content"
            android:text="我"
            android:textColor="#b6b3b3" />
    </LinearLayout>
</LinearLayout>
```

通过代码可以看出，一个 LinearLayout 内部包装了四个子 LinearLayout，外部的 LinearLayout 采用水平布局（设置了 orientation 属性为 horizontal），四个子 LinearLayout 采用了垂直布局（每一个子 LinearLayout 的属性 orientation 都为 vertical）。每一个 LinearLayout 内都添加了一个 ImageView（图片控件）和一个 TextView（文本控件）。

除了上面讲过的属性以外，还有一些属性没有介绍，通过表 3.2 所示来说明它们的用法。

表 3.2　LinearLayout 属性

属 性 名	说　　明
background	添加背景
gravity	设置子组件在父组件中的位置
weight	权重属性
src	为 ImageView 添加引用
textColor	为 TextView 文本设置颜色

上述代码为父 LinearLayout 设置了 background 属性，其值为 #ffffffff（纯白色）；为子 LinearLayout 设置了 gravity 属性并设置其值为 center，子 LinearLayout 将居中显示。设置 weight 属性可以平分整个屏幕的宽，这个属性的用法在后面的章节会详细介绍；为 ImageView 添加了 src 属性来设置图片背景，这个值的形式为"@drawable/图片名"；为 TextView 添加了 textColor 属性来设置字体颜色，其值为 #b6b3b3（浅灰色）。

可以通过 Android Studio 的预览功能查看效果如图 3.17 所示。

可以看出，父 LinearLayout 的背景为纯白色，每个子 LinearLayout 也都是居中显示并四等分整个屏幕的宽，每个 LinearLayout 上方的 ImageView 控件都添加了不同的图片，下方的 TextView 都设置了不同的文本和相同的文本颜色（浅灰色）。

图 3.17　Android 布局 LinearLayout 实例

3.4　Android 线性布局的重要属性

3.3 节的实例中用到了两个属性 gravity 和 layout_weight，这两个属性在 Android 开发中会经常用到，用法也比较复杂，下面讲解这两个属性的用法。

3.4.1　gravity 属性

Android 中的 gravity 属性有两种形式：layout_gravity 和 gravity，这两种有什么区别呢？从字面意思上就可以大概理解，第一个 layout_gravity 控制控件在父布局中的位置（和 margin 比较类似），gravity 可以控制控件中内容的显示位置（和 padding 比较类似）。下面还会通过实例来比较这两个属性的效果。除了上面用到的属性值 center 之外，还提供了如

表 3.3 所示中常用属性值供开发者调用（一次设置多个属性值用"|"隔开）：

表 3.3 gravity 属性

属 性 值	说　　明
right	置于右侧（线性布局设置垂直方式有效）
left	置于左侧（线性布局设置垂直方式有效）
top	置于顶部（线性布局设置水平方式有效）
bottom	置于底部（线性布局设置水平方式有效）
center_vertical	垂直方向居中显示（线性布局设置水平方式有效）
center_horizontal	水平方向居中显示（线性布局设置垂直方式有效）
center	水平和垂直方向居中显示

首先我们看一下 layout_gravity 的用法（activity_main.xml）：

```xml
<?xml version="1.0" encoding="utf-8"?>
<LinearLayout xmlns:android="http://schemas.android.com/apk/res/android"
    android:layout_width="match_parent"
    android:layout_height="match_parent"
    android:orientation="vertical">

    <TextView
        android:layout_width="wrap_content"
        android:layout_height="wrap_content"
        android:layout_gravity="center_horizontal"
        android:text="center_horizontal"
        android:textSize="30dp" />

    <TextView
        android:layout_width="wrap_content"
        android:layout_height="wrap_content"
        android:layout_gravity="right"
        android:text="right"
        android:textSize="30dp" />

    <TextView
        android:layout_width="wrap_content"
        android:layout_height="wrap_content"
        android:layout_gravity="left"
        android:text="left"
        android:textSize="30dp" />

</LinearLayout>
```

上述代码设置了 Linearlayout 的 orientation 属性值为 vertical（垂直布局），添加了三个 TextView 控件，并分别为这三个 TextView 添加了 layout_gravity 属性，其值分别为 center_horizontal（水平居中）、right（居右）和 left（居左），这时看一下预览窗口中的显示如图 3.18 所示。

下面修改 orientation 属性值为 horizontal，然后看一下另外几个属性值的用法，修改 activity_main.xml 如下：

```xml
<?xml version="1.0" encoding="utf-8"?>
<LinearLayout xmlns:android="http://schemas.android.com/apk/res/android"
    android:layout_width="match_parent"
    android:layout_height="match_parent"
    android:orientation="horizontal">

    <TextView
        android:layout_width="wrap_content"
        android:layout_height="wrap_content"
        android:layout_gravity="center_vertical"
        android:text="center_vertical"
        android:textSize="30dp" />

    <TextView
        android:layout_width="wrap_content"
        android:layout_height="wrap_content"
        android:layout_gravity="bottom"
        android:text="bottom"
        android:textSize="30dp" />

    <TextView
        android:layout_width="wrap_content"
        android:layout_height="wrap_content"
        android:layout_gravity="top"
        android:text="top"
        android:textSize="30dp" />

</LinearLayout>
```

这里同样添加了三个 TextView 并分别设置了其 layout_gravity 属性值为 center_vertical（垂直居中）、bottom（底部）和 top（顶部），这时查看效果如图 3.19 所示。

图 3.18　layout_gravity 属性示意图一　　　　图 3.19　layout_gravity 属性示意图二

可以看出，这时 center_vertical 将垂直居中，top 将位于界面的顶部，bottom 将位于界面的底部。

下面来看 gravity 属性的用法，代码如下：

```xml
<?xml version="1.0" encoding="utf-8"?>
<LinearLayout xmlns:android="http://schemas.android.com/apk/res/android"
    android:layout_width="match_parent"
    android:layout_height="match_parent"
    android:orientation="vertical">

    <TextView
        android:layout_width="150dp"
        android:layout_height="150dp"
        android:layout_gravity="center_horizontal"
        android:background="#41e67b"
        android:gravity="center_horizontal|center_vertical"
        android:text="center"
        android:textSize="30dp" />

    <TextView
        android:layout_width="120dp"
        android:layout_height="120dp"
        android:layout_gravity="left"
        android:background="#355fa1"
        android:gravity="right|bottom"
        android:text="right|bottom"
        android:textSize="30dp" />

    <TextView
        android:layout_width="150dp"
        android:layout_height="150dp"
        android:layout_gravity="right"
        android:background="#afb639"
        android:gravity="top|right"
        android:text="top|right"
        android:textSize="30dp" />

</LinearLayout>
```

为了演示方便，将各个 TextView 的宽和高都设置得足够大并为每个 TextView 都添加了 background 属性。第一个 TextView 添加了两个 gravity 属性值，中间用"|"符号隔开，这两个属性值（center_horizontal 和 center_vertical）和一个 center 是一样的效果；第二个 TextView 为 gravity 添加了两个属性值 right|bottom 即右下角；第三个 TextView 设置 gravity 属性值为 top|right 即右上角。查看右侧的预览窗口，如图 3.20 所示。

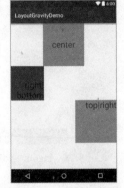

图 3.20 gravity 属性示意图

3.4.2 layout_weight 属性

layout_weight 在分配屏幕的宽高上有很大的用处，它的用法很灵活，结合不同的宽高值和 weight 值可以实现不同的效果和要求。

1. layout_width="match_parent"

首先看一下设置宽为 match_parent 时，layout_weight 不同值时的效果，代码如下：

```xml
<?xml version="1.0" encoding="utf-8"?>
<LinearLayout xmlns:android="http://schemas.android.com/apk/res/android"
    android:layout_width="match_parent"
    android:layout_height="match_parent"
    android:orientation="horizontal">

    <TextView
        android:layout_width="match_parent"
        android:layout_height="wrap_content"
        android:layout_weight="5"
        android:background="#cc5858"
        android:text="layout_weight=5"
        android:textSize="20dp" />

    <TextView
        android:layout_width="match_parent"
        android:layout_height="wrap_content"
        android:layout_weight="1"
        android:background="#5baf54"
        android:gravity="center_horizontal"
        android:paddingBottom="15dp"
        android:text="layout_weight=1"
        android:textSize="20dp" />
</LinearLayout>
```

上述代码中 LinearLayout 的 orientation 设置成了 horizontal 水平布局，在 LinearLayout 中添加了两个 TextView 并设置其宽的属性为 match_parent（若此时不添加 layout_weight 属性，则第一个 TextView 将会覆盖第二个 TextView）。为第一个 TextView 设置了 layout_weight 属性值为 5，为第二个 TextView 设置了 layout_weight 属性值为 1，这两个值具体有什么作用可以查看预览窗口，如图 3.21 所示。

为了演示方便，这里为 TextView 添加了 background 属性，可以看出 layout_weight 为 5 时反而宽度很小，layout_weight 为 1 时宽度很大，两个 TextView 的比例基本上是 1:5。

修改第一个 TextView 的 layout_weight 值为 10，再次查看如图 3.22 所示。第一个 TextView 的宽被压缩得更小了，当第一个 TextView 的 layout_weight 为 100 时，如图 3.23 所示。

图 3.21　layout_weight 属性示意图一　　图 3.22　layout_weight 属性示意图二　　图 3.23　layout_weight 属性示意图三

可以看出第一个 TextView 基本被压缩地隐藏了。

2. layout_width="wrap_content"

修改 TextView 的宽为 wrap_content 时再次看一下效果，修改代码如下：

```xml
<?xml version="1.0" encoding="utf-8"?>
<LinearLayout xmlns:android="http://schemas.android.com/apk/res/android"
    android:layout_width="match_parent"
    android:layout_height="match_parent"
    android:orientation="horizontal">

    <TextView
        android:layout_width="wrap_content"
        android:layout_height="wrap_content"
        android:layout_weight="5"
        android:background="#cc5858"
        android:text="layout_weight=5"
        android:textSize="20dp" />

    <TextView
        android:layout_width="wrap_content"
        android:layout_height="wrap_content"
        android:layout_weight="1"
        android:background="#5baf54"
        android:gravity="center_horizontal"
        android:paddingBottom="15dp"
        android:text="layout_weight=1"
        android:textSize="20dp" />
</LinearLayout>
```

再次查看预览窗口，如图 3.24 所示。可以看出，和设置 match_parent 相反，设置为 wrap_content 时，layout_weight 的值越大占据的宽越大，但是并没有按照 5:1 显示。再次修改第一个 TextView 的 layout_weight 属性值为 10，预览图片如图 3.25 所示。第一个 TextView 的宽仅仅增加了一点，第二个 TextView 仍然是一行包裹显示。也就是说不管第一个 TextView 的 layout_weight 值有多大，第二个 TextView 都会包裹内容，不会被压缩到消失。

3. layout_width="0dp"

设置 layout_width 为 0dp 时才是正确的 layout_weight 属性使用方法，因为 SDK 中对 layout_weight 的使用方法有如下解释：

```
In order to improve the layout efficiency when you specify the weight,
you should change the width of the EditText to be zero (0dp). Setting the
width to zero improves layout performance because using "wrap_content"as
the width requires the system to calculate a width that is ultimately
irrelevant because the weight value requires another width calculation to
fill the remaining space.
```

也就是说，在某个方向上使用 layout_weight 属性，推荐将这个方向上的 width 设置成 0dp，系统将会采用另一种算法来计算控件的控件占比，这时 layout_weight 属性值和占据的"宽度"将成正比例。

修改 activity_layout.xml 代码如下：

```xml
<?xml version="1.0" encoding="utf-8"?>
<LinearLayout xmlns:android="http://schemas.android.com/apk/res/android"
    android:layout_width="match_parent"
    android:layout_height="match_parent"
    android:orientation="horizontal">

    <TextView
        android:layout_width="0dp"
        android:layout_height="wrap_content"
        android:layout_weight="2"
        android:background="#cc5858"
        android:padding="10dp"
        android:text="layout_weight=2"
        android:textSize="20dp" />

    <TextView
        android:layout_width="0dp"
        android:layout_height="wrap_content"
        android:layout_weight="1"
        android:background="#5baf54"
        android:gravity="center_horizontal"
        android:padding="10dp"
        android:text="layout_weight=1"
        android:textSize="20dp" />
</LinearLayout>
```

因为布局是水平布局，所以其方向上的 width 就是 layout_width，设置 layout_width 为 0dp，查看预览窗口如图 3.26 所示。

图 3.24　layout_weight 属性示意图四

图 3.25　layout_weight 属性示意图五

图 3.26　layout_weight 属性示意图六

可以看出，layout_weight 为 2 的 TextView 所占据的宽度是 layout_weight 为 1 的 TextView 所占据宽度的两倍，因此，推荐在开发时使用 0dp。

3.4.3 weightSum 属性

上面讲解了 layout_weight 属性的使用，Android 还提供了一个 weightSum 属性供开发者调用。通过名字直观分析，它应该是所有 layout_weight 的和，此属性将在父布局中使用。

下面通过一个实例看一下 weightSum 的用法：

```xml
<?xml version="1.0" encoding="utf-8"?>
<LinearLayout xmlns:android="http://schemas.android.com/apk/res/android"
    android:layout_width="match_parent"
    android:layout_height="match_parent"
    android:gravity="center"
    android:orientation="horizontal"
    android:weightSum="2">

    <TextView
        android:layout_width="0dp"
        android:textSize="28dp"
        android:gravity="center"
        android:textColor="#ffffff"
        android:background="#1d6e09"
        android:layout_height="wrap_content"
        android:layout_weight="1"
        android:text="Hello World!" />
</LinearLayout>
```

上述代码中 weightSum 放在父布局中并设置其值为 2，这时可以认为整个宽为 2，在子控件 TextView 中设置 layout_weight 为 1 并设置其 layout_width 为 0dp，可以认为 TextView 占据了整个宽的一半，如图 3.27 所示。

可以看出，TextView 居中并占据整个宽的一半。两个控件时同样也可以按照比例占据屏幕宽的一半，修改代码如下：

```xml
<?xml version="1.0" encoding="utf-8"?>
<LinearLayout xmlns:android="http://schemas.android.com/apk/res/android"
    android:layout_width="match_parent"
    android:layout_height="match_parent"
    android:gravity="center"
    android:orientation="horizontal"
    android:weightSum="6">

    <TextView
        android:layout_width="0dp"
        android:layout_height="wrap_content"
        android:layout_weight="2"
        android:background="#759c6c"
        android:gravity="center"
        android:text="2"
        android:textColor="#ffffff"
        android:textSize="28dp" />

    <TextView
```

```
        android:layout_width="0dp"
        android:layout_height="wrap_content"
        android:layout_weight="1"
        android:background="#b33174"
        android:gravity="center"
        android:text="1"
        android:textColor="#ffffff"
        android:textSize="28dp" />
</LinearLayout>
```

上述代码设置父布局的 weightSum 为 6，将整个屏幕的宽分成 6 份，将第一个 TextView 的 layout_weight 属性值设为 2，它将占据 2 份屏幕的宽，将第二个 TextView 的 layout_weight 属性值设为 1，它将占据 1 份屏幕的宽，查看预览窗口如图 3.28 所示。

图 3.27 weightSum 属性示意图一

图 3.28 weightSum 属性示意图二

可以看出，第一个 TextView 是第二个 TextView 的宽的两倍，这两个 TextView 占据整个屏幕的一半。

3.5 Android 布局之相对布局——RelativeLayout

RelativeLayout 继承于 android.widget.ViewGroup，按照子元素之间的位置关系完成布局，作为 Android 系统五大布局中最灵活也是最常用的一种布局方式，非常适合于一些比较复杂的界面设计。

RelativeLayout 常用的位置属性如表 3.4 所示。

表 3.4 RelativeLayout 常用的位置属性

属性	说明
android:layout_toLeftOf	该控件位于引用控件的左方
android:layout_toRightOf	该控件位于引用控件的右方
android:layout_above	该控件位于引用控件的上方
android:layout_below	该控件位于引用控件的下方
android:layout_centerInParent	该控件是否相对于父组件居中
android:layout_centerHorizontal	该控件是否横向居中
android:layout_centerVertical	该控件是否垂直居中

续表

属　　性	说　　明
android:layout_alignParentLeft	该控件是否对齐父组件的左端
android:layout_alignParentRight	该控件是否齐其父组件的右端
android:layout_alignParentTop	该控件是否对齐父组件的顶部
android:layout_alignParentBottom	该控件是否对齐父组件的底部

下面选择部分属性进行实例讲解，首先通过实例看一下前面属性 layout_toLeftOf、layout_toRightOf、layout_above 和 layout_centerInParent 的用法：

```xml
<?xml version="1.0" encoding="utf-8"?>
<RelativeLayout xmlns:android="http://schemas.android.com/apk/res/android"
    android:layout_width="match_parent"
    android:layout_height="match_parent">

    <Button
        android:id="@+id/btn1"
        android:layout_width="wrap_content"
        android:layout_height="wrap_content"
        android:layout_centerInParent="true"
        android:text="Button1"/>

    <Button
        android:id="@+id/btn2"
        android:layout_width="wrap_content"
        android:layout_height="wrap_content"
        android:layout_above="@+id/btn1"
        android:layout_toLeftOf="@+id/btn1"
        android:text="Button2"/>

    <Button
        android:id="@+id/btn3"
        android:layout_width="wrap_content"
        android:layout_height="wrap_content"
        android:layout_above="@+id/btn1"
        android:layout_toRightOf="@+id/btn1"
        android:text="Button3"/>

    <Button
        android:id="@+id/btn4"
        android:layout_width="wrap_content"
        android:layout_height="wrap_content"
        android:layout_above="@+id/btn2"
        android:layout_toLeftOf="@+id/btn3"
        android:layout_toRightOf="@+id/btn2"
        android:text="Button4"/>

</RelativeLayout>
```

上述代码为 Button1 添加了 layout_centerInParent 属性，并设置其值为 true，Button1 将置于父控件 RelativeLayout 的正中；为 Button2 添加了属性 layout_above 并设置其值为"@+id/btn1"，也就是 Button2 将位于 Button1 的上方，同时为 Button2 添加了属性 layout_ toLeftOf 并设置其值为"@+id/btn1",Button2 将位于 Button1 的左边;为 Button3 添加了属性 layout_above 并设置其值为"@+id/btn1",也就是 Button3 将位于 Button1 的上方,同时为 Button3 添加了属性 layout_toRightOf 并设置其值为"@+id/btn1",Button3 将位于 Button1 的右边;为 Button4 添加了三个位置属性,Button4 将位于 Button2 的上方,位于 Button2 的右边,位于 Button3 的左边。

查看预览窗口如图 3.29 所示。

下面通过实例来看一下属性 layout_alignParentLeft、layout_alignParentRight、layout_alignParentTop 和 layout_alignParentBottom 的用法:

```xml
<?xml version="1.0" encoding="utf-8"?>
<RelativeLayout xmlns:android="http://schemas.android.com/apk/res/android"
    android:layout_width="match_parent"
    android:layout_height="match_parent">

    <TextView
        android:layout_width="wrap_content"
        android:layout_height="wrap_content"
        android:layout_alignParentBottom="true"
        android:padding="5dp"
        android:text="TextView1"
        android:textSize="26dp" />
    <TextView
        android:layout_width="wrap_content"
        android:layout_height="wrap_content"
        android:layout_alignParentLeft="true"
        android:padding="5dp"
        android:text="TextView2"
        android:textSize="26dp" />

    <TextView
        android:layout_width="wrap_content"
        android:layout_height="wrap_content"
        android:layout_alignParentBottom="true"
        android:layout_alignParentRight="true"
        android:padding="5dp"
        android:text="TextView3"
        android:textSize="26dp" />

    <TextView
        android:layout_width="wrap_content"
        android:layout_height="wrap_content"
        android:layout_alignParentRight="true"
        android:padding="5dp"
        android:text="TextView4"
        android:textSize="26dp" />
</RelativeLayout>
```

上述代码为 TextView1 添加了 layout_alignParentBottom 属性,值为 true,控件将置于父布局底部(默认在左边);为 TextView2 添加了 layout_alignParentLeft 属性,值为 true,控件将置于父布局的左边;为 TextView3 添加了 layout_alignParentButton 属性,值为 true,并添加了 layout_alignParentRight 属性,值为 true,控件将置于父布局的右下角;为

TextView4 添加了 layout_alignParentRight 属性，值为 true，控件将置于父布局的右边。

查看预览窗口如图 3.30 所示。

图 3.29　RelativeLayout 属性示意图一　　　　图 3.30　RelativeLayout 属性示意图二

可以看出，TextView1 位于左下角，TextView2 位于左上角，TextView3 位于右下角，TextView4 位于右上角。

3.6　Android 布局之帧布局——FrameLayout

FrameLayout 是比较简单的布局方式，所有的控件层叠显示，默认放在屏幕的左上角，最先添加的控件放在最底层，后添加的控件在先添加的控件上面。

下面通过一个实例看一下这个 FrameLayout 的基础用法：

```
<FrameLayout xmlns:android="http://schemas.android.com/apk/res/android"
    android:id="@+id/FrameLayout1"
    android:layout_width="match_parent"
    android:layout_height="match_parent">

    <TextView
        android:layout_width="200dp"
        android:layout_height="200dp"
        android:background="#FF6143" />

    <TextView
        android:layout_width="150dp"
        android:layout_height="150dp"
        android:background="#7BFE00" />

    <TextView
        android:layout_width="100dp"
        android:layout_height="100dp"
        android:background="#FFFF00" />

</FrameLayout>
```

上述代码一共在 FrameLayout 中添加了三个 TextView 控件：第一个 TextView 的尺寸最大，放在最底层；第二个 TextView 在第一个 TextView 的上面；最后一个 TextView 的尺寸最小，放在所有 TextView 的最上方。为了方便观察，这里为每个 TextView 设置了不同的背景值，查看预览窗口如图 3.31 所示。

可以看出，所有 TextView 都层叠地堆在屏幕的左上角。当然也可以添加 layout_gravity 属性修改 FrameLayout 的默认左上显示，代码如下：

```xml
<FrameLayout xmlns:android="http://schemas.android.com/apk/res/android"
    android:id="@+id/FrameLayout1"
    android:layout_width="match_parent"
    android:layout_height="match_parent">

    <TextView
        android:layout_width="200dp"
        android:layout_height="200dp"
        android:layout_gravity="center"
        android:background="#FF6143" />

    <TextView
        android:layout_width="150dp"
        android:layout_height="150dp"
        android:layout_gravity="center"
        android:background="#7BFE00" />

    <TextView
        android:layout_width="100dp"
        android:layout_height="100dp"
        android:layout_gravity="center"
        android:background="#FFFF00" />

</FrameLayout>
```

上述代码为每个 TextView 都添加了 layout_gravity 属性并设置其值为 center，查看预览窗口如图 3.32 所示。

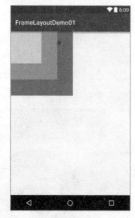

图 3.31　FrameLayout 布局示意图一

图 3.32　FrameLayout 布局示意图二

可以看出，所有的 TextView 都居中显示。

相对 LinearLayout 和 RelativeLayout 而言，FrameLayout 布局在开发中不是很常用，

在一些需要重叠显示的场景下才会使用。下面介绍一个经典的实例，代码如下：

```xml
<?xml version="1.0" encoding="utf-8"?>
<LinearLayout xmlns:android="http://schemas.android.com/apk/res/android"
    android:layout_width="match_parent"
    android:layout_height="match_parent"
    android:layout_marginTop="30dp"
    android:gravity="center"
    android:orientation="vertical">
    <FrameLayout
        android:layout_width="250dp"
        android:layout_height="250dp"
        android:layout_margin="10dp">

        <ImageView
            android:layout_width="wrap_content"
            android:layout_height="wrap_content"
            android:layout_gravity="center"
            android:src="@drawable/game_disc" />

        <ImageView
            android:layout_width="wrap_content"
            android:layout_height="wrap_content"
            android:layout_gravity="center"
            android:src="@drawable/game_disc_light" />

        <ImageButton
            android:layout_width="wrap_content"
            android:layout_height="wrap_content"
            android:layout_gravity="center"
            android:background="@drawable/play_button_icon" />

        <ImageView
            android:layout_width="50sp"
            android:layout_height="140sp"
            android:layout_gravity="right"
            android:src="@drawable/index_pin" /><!-- 拨杆 -->
    </FrameLayout>
</LinearLayout>
```

上述代码在 LinearLayout 中嵌套了一个 FrameLayout，LinearLayout 添加了 gravity 属性并设置其值为 true，其子布局（FrameLayout）将居中显示。在 FrameLayout 中添加了三个 ImageView 和一个 ImageButton，前三个控件都设置了其 layout_gravity 属性，其值为 cente，最后一个 ImageView 的控件添加了 layout_gravity 属性其值为 right。查看预览窗口如图 3.33 所示。

可以看出，本实例模仿了一个唱片播放器，三个 ImageView 和一个 ImageButton 层叠显示。

图 3.33　FrameLayout 布局实例示意图

3.7 Android 布局优化

作为一个有追求的程序员，功能实现应该是最低要求，实现基本功能之后，性能优化才应该是一个持续而永恒的话题。现如今的 APP 布局都十分复杂，需要多层多级嵌套，页面加载时，逐层渲染布局，层数背景越多自然渲染时间就会越久，占用的 GPU 和 CPU 资源越多，给用户的感觉也就是"越卡"，因此减少不必要的嵌套应该是布局优化的第一步。

3.7.1 过度绘制

如何查看是否是"过度绘制（Overdraw）"呢？在"开发人员选项"菜单中有一个"调试 GPU 过度绘制"的开关，如图 3.34 所示。选择这个开关，这时会弹出选择对话框，如图 3.35 所示。

图 3.34　Android 过度绘制开关一　　　图 3.35　Android 过度绘制开关二

选择"显示过度绘制区域"选项，即可开启过度绘制的检测。官网上对开启"过度绘制"开关后的颜色含义做了介绍，如图 3.36 所示。

即灰色表示无过度绘制，浅蓝色表示一层过度绘制，绿色表示两层过度绘制，粉红色表示三层过度绘制，红色表示四层或四层以上过度绘制。一般来讲，在开发中应禁止四层以上过度绘制。

下面通过一个实例模仿一些过度绘制，代码如下：

```xml
<?xml version="1.0" encoding="utf-8"?>
<LinearLayout xmlns:android="http://schemas.android.com/apk/res/android"
    android:id="@+id/activity_main"
    android:layout_width="match_parent"
    android:layout_height="match_parent"
    android:background="#ffffff"
    android:orientation="vertical">

    <LinearLayout
```

```xml
        android:layout_width="match_parent"
        android:layout_height="40dp"
        android:background="#ffffff">

        <LinearLayout
            android:layout_width="match_parent"
            android:layout_height="40dp"
            android:background="#ffffff">

            <LinearLayout
                android:layout_width="match_parent"
                android:layout_height="40dp"
                android:background="#ffffff">

                <TextView
                    android:layout_width="match_parent"
                    android:layout_height="match_parent"
                    android:gravity="center"
                    android:text="四层布局" />

            </LinearLayout>
        </LinearLayout>
    </LinearLayout>

    <LinearLayout
        android:layout_width="match_parent"
        android:layout_height="40dp"
        android:background="#ffffff">

        <LinearLayout
            android:layout_width="match_parent"
            android:layout_height="40dp"
            android:background="#ffffff">

            <TextView
                android:layout_width="match_parent"
                android:layout_height="match_parent"
                android:gravity="center"
                android:text="三层布局" />

        </LinearLayout>
    </LinearLayout>

    <LinearLayout
        android:layout_width="match_parent"
        android:layout_height="40dp"
        android:background="#ffffff">

        <TextView
            android:layout_width="match_parent"
            android:layout_height="match_parent"
            android:gravity="center"
```

```
        android:text="两层布局" />
    </LinearLayout>

    <TextView
        android:layout_width="match_parent"
        android:layout_height="match_parent"
        android:gravity="center"
        android:text="一层布局" />

</LinearLayout>
```

在开启"显示过度绘制"开关的真机中运行,结果如图 3.37 所示。

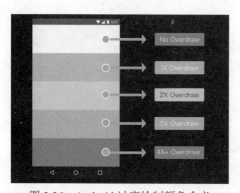

图 3.36　Android 过度绘制颜色含义

图 3.37　Android 过度绘制实例示意图

可以看出,由下到上,过度绘制越来越严重,根据这些颜色的提示,程序员即可快速地定位可能出现过度绘制的区域,然后采取相应操作减少过度绘制的层数。

3.7.2　布局优化之 include 标签

有开发经验的读者都知道,可复用性是代码质量评价的一项重要指标,代码复用可以提高开发效率、提高代码的可维护性并减少一些低级错误的发生。因此,在开发中要尽量提高代码的可复用性。我们都知道,开发中经常会抽取一些常用的 Java 代码做成工具类,那么布局文件可不可以复用呢?当然是可以的,将一些共用的组合布局抽取出来,可以增强 UI 风格的一致性并提高可维护性。

这里以顶部 Tab 导航条为例。导航条中包括一个返回箭头和当前页面的 title,假设 APP 中每一个页面中都有这样一个导航条,那么我们就可以抽取这个导航条为一个单独的布局,代码如下(top.xml):

```
<?xml version="1.0" encoding="utf-8"?>
<LinearLayout xmlns:android="http://schemas.android.com/apk/res/android"
    android:id="@+id/top"
    android:layout_width="match_parent"
    android:layout_height="wrap_content"
    android:orientation="horizontal">

    <ImageView
```

```
        android:layout_width="30dp"
        android:layout_height="30dp"
        android:layout_margin="5dp"
        android:src="@drawable/left" />

    <TextView
        android:layout_width="wrap_content"
        android:layout_height="30dp"
        android:layout_margin="5dp"
        android:gravity="center"
        android:text="title"
        android:textSize="24sp" />

</LinearLayout>
```

在预览窗口中查看如图 3.38 所示。

在 activity_main.xml 中引入这个布局,代码如下:

```
<?xml version="1.0" encoding="utf-8"?>
<RelativeLayout xmlns:android="http://schemas.android.com/apk/res/android"
    android:id="@+id/activity_main"
    android:layout_width="match_parent"
    android:layout_height="match_parent">

    <include layout="@layout/top" />

    <TextView
        android:layout_width="match_parent"
        android:layout_height="match_parent"
        android:layout_below="@+id/top"
        android:gravity="center"
        android:text="activity_main"
        android:textSize="30sp" />
</RelativeLayout>
```

这里使用"<include/>"标签将一个布局引入到另一个布局文件中,这时引入的子布局就真实存在于这个布局中了。因此,下面就可以设置 TextView 的 layout_below 属性值为"@+id/top"了。查看预览窗口,如图 3.39 所示。

可以看出,top.xml 已成功引入到 activity_main.xml 布局文件中并显示出来。同样可以使用 include 标签将 top 布局引入到其他布局中,代码如下(activity_second.xml):

```
<?xml version="1.0" encoding="utf-8"?>
<LinearLayout xmlns:android="http://schemas.android.com/apk/res/android"
    android:layout_width="match_parent"
    android:layout_height="match_parent"
    android:orientation="vertical">

    <include layout="@layout/top" />

    <TextView
        android:layout_width="match_parent"
```

```
            android:layout_height="match_parent"
            android:gravity="center"
            android:text="activity_second"
            android:textSize="30sp" />

</LinearLayout>
```

查看预览窗口如图 3.40 所示。

图 3.38　Android 布局优化
实例示意图一

图 3.39　Android 布局优化
实例示意图二

图 3.40　Android 布局优化
实例示意图三

同样，top.xml 引入到了 activity_second.xml 中并被显示出来，可以看出，activity_main 和 activity_second 有了统一的 UI 风格。然而，这里还没有彻底体现出布局复用的最大的优点，那就是布局的可维护性大大提高。设想一下如下场景，若一个 APP 中有几十个布局，每个布局都有独立的顶部导航栏，此时若有需求需要将顶部导航栏的图片都更换一下，那么就必须一个布局一个布局地进行修改，效率低下，还极易引入错误。若每个页面都采用 include 的方式，则只需要修改被 include 方式导入的文件中的图片即可，简单且安全。

第 4 章 Android 基础控件操作实战

控件可以认为是为了方便开发而封装的功能集合体，Android APP 即可被认为是各类控件有机的组合。因此，对于初学者来说，控件的操作就是开发的基础，是基本功、硬把式，要充分熟悉基本控件的基本用法。本章首先介绍常用控件的属性及方法，通过充满创意而又切实实用的实例循序渐进地讲解这些属性及方法的用法，最后通过扫一扫看动态图的方式，形象地展示这些控件的功能及运行效果。

4.1 炫酷之星——TextView 控件

4.1.1 常用属性介绍

TextView 是 Android 中最常用的控件，主要承担文本显示的工作，任何 APP 都不可避免地会用到它。同时，TextView 的属性和方法在所有控件中也是最为丰富的，其常用属性和方法，如表 4.1 所示。

表 4.1 TextView 常用属性和方法

配置属性	相关方法	说 明
android:text	setText(CharSequence,TextView.BufferType)	最常用的属性，设置组件显示文字
android:textColor	setTextColor(int)	设置文字颜色
android:textSize	setTextSize(int,float)	设置字体大小
android:textStyle	setTypeface(Typeface)	设置字体样式
android:gravity	setGravity(int)	文字显示位置
android:width	setWidth(int)	设置组件宽度
android:singleLine	setTransformationMethod(TransformationMethod)	是否单行显示
android:maxLines	setMaxLines(int)	最多行数
android:ellipsize	setEllipsize(TextUtils.TruncateAt)	文字长度超过组件宽度时可以选择的显示方式
android:drawableRight	setCompoundDrawablesWithIntrinsicBounds(int,int,int,int)	组件中插入图片
android:autoLink	setAutoLinkMask(int)	设置链接

4.1.2 TextView 实战演练

表格能展现的东西往往生硬，下面通过一个实例来展示表 4.1 中部分属性的用法，代码如下：

```
<?xml version="1.0" encoding="utf-8"?>
<LinearLayout xmlns:android="http://schemas.android.com/apk/res/android"
    android:layout_width="match_parent"
    android:layout_height="match_parent"
```

```
        android:orientation="vertical">

    <TextView
        android:id="@+id/textView"
        android:layout_width="match_parent"
        android:layout_height="wrap_content"
        android:drawableRight="@android:drawable/ic_lock_lock"
        android:gravity="center"
        android:singleLine="true"
        android:text="This is test String... This is test String..."
        android:textColor="#186806"
        android:textSize="24sp"
        android:textStyle="bold" />
</LinearLayout>
```

TextView 标签中"android:"后面的单词表示属性名,"="后面用引号包裹的部分表示属性值,通过这种"键 - 值"对的关系设定控件的一个个属性。这里为 TextView 添加的属性有:

- id 属性:这个属性为控件指定了唯一的 id 号,通过这个 id 就可以在项目中的任何地方找到这个控件(可以认为是控件名)。一般来讲,这个属性是必须的,格式比较固定"@+id/用户自定义名"。注意,设置这个属性值时要保证其唯一性。
- layout_width 和 layout_height 属性:宽高属性,定义一个控件的大小,在 Android Studio 中,每创建一个控件,这两个属性是会自动生成的,这也表明了它们是定义一个控件所必需的。常用的值有两个:match_parent(占满一行)、wrap_content(包裹内容)。当然用户也可以自定义其大小,考虑到 Android 设备分辨率的多样性,一般采用 dp 作为单位。
- drawableRight 属性:这个属性可以在 TextView 中插入图片,后面的 Right 后缀表明插入的位置(右边),当然要在左边插入要怎么做,相信读者也应该明白了。引号中是属性对应的值,如"@android:"表示这个图片是 Android 系统内置的图片,若要引用项目 drawable 文件夹下的图片,直接引用"@drawable"即可。
- gravity 属性:这个属性一般用于控制控件内部内容在控件中的什么位置显示。在 Android Studio 中按下快捷键:Ctrl+Alt+Space,即可弹出 gravity 的所有属性值,如图 4.1 所示。
- singleLine 属性:中文译为单行模式,主要有 true 和 false 两个值。true 表示单行模式,false 表示多行模式。不设置这个属性时默认为 false,即多行模式。
- text 属性:这是 TextView 的核心属性,也就是 TextView 的显示内容。

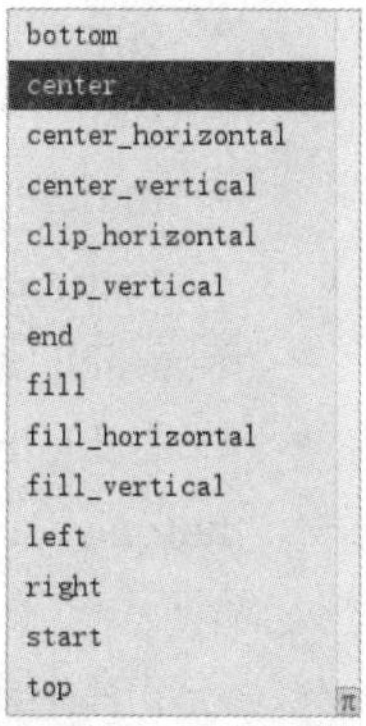

图 4.1 gravity 属性值

- textColor 属性:设置字体的颜色,其值是十六位的数字,也可以使用 color 文件中保存的颜色值。

- textSize 属性：这个属性可以设置显示文本的字体大小，考虑到 Android 设备屏幕分辨率的多样性，一般采用 sp 作为单位。
- textStyle 属性：这个属性用于设置字体的样式。Android 提供了三个值供开发者选用：bold（加粗）、italic（斜体）、normal（正常），其中 normal 为默认值。

在模拟器中运行这个项目，如图 4.2 所示。

可以看出，由于设置了单行模式，未能显示的文本以省略号的形式展示。读者可以考虑一下，对于展示空间有限而展示内容过多的场景怎么实现呢？生活当中不乏这样的事情，出门逛街时可以看到各式各样的 LED 显示屏，一般 LED 显示屏的宽度都是有限的，若要显示较长的文本信息就必须让文本滚动起来，循环播放。Android 中可不可以实现这样的效果呢？

下面通过一个实例来进行研究，代码如下：

```
<TextView
    android:id="@+id/tv_marquee"
    android:layout_width="match_parent"
    android:layout_height="wrap_content"
    android:textColor="@android:color/black"
    android:ellipsize="marquee"
    android:focusable="true"
    android:focusableInTouchMode="true"
    android:marqueeRepeatLimit="marquee_forever"
    android:scrollHorizontally="true"
    android:singleLine="true"
    android:text=" 这是跑马灯的效果这是跑马灯的效果这是跑马灯的效果这是跑马灯的效果 ">
</TextView>
```

这里用到了几个属性：

- ellipsize 属性：主要用于解决显示文本长度长于控件宽度的场景。系统提供了几个值：
  - end——省略号显示位置，省略号在尾部；
  - start——省略号在头部；
  - middle——省略号在中部；
  - marquee——滚动方式显示（本例采用）。
- focusable 属性：控件获得焦点。
- focusableInTouchMode 属性：针对触摸屏获得当前焦点。
- marqueeRepeatLimit 属性：这里设置其值为 marquee_forever，滚动播放无限制次数循环。
- scrollHorizontally 属性：水平方式显示。

注意，为了获得滚动播放的效果，设置文本的长度要长于控件宽度。由于是动态显示，静态图片无法展示其效果，查看动态图，请扫描图 4.3 所示的二维码。

TextView 除了可以显示文本外，还可以对特殊文本进行识别，这些特殊文本包括网址、电话、邮箱等，只需要设置 autoLink 属性即可。代码如下：

第4章　Android基础控件操作实战　◆　067

图 4.2　TextView 运行实例

图 4.3　TextView 运行实例二维码

```xml
<?xml version="1.0" encoding="utf-8"?>
<LinearLayout xmlns:android="http://schemas.android.com/apk/res/android"
    android:layout_width="match_parent"
    android:layout_height="match_parent"
    android:orientation="vertical">

    <TextView
        android:layout_width="match_parent"
        android:layout_height="wrap_content"
        android:autoLink="phone"
        android:gravity="center_horizontal"
        android:padding="10dp"
        android:text="phone: 13057655618"
        android:textSize="18sp" />

    <TextView
        android:layout_width="match_parent"
        android:layout_height="wrap_content"
        android:autoLink="web"
        android:gravity="center_horizontal"
        android:padding="10dp"
        android:text="web: www.baidu.com"
        android:textSize="18sp" />

    <TextView
        android:layout_width="match_parent"
        android:layout_height="wrap_content"
        android:autoLink="email|phone"
        android:gravity="center_horizontal"
        android:padding="10dp"
        android:text="email: 291214603@qq.com"
        android:textSize="18sp" />
</LinearLayout>
```

上述代码中每一个 TextView 都设置了对应的 autoLink 属性。Android 主要提供了六个 autoLink 的属性值供开发者调用，主要包括：web（匹配网址）、phone（匹配电话）、email（匹配邮箱）、map（匹配地图）、all（匹配所有）、none。运行实例如图 4.4 所示，单击 phone 后面的电话会跳转到拨号界面，单击 web 后面的链接会跳转到浏览器并打开百度，单击 email 后面的链接，若手机安装了邮箱应用将会跳转到邮箱应用。查看动态图，请扫描图

4.5 所示的二维码。

图 4.4 TextView autoLink 属性实例

图 4.5 TextView autoLink 属性实例二维码

除了通过属性进行设置，还可以在代码中通过方法进行设置。

新建项目，其主布局文件代码如下：

```xml
<?xml version="1.0" encoding="utf-8"?>
<RelativeLayout xmlns:android="http://schemas.android.com/apk/res/android"
    android:id="@+id/activity_main"
    android:layout_width="match_parent"
    android:layout_height="match_parent">

    <TextView
        android:id="@+id/test"
        android:layout_width="match_parent"
        android:layout_height="wrap_content"
        android:gravity="center_horizontal"
        android:padding="10dp"
        android:text="web: www.baidu.com ;"
        android:textSize="18sp" />
</RelativeLayout>
```

为了方便在代码中引用这个 TextView，为其设置了 id 属性。

MainActivity.java 代码如下：

```java
public class MainActivity extends AppCompatActivity {
    private TextView mTextView;

    @Override
    protected void onCreate(Bundle savedInstanceState) {
        super.onCreate(savedInstanceState);
        setContentView(R.layout.activity_main);
        mTextView=(TextView)findViewById(R.id.test);
        mTextView.setAutoLinkMask(Linkify.WEB_URLS);// 识别 URL 类型文本
        mTextView.setMovementMethod(LinkMovementMethod.getInstance());
    }
}
```

调用 TextView 的 setAutoLinkMask 方法传入 Linkify.WEB_URLS 常量用于匹配网址类型，要想链接可单击并跳转，还需要调用 textView.setMovementMethod 方法。Android 提供了 LinkMovementMethod 类实现对于文本内容中超链接的遍历，并且支持对于超链接的单击事件，运行实例并单击链接，如图 4.6 所示。选择打开这个链接的浏览器（Chrome），这时会跳转到百度首页，如图 4.7 所示。

图 4.6　TextView 动态设置 autoLink　　　　图 4.7　TextView 动态设置 autoLink 跳转

4.2　用户之窗——EditText 控件

4.2.1　常用属性介绍

EditText 常用于获取用户输入的内容，是用户和系统交互的窗户。继承结构如下：
public class
EditText
extends TextView
java.lang.Object
　　□　android.view.View
　　　　□　android.widget.TextView
　　　　　　□　android.widget.EditText

由继承结构可以看出，EditText 继承自 TextView，因此 TextView 中的一些属性和方法也可以在 EditText 中使用。EditText 的常用属性如表 4.2 所示。

表 4.2　EditText 的常用属性

属　　性	说　　明
password	密文显示
numeric	只能输入数字
maxLength	最长输入
editable	是否可编辑
selectAllOnFocus	默认全部文本选中获得焦点
hint	提示文本，输入文本后自动隐藏

4.2.2 EditText 实战演练

下面通过一段代码来演示表 4.2 中的属性：

```xml
<?xml version="1.0" encoding="utf-8"?>
<RelativeLayout xmlns:android="http://schemas.android.com/apk/res/android"
    android:layout_width="match_parent"
    android:layout_height="match_parent">

    <TextView
        android:id="@+id/tv_num"
        android:layout_width="wrap_content"
        android:layout_height="wrap_content"
        android:padding="10dp"
        android:text="电话号码："
        android:textSize="18sp" />

    <EditText
        android:id="@+id/et_num"
        android:layout_width="wrap_content"
        android:layout_height="wrap_content"
        android:layout_toRightOf="@+id/tv_num"
        android:hint="这里是提示内容，只能输入整数"
        android:numeric="integer"
        android:selectAllOnFocus="true" />

    <TextView
        android:id="@+id/tv_password"
        android:layout_width="wrap_content"
        android:layout_height="wrap_content"
        android:layout_alignParentLeft="true"
        android:layout_below="@+id/et_num"
        android:layout_marginTop="41dp"
        android:padding="5dp"
        android:text="密码："
        android:textSize="18sp" />

    <EditText
        android:id="@+id/edit_password"
        android:layout_width="wrap_content"
        android:layout_height="wrap_content"
        android:layout_alignBottom="@+id/tv_password"
        android:layout_toRightOf="@+id/tv_password"
        android:drawableRight="@android:drawable/ic_lock_lock"
        android:hint="密文显示输入，最长 8 位"
        android:maxLength="8"
        android:password="true" />
```

```xml
<TextView
    android:id="@+id/tv_unedit"
    android:layout_width="wrap_content"
    android:layout_height="wrap_content"
    android:layout_alignParentLeft="true"
    android:layout_below="@+id/tv_password"
    android:layout_marginTop="51dp"
    android:padding="5dp"
    android:text=" 不可编辑："
    android:textSize="18sp" />

<EditText
    android:id="@+id/edit_false"
    android:layout_width="wrap_content"
    android:layout_height="wrap_content"
    android:layout_alignBottom="@+id/tv_unedit"
    android:layout_alignLeft="@+id/et_num"
    android:editable="false"
    android:hint=" 这里是提示内容，内容不可编辑 " />
</RelativeLayout>
```

这里引入了三个 EditText 控件：第一个 EditText 控件引入 numeric 属性并设置其值为 integer，表示允许输入的文本类型为整数；第二个 EditText 控件引入 password 属性并设置其值为 true，表示文本密文显示，设置其 maxLength 属性值为 8，表示最多输入 8 位；第三个 EditText 控件设置 editable 属性为 false，表示这个 EditText 不可编辑。

运行实例结果如图 4.8 所示。可以看出，第一个 EditText 控件中的内容默认全选，并且单击时弹出默认的数字键盘；第二个 EditText 的内容密文显示，第三个 EditText 不可编辑。此外，为了提高用户体验，为 EditText 添加了 hint 属性，提示当前 EditText 要输入的内容。查看动态图，请扫描图 4.9 中的二维码。

图 4.8　EditText 基础实例

图 4.9　EditText 基础实例二维码

上面演示了 EditText 的常用属性，下面通过一个较实用的例子来学习 EditText 的其他用法。用过QQ的同学都会关注到它的一个比较人性化的功能,那就是在聊天时可以看到对方是否正在输入,这里结合EditText的输入监听来模拟这个小功能。

布局文件如下:

```xml
<?xml version="1.0" encoding="utf-8"?>
<RelativeLayout xmlns:android="http://schemas.android.com/apk/res/android"
    android:layout_width="match_parent"
    android:layout_height="match_parent">

    <TextView
        android:id="@+id/textView"
        android:layout_width="match_parent"
        android:layout_height="wrap_content"
        android:padding="10dp"
        android:textSize="16sp" />

    <EditText
        android:id="@+id/edit"
        android:layout_width="match_parent"
        android:layout_height="wrap_content"
        android:layout_below="@+id/textView"
        android:gravity="top"
        android:inputType="textMultiLine"
        android:minHeight="100dp"
        android:scrollbars="vertical"
        android:text="Hello World!" />

</RelativeLayout>
```

上述代码中TextView控件用于显示当前用户是否正在输入,EditText作为模拟的输入框,这里为EditText添加了几个属性:

- inputType属性:即输入类型属性,其值为textMultiLine,表示允许多行输入。
- miniHeight属性:定义EditText的最小高度。
- scrollbars属性:设置这个属性可以表明EditText含有滚动条,设置其值为vertical表示滚动条为垂直的,除了这个值之外,还提供了一个horizontal(水平滚动条)供开发者调用。

MainActivity.java代码如下:

```java
public class MainActivity extends AppCompatActivity {
    private EditText mEditText;
    private TextView mTextView;
    private TimeCount mtTimeCount;

    @Override
    protected void onCreate(Bundle savedInstanceState) {
        super.onCreate(savedInstanceState);
        setContentView(R.layout.activity_main);
        mEditText=(EditText)findViewById(R.id.edit);
        mTextView=(TextView)findViewById(R.id.textView);
        mtTimeCount = new TimeCount(2 * 1000, 1000);
```

```java
        mEditText.addTextChangedListener(new TextWatcher() {
            @Override
            public void beforeTextChanged(CharSequence s,
                             int start, int count, int after){

            }

            @Override
            public void onTextChanged(CharSequence s,
                             int start, int before, int count) {
                mTextView.setText("正在输入 ...");
                mTextView.setTextColor(Color.GREEN);
            }

            @Override
            public void afterTextChanged(Editable s) {
                mtTimeCount.cancel();
                mtTimeCount.start();
            }
        });
    }

    class TimeCount extends CountDownTimer {
        // 构造方法
        public TimeCount(long totalTime, long interval) {
            super(totalTime, interval);
        }
        @Override
        public void onTick(long millisUntilFinished) {// 覆写方法 - 计时中

        }
        @Override
        public void onFinish() {// 覆写方法,计时结束
            mTextView.setText("");
        }
    }
}
```

这里主要有两个重点：为 EditText 添加文本变化监听；设置一个倒计时类,用于用户输入结束的计时,即用户停止输入 2s 后认为真正停止输入。

为 EditText 控件添加文本变化的监听,采用匿名内部类的形式实现,实现监听接口需要覆写三个方法,分别是：

- beforeTextChanged：文本内容变化前回调。
- onTextChanged：文本变化中回调。
- afterTextChanged：文本变化后回调。

这里主要关注的是文本变化中和文本变化后,文本变化中设置 TextView 的显示为"正在输入 ...",文本变化结束后调用倒计时类的 cancel 和 start 方法,开始计时,计时结束后设置 TextView 为空。

对于倒计时类，这里采用内部类的形式，自定义倒计时类继承至 CountDownTimer 类，CountDownTimer 类是一个抽象类，继承它要实现它的两个抽象方法，即：
- onTick 方法：根据实例化时传入的计时间隔回调，单位为毫秒，例如传入 1000 表示一秒钟回调一次，其方法中的参数 millsUntilFinished 表示离倒计时结束还剩余的时间。
- OnFinish 方法：倒计时结束后回调，这里将 TextView 的显示内容置空。

运行实例，如图 4.10 所示，查看动态图效果，请扫描图 4.11 中的二维码。

图 4.10　EditText 模拟"正在输入"
提示功能一

图 4.11　EditText 模拟"正在输入"
提示功能一二维码

对于这种有默认文本的 EditText，初次运行时可以看出光标在文本的开始位置，这样用户体验会不太好，可以通过调用 EditText 的 setSelection 方法可以将光标移动到文本的末尾，代码如下：

```
mEditText.setSelection(mEditText.getText().length());
```

setSeleciton 方法需要传入一个 int 值，即光标要插入的位置，这里传入了文本的长度，将光标移动到文本的最后。

再次运行实例，如图 4.12 所示。

可以看到，光标移动到了文本的最后。

4.2.3　EditText 实战进阶

文本框是人机交互的重要窗口，经常用来输入一些信息，用户名和密码的输入是最常用的场景，若用户名或密码输入错误时一个一个字符地删除通常比较费时费力，因此很多 APP 都添加了一键清空的功能。下面通过一个实例来看一下如何实现这个功能。

图 4.12　EditText 模拟"正在输入"
提示功能二

布局文件代码如下：

```xml
<RelativeLayout xmlns:android="http://schemas.android.com/apk/res/android"
    android:layout_width="match_parent"
    android:layout_height="match_parent">

    <TextView
        android:id="@+id/title"
        android:layout_width="match_parent"
        android:layout_height="wrap_content"
        android:layout_alignParentTop="true"
        android:gravity="center"
        android:text="登    录"
        android:textSize="25sp" />

    <com.example.administrator.deleteedittext.DeleteEditText
        android:id="@+id/det_test"
        android:layout_width="match_parent"
        android:layout_height="wrap_content"
        android:layout_below="@+id/title"
        android:drawableLeft="@drawable/user_account"
        android:drawableRight="@drawable/user_delete"
        android:hint="请输入账号名" />

    <com.example.administrator.deleteedittext.DeleteEditText
        android:id="@+id/user_password_input"
        android:layout_width="match_parent"
        android:layout_height="wrap_content"
        android:layout_below="@+id/det_test"
        android:layout_marginTop="10dp"
        android:drawableLeft="@drawable/user_password"
        android:drawableRight="@drawable/user_delete"
        android:hint="请输入密码"
        android:inputType="textPassword"
        android:singleLine="true" />
</RelativeLayout>
```

上述代码采用了相对布局的方式，为 TextView 添加了 layout_alignParentTop 属性并设置属性值为 true，这时 TextView 将在最顶部；采用"包.类"的方式引入自定义的 View 并设置了相对应的 drawableLeft 和 drawableRight 在 EditText 内部左右两边显示图片；设置了 hint 属性，用于提示用户输入，提升用户体验；为密码文本框添加了 inputType 属性，并设置其值为 textPassword，密文显示输入。

自定义 View 的代码如下：

```java
public class DeleteEditText extends EditText {
    private Drawable mRightDrawable;
    // 控件是否获得焦点标志位
    boolean mIsHasFocus;

    // 构造方法1
    public DeleteEditText(Context context) {
```

```java
        super(context);
        init();
    }

    // 构造方法2
    public DeleteEditText(Context context, AttributeSet attrs) {
        super(context, attrs);
        init();
    }

    // 构造方法3
    public DeleteEditText(Context context, AttributeSet attrs, int defStyle) {
        super(context, attrs, defStyle);
        init();
    }

    private void init() {
        // 本方法获取控件上下左右四个方位插入的图片
        Drawable drawables[] = this.getCompoundDrawables();
        mRightDrawable = drawables[2];
        // 添加文本改变监听
        this.addTextChangedListener(new TextWatcherImpl());
        // 添加触摸改变监听
        this.setOnFocusChangeListener(new OnFocusChangeImpl());
        // 初始设置所有右边图片不可见
        setClearDrawableVisible(false);
    }

    private class OnFocusChangeImpl implements OnFocusChangeListener {
        @Override
        public void onFocusChange(View v, boolean hasFocus) {
            mIsHasFocus = hasFocus;
            if (mIsHasFocus) {
                // 如果获取焦点并且判断输入内容不为空则显示删除图标
                boolean isNoNull = getText().toString().length() >= 1;
                setClearDrawableVisible(isNoNull);
            } else {
                // 否则隐藏删除图标
                setClearDrawableVisible(false);
            }
        }
    }

    // 本方法控制右边图片的显示与否
    private void setClearDrawableVisible(boolean isNoNull) {
        Drawable rightDrawable;
        if (isNoNull) {
            rightDrawable = mRightDrawable;
        } else {
```

```
            rightDrawable = null;
    }
    // 使用代码设置该控件 left, top, right 和 bottom 处的图标
    setCompoundDrawables(getCompoundDrawables()[0],
            getCompoundDrawables()[1], rightDrawable,
            getCompoundDrawables()[3]);
}

private class TextWatcherImpl implements TextWatcher {

    // 内容输入后
    @Override
    public void afterTextChanged(Editable s) {
        boolean isNoNull = getText().toString().length() >= 1;
        setClearDrawableVisible(isNoNull);
    }

    // 内容输入前
    @Override
    public void beforeTextChanged(CharSequence s, int start, int count, int after) {

    }

    // 内容输入中
    @Override
    public void onTextChanged(CharSequence s, int start, int before, int count) {

    }
}

@Override
public boolean onTouchEvent(MotionEvent event) {
    switch (event.getAction()) {
        // 抬手指事件
        case MotionEvent.ACTION_UP:
            // 删除图片右侧到 EditText 控件最左侧距离
            int length1 = getWidth() - getPaddingRight();
            // 删除图片左侧到 EditText 控件最左侧距离
            int length2 = getWidth() - getTotalPaddingRight();
            // 判断单击位置是否在图片上
            boolean isClean = (event.getX() > length2)
                    && (event.getX() < length1);
            if (isClean) {
                setText("");
            }
            break;
        default:
```

```
                break;
        }
        return super.onTouchEvent(event);
    }
}
```

先说一下本功能的主要思路,这里有两个关键点:
- 判断何时显示删除图标:主要通过两个监听实现。第一个是焦点变化的监听,通过 setOnFocusChangeListener(new OnFocusChangeImpl()) 方法添加监听,当焦点改变时,判断当前文本框的内容不为空时显示删除图标;第二个是文本变化监听,通过 addTextChangedListener(new TextWatcherImpl()) 添加,同样也是判断若当前文本框的内容不为空时显示删除图标。
- 监听删除图标的单击事件:对于删除图标的单击事件,这里采取了间接的方式,通过判断用户触摸屏幕的坐标进行判断,若触摸坐标和删除图标的位置坐标一致,则认为单击了删除图片,对应清空文本框内容。

运行实例,如图 4.13 所示。可以看出,当文本框中有内容,并且获得焦点时,最左侧显示删除图标,单击删除图标即可删除当前文本框的内容。查看动态图,请扫描图 4.14 中的二维码。

图 4.13　自定义 EditText 实现一键删除功能　　图 4.14　自定义 EditText 实现一键删除功能二维码

4.3　交互之王——Button 控件

Button 是 Android 应用中最常用到的控件之一,主要用来响应用户单击事件。每个应用中都包含了多个 Button 响应和解决用户各种单击交互事件,因此,说它为交互之王一点儿都不过分。通过 Button 的源码 public class Button extends TextView 可以看出 Button 控件继承自 TextView,自然很多 TextView 的属性和方法在 Button 中也同样适用。

Button 是用来响应单击事件的,具体怎么响应呢?开发者需要为 Button 设置单击事件监听,实现监听的方式很多,下面讲解比较主流的几种做法。

4.3.1　Button 单击事件响应

方式一:通过 onClick 属性

做过 .NET 的同学应该熟悉 onClick 属性,当然,除此之外多种开发平台都有这样一个属性，设置这个属性并在代码中添加相应的方法即可实现对用户单击事件的监听。

布局文件代码如下：

```xml
<?xml version="1.0" encoding="utf-8"?>
<RelativeLayout xmlns:android="http://schemas.android.com/apk/res/android"
    android:layout_width="match_parent"
    android:layout_height="match_parent">

    <Button
        android:onClick="onClick"
        android:layout_width="match_parent"
        android:layout_height="wrap_content"
        android:text=" 通过 onClick 属性实现单击事件监听 " />
</RelativeLayout>
```

上述代码在相对布局中添加了一个 Button 控件，为其添加了 onClick 属性并设置其值为 onClick。

MainActivity.java 代码如下：

```java
public class MainActivity extends AppCompatActivity {

    @Override
    protected void onCreate(Bundle savedInstanceState) {
        super.onCreate(savedInstanceState);
        setContentView(R.layout.activity_main);
    }

    public void onClick(View view){
        Toast.makeText(MainActivity.this, "Onclick", Toast.LENGTH_
        SHORT).show();
    }
}
```

要注意的是，在代码中添加方法的方法名和 onClick 的属性值必须一致。这个方法的写法比较固定，即 public void onClick 的属性值（View view）。通过 Toast 显示单击事件是否成功。运行实例，如图 4.15 所示。

单击 Button 按钮，Toast 信息成功弹出，监听添加成功。这是一个 Button 的情形，若有多个 Button 时怎么区别是哪一个 Button 被单击了呢？有两种方式：一是为每一个 Button 设置不同的 onClick 属性值，然后在代码中分别添加不同的方法；二是设置相同的 onClick 属性值，通过 Button 的 id 分辨不同的 Button。从代码设计的优雅角度来讲，宜选择后者。下面通过一个实例来看一下如何应对多个 Button 的情形。

布局代码如下：

图 4.15　Button onClick 实现监听一

```xml
<?xml version="1.0" encoding="utf-8"?>
<LinearLayout xmlns:android="http://schemas.android.com/apk/res/android"
    android:layout_width="match_parent"
    android:orientation="vertical"
    android:layout_height="match_parent">

    <Button
        android:id="@+id/btn1"
        android:onClick="onClick"
        android:layout_width="match_parent"
        android:layout_height="wrap_content"
        android:text="ButtonOne" />
    <Button
        android:id="@+id/btn2"
        android:onClick="onClick"
        android:layout_width="match_parent"
        android:layout_height="wrap_content"
        android:text="ButtonTwo" />
</LinearLayout>
```

上述代码添加了两个 Button 控件，为每一个控件添加了 onClick 属性且属性值相同，为了区别不同的 Button，这里为每一个 Button 都设置了不同的 id。

MainActivity.java 代码如下：

```java
public class MainActivity extends AppCompatActivity {

    @Override
    protected void onCreate(Bundle savedInstanceState) {
        super.onCreate(savedInstanceState);
        setContentView(R.layout.activity_main);
    }

    public void onClick(View view) {
        switch (view.getId()) {
            case R.id.btn1:
                Toast.makeText(MainActivity.this, "btn1", Toast.LENGTH_
                SHORT).show();
                break;
            case R.id.btn2:
                Toast.makeText(MainActivity.this, "btn2", Toast.LENGTH_
                SHORT).show();
                break;
        }
    }
}
```

可以看出，这里添加了一个 onClick 方法，通过传入的参数 view 结合其 getId 方法判断是哪个 Button 的单击事件。运行实例并单击第一个 Button，结果如图 4.16 所示。单击第二个 Button，结果如图 4.17 所示。

单击 ButtonOne 弹出内容为 btn1 的 Toast，单击 ButtonTwo 弹出内容为 btn2 的 Toast。

图 4.16　Button onClick 实现监听二　　　　图 4.17　Button onClick 实现监听三

方式二：实现 OnClickListener 接口

上面的实例还可以通过实现 OnClickListner 接口来实现，这时就可以去掉 onClick 属性了，此时主布局文件（activity_main.xml）代码如下：

```xml
<?xml version="1.0" encoding="utf-8"?>
<LinearLayout xmlns:android="http://schemas.android.com/apk/res/android"
    android:layout_width="match_parent"
    android:orientation="vertical"
    android:layout_height="match_parent">

    <Button
        android:id="@+id/btn1"
        android:layout_width="match_parent"
        android:layout_height="wrap_content"
        android:text="ButtonOne" />
    <Button
        android:id="@+id/btn2"
        android:layout_width="match_parent"
        android:layout_height="wrap_content"
        android:text="ButtonTwo" />
</LinearLayout>
```

MainActivity.java 代码如下：

```java
public class MainActivity extends AppCompatActivity implements View.OnClickListener {
    private Button mButton1, mButton2;

    @Override
    protected void onCreate(Bundle savedInstanceState) {
        super.onCreate(savedInstanceState);
        setContentView(R.layout.activity_main);
        initViews();// 初始化控件
```

```java
    }

    private void initViews() {
        mButton1 = (Button) findViewById(R.id.btn1);
        mButton2 = (Button) findViewById(R.id.btn2);
        mButton1.setOnClickListener(this);// 注册监听
        mButton2.setOnClickListener(this);// 注册监听
    }

    @Override
    public void onClick(View v) {// 覆写 onClick 方法
        switch (v.getId()) {
            case R.id.btn1:
                Toast.makeText(MainActivity.this, "btn1", Toast.LENGTH_
                SHORT).show();
                break;
            case R.id.btn2:
                Toast.makeText(MainActivity.this, "btn2", Toast.LENGTH_
                SHORT).show();
                break;
        }
    }
}
```

这里 MainActivity 实现了 OnClickListener 接口，其 Android 源码如下：

```java
/**
 * Interface definition for a callback to be invoked when a view is clicked.
 */
public interface OnClickListener {
    /**
     * Called when a view has been clicked.
     *
     * @param v The view that was clicked.
     */
    void onClick(View v);
}
```

可以看出，这个接口中有一个抽象方法，实现 OnClickListener 接口时自然要覆写这个 onClick 方法，接口中的参数 v 即为被单击的 View 对象，通过 View 的 getId 方法获取这个 View 对象的 id，这个 id 即可用于判断不同的 View 控件。

总结一下，实现 Button 的单击事件监听需要三步：

- 实现 OnClickListenr 接口。
- 覆写其 onClick 方法并处理单击逻辑。
- 注册监听事件（初学者经常会忘记，要注意）。

方式三：内部类方式

布局代码和方式二一样，这里就不再介绍。

MainActivity.java 代码如下：

```java
public class MainActivity extends AppCompatActivity {
    private Button mButton1, mButton2;

    @Override
    protected void onCreate(Bundle savedInstanceState) {
        super.onCreate(savedInstanceState);
        setContentView(R.layout.activity_main);
        initViews();// 初始化控件
        mButton1.setOnClickListener(new View.OnClickListener() {
            @Override
            public void onClick(View v) {
                Toast.makeText(MainActivity.this, "btn1", Toast.LENGTH_
                SHORT).show();
            }
        });
        mButton2.setOnClickListener(new View.OnClickListener() {
            @Override
            public void onClick(View v) {
                Toast.makeText(MainActivity.this, "btn2", Toast.LENGTH_
                SHORT).show();
            }
        });
    }

    private void initViews() {
        mButton1 = (Button) findViewById(R.id.btn1);
        mButton2 = (Button) findViewById(R.id.btn2);
    }
}
```

initViews 方法用于初始化控件，通过 findViewById 方法并传入控件 id 来初始化控件，然后调用 View 的 setOnClickListener 方法设置事件监听。调用这个方法时需要传入 OnClickListener 对象，这里通过 new 的方式实例化，实例化这个接口时同时覆写了其 onClick 方法，这种方式称为匿名内部类方式。

除了匿名内部类的方式之外还可以通过内部类的方式来实现，代码如下：

```java
public class MainActivity extends AppCompatActivity {
    private Button mButton1, mButton2;

    @Override
    protected void onCreate(Bundle savedInstanceState) {
        super.onCreate(savedInstanceState);
        setContentView(R.layout.activity_main);
        initViews();
        mButton1.setOnClickListener(new OnClickListenerImpl());
        mButton2.setOnClickListener(new OnClickListenerImpl());
    }
    private void initViews() {
        mButton1 = (Button) findViewById(R.id.btn1);
        mButton2 = (Button) findViewById(R.id.btn2);
    }
```

```java
        // 创建内部类 OnClickListenerImpl 实现 OnClickListener 接口
        private class OnClickListenerImpl implements View.OnClickListener {
            @Override
            public void onClick(View v) {
                switch (v.getId()){
                    case R.id.btn1:
                        Toast.makeText(MainActivity.this, "btn1", Toast.
                        LENGTH_SHORT).show();
                        break;
                    case R.id.btn2:
                        Toast.makeText(MainActivity.this, "btn2", Toast.
                        LENGTH_SHORT).show();
                        break;
                }
            }
        }
    }
```

上述代码创建了内部类 OnClickListenerImpl 实现 OnClickListener 接口，覆写其 onClick 方法，同样通过 View 的 getId 方法获取控件的 id，通过 id 判断是哪个 View 被单击，在注册监听时传入 OnClickListenerImpl 对象。

4.3.2 clickable 属性设置无效分析

不知道开发者们有没有遇到这样的问题：明明将 onClick 属性设置成了 false，为什么单击时还可以响应单击事件呢？下面我们来做个实验。

主布局文件（activity_main.xml）代码如下：

```xml
<?xml version="1.0" encoding="utf-8"?>
<LinearLayout xmlns:android="http://schemas.android.com/apk/res/android"
    android:layout_width="match_parent"
    android:layout_height="match_parent">

    <Button
        android:id="@+id/btn1"
        android:layout_width="match_parent"
        android:layout_height="wrap_content"
        android:clickable="false"
        android:text="点我点我" />
</LinearLayout>
```

这里 Button 控件设置了 clickable 属性并设置其值为 false。

MainActivity.java 代码如下：

```java
public class MainActivity extends Activity {
    private Button mButton1;

    @Override
    protected void onCreate(Bundle savedInstanceState) {
        super.onCreate(savedInstanceState);
        setContentView(R.layout.activity_main);
        mButton1 = (Button) findViewById(R.id.btn1);
```

```
        mButton1.setOnClickListener(new View.OnClickListener() {
            @Override
            public void onClick(View v) {
                Toast.makeText(MainActivity.this, "点我干嘛", Toast.
                LENGTH_SHORT).show();
            }
        });
    }
}
```

上述代码通过匿名内部类的方式实现了单击事件监听，运行实例，如图 4.18 所示。

图 4.18　Button clickable 属性无效分析

可以看出打出了 Toast 信息，这时读者可能会迷糊了，为什么明明设置了 clickable 属性并且设置了其值为 false，单击事件还是响应了？这是为什么呢？从源码的角度理解，查看源码如下：

```
/**
 * Register a callback to be invoked when this view is clicked. If this
 * view is not clickable, it becomes clickable.
 *
 * @param l The callback that will run
 *
 * @see #setClickable(boolean)
 */
public void setOnClickListener(@Nullable OnClickListener l) {
    if (!isClickable()) {
        setClickable(true);
    }
    getListenerInfo().mOnClickListener = l;
}
```

从源码中可以看出，在设置事件监听时，setOnClickListener 方法里首先会调用 isClickable 方法判断当前 View 是否可单击，若当前 View 不可单击，则调用 setClickable

方法设置其为 true，这时也就解释了为什么在布局中设置属性无效了，都是源码干的好事（代码里设置优先级高）。

4.3.3　Button 实战进阶

一般 APP 都有用户注册功能，用户注册需要获取手机验证码，单击"获取验证码"按钮时短信平台通过短信的形式发送到用户手机。为了防止用户多次单击并提高用户体验，一次按下时，Button 将变得不可单击并且在 Button 上将显示下一次可获取验证码的剩余时间。下面通过实例来模拟这个功能。

布局文件代码如下：

```xml
<?xml version="1.0" encoding="utf-8"?>
<RelativeLayout xmlns:android="http://schemas.android.com/apk/res/android"
    android:layout_width="match_parent"
    android:layout_height="match_parent">

    <EditText
        android:layout_width="match_parent"
        android:layout_height="wrap_content"
        android:layout_marginTop="20dp"
        android:layout_toLeftOf="@+id/btn" />

    <Button
        android:id="@+id/btn"
        android:layout_width="100dp"
        android:layout_height="wrap_content"
        android:layout_alignParentRight="true"
        android:layout_marginTop="20dp"
        android:background="@android:color/transparent"
        android:text=" 获取验证码 " />
</RelativeLayout>
```

上述代码采用了相对布局的方式，添加了一个 EditText 控件，设置其 layout_toLeftOf 属性为" @+id/btn"，也就是 EditText 在 Button 的左边。添加了一个 Button 控件，设置其 layout_alignParentRight 属性并设置其值为 true，Button 控件位于布局的右边，设置了其 background 属性并设置其值为" @android:color/transparent"，" @android:"开头说明了这个 color 值来自 Android 系统内部，按住 Ctrl 键并单击这个值跳到 Android 内部的 color 文件，可以查看其值为 #00000000，也就是背景透明。

MainActivity.java 代码如下：

```java
public class MainActivity extends Activity {
    private Button mButtonClick;
    private TimeCount mTimeCount;

    @Override
    protected void onCreate(Bundle savedInstanceState) {
        super.onCreate(savedInstanceState);
        setContentView(R.layout.activity_main);
        mButtonClick = (Button) findViewById(R.id.btn);
```

```java
        mTimeCount = new TimeCount(60 * 1000, 1000);// 实例化 TimeCount 类
        mButtonClick.setOnClickListener(new View.OnClickListener() {
            @Override
            public void onClick(View v) {
                mTimeCount.start();// 调用 start 方法开始倒计时
            }
        });
    }

    class TimeCount extends CountDownTimer {
        // 构造方法
        public TimeCount(long totalTime, long interval) {
            super(totalTime, interval);
        }

        @Override
        public void onTick(long millisUntilFinished) {// 覆写方法 - 计时中
            mButtonClick.setEnabled(false);// 按钮不可单击
            // 参数 millisUntilFinished 表示剩余时间
            mButtonClick.setText(millisUntilFinished / 1000 + "秒");
        }

        @Override
        public void onFinish() {                      // 覆写方法，即时结束
            mButtonClick.setEnabled(true);            // 恢复按钮可单击
            mButtonClick.setText("重新获取");          // 修改按钮提示文字
        }
    }
}
```

上述代码通过匿名内部类的方式为 Button 设置了单击事件监听，单击 Button 时调用了 TimeCount 类（继承自 CountDownTimer 类）的 start 方法，开始倒计时。

对于倒计时，这里还是用到了 CountDownTimer 类，它是一个抽象类，这里自定义了一个 TimeCount 类来继承这个抽象类，TimeCount 类实现倒计时功能，覆写 onTick 方法可以实现定期通知的操作，覆写 onFinish 方法，执行定时结束后的相关操作。

对于 CountDownTimer 类的使用，API 中给出了参考实例：

```java
new CountDownTimer(30000, 1000) {
    public void onTick(long millisUntilFinished) {
        mTextField.setText("seconds remaining: " + millisUntilFinished
            / 1000);
    }
    public void onFinish() {
        mTextField.setText("done!");
    }
}.start();
```

CountDownTimer 类实例化时需要传递两个参数：第一个参数是倒计时总时长；第二个参数是时间间隔。从样例代码中也可以看出，实现该抽象类同时覆写了两个方法 onTick 方法（根据实例化 CountDownTimer 类时传入的时间间隔，定期调用的方法，样例中利用

它实现了信息更新的操作)和 onFinish 方法(倒计时结束时调用的方法)。最后调用其 start 方法,开始倒计时。从 API 文档中还可以看出 start 方法考虑了线程安全问题。

运行实例并单击"获取验证码"按钮,如图 4.19 所示,倒计时结束后如图 4.20 所示。查看动态图,请扫描图 4.21 中的二维码。

图 4.19 Button 获取验证码倒计时一

图 4.20 Button 获取验证码倒计时二

图 4.21 Button 获取验证码倒计时二维码

4.4 执行中的指示器——ProgressBar

ProgressBar 可以作为一些操作过程中的进度指示器,操作进度实时反馈给用户。在一些特殊情况下可能会使用第二进度条用以辅助显示,例如在观看视频时,第一进度条可以显示当前播放进度,而第二进度条则用以显示缓冲进度,用户体验更佳。

同时,对于某些不确定进度的情况,例如进行网络连接时,可以使用转圈的动画作为一个进度指示器,提示用户此时正在加载操作。

4.4.1 ProgressBar 样例

对于如何使用 ProgressBar,API 文档也给了一个代码样例:

```
public class MyActivity extends Activity {
    private static final int PROGRESS = 0x1;
    private ProgressBar mProgress;
    private int mProgressStatus = 0;
    private Handler mHandler = new Handler();
    protected void onCreate(Bundle icicle) {
        super.onCreate(icicle);

        setContentView(R.layout.progressbar_activity);
        mProgress = (ProgressBar) findViewById(R.id.progress_bar);
        // Start lengthy operation in a background thread
        new Thread(new Runnable() {
            public void run() {
```

```
                while (mProgressStatus < 100) {
                    mProgressStatus = doWork();
                    // Update the progress bar
                    mHandler.post(new Runnable() {
                        public void run() {
                            mProgress.setProgress(mProgressStatus);
                        }
                    });
                }
            }
        }).start();
    }
}
```

从示例代码中可以看出，doWork 方法属于耗时操作，因此这里新开了一个线程（耗时操作不能在主线程 UI 中运行，否则可能会造成 ANR，即应用程序无响应的情况，这里应该注意）。对于将耗时操作的进度反馈到 UI 线程，也有较多方法（后面课程会详细讲解），这里采用了 Handler 的 post 方法，将实时操作进度反馈到 UI 线程中的 ProgressBar 中。最后不要忘记调用它的 start 方法，启动线程。

在布局文件中引入一个 ProgressBar 可以使用一个 ProgressBar 标签，还需要一些属性修饰。下面对常用属性进行介绍，如表 4.3 所示。

表 4.3 ProgressBar 的常用属性

属 性	说 明
android:max	进度条最大值
android:maxHeight	进度条最大高度
android:progress	进度条初始值，介于 0 到最大值之间
android:secondaryProgress	第二进度条

Android 提供了几种原生的进度条样式，可以通过 style 属性在布局文件中设置进度条的样式。

- Widget.ProgressBar.Horizontal，水平进度条样式。
- Widget.ProgressBar.Small，小进度条。
- Widget.ProgressBar.Large，大进度条。
- Widget.ProgressBar.Inverse，不断跳变并旋转的进度条。
- Widget.ProgressBar.Small.Inverse，小的不断跳变并旋转的进度条。
- Widget.ProgressBar.Large.Inverse，大的不变跳变并旋转的进度条。

下面通过一个小实例来对上面进度条样式和属性进行学习。

4.4.2 ProgressBar 基础用法

布局代码如下：

```
<?xml version="1.0" encoding="utf-8"?>
<LinearLayout xmlns:android="http://schemas.android.com/apk/res/android"
    android:layout_width="match_parent"
    android:layout_height="match_parent"
```

```xml
    android:orientation="vertical">
    <!-- 水平进度条 -->
    <ProgressBar
        style="@android:style/Widget.ProgressBar.Horizontal"
        android:layout_width="match_parent"
        android:layout_height="wrap_content"
        android:max="100"
        android:progress="50"
        android:secondaryProgress="60" />
    <!-- 小进度条 -->
    <ProgressBar
        style="@android:style/Widget.ProgressBar.Small"
        android:layout_width="match_parent"
        android:layout_height="wrap_content" />
    <!-- 大进度条 -->
    <ProgressBar
        style="@android:style/Widget.ProgressBar.Large"
        android:layout_width="match_parent"
        android:layout_height="wrap_content" />
    <ProgressBar
        style="@android:style/Widget.ProgressBar.Inverse"
        android:layout_width="match_parent"
        android:layout_height="wrap_content" />
    <ProgressBar
        style="@android:style/Widget.ProgressBar.Large.Inverse"
        android:layout_width="match_parent"
        android:layout_height="wrap_content" />
    <ProgressBar
        style="@android:style/Widget.ProgressBar.Small.Inverse"
        android:layout_width="match_parent"
        android:layout_height="wrap_content" />
</LinearLayout>
```

上述布局文件中定义了六种样式的进度条，第一个进度条的 style 属性设置值为 "@android:style/Widget.ProgressBar.Horizontal"，也就是调用了 Android 内置的水平进度条样式，并对第一个进度条设置了 max（最大值）、progress（当前进度）、secondaryProgress（第二进度）等属性；其余五个进度条也分别设置了不同的样式，这里不再一一介绍。下面运行实例观察一下不同样式进度条的外观差异，如图 4.22 所示。

除了在布局中显示进度条之外，还可以在标题栏中显示进度条。

新建一个项目，主布局文件代码如下：

```xml
<?xml version="1.0" encoding="utf-8"?>
<LinearLayout xmlns:android="http://schemas.android.com/apk/res/android"
    android:layout_width="match_parent"
    android:layout_height="match_parent"
    android:orientation="vertical">

    <Button
```

```xml
        android:id="@+id/btn_show"
        android:layout_width="match_parent"
        android:layout_height="wrap_content"
        android:onClick="show"
        android:text=" 显示标题栏进度条 " />

    <Button
        android:id="@+id/btn_dismiss"
        android:layout_width="match_parent"
        android:layout_height="wrap_content"
        android:onClick="dismiss"
        android:text=" 隐藏标题栏进度条 " />
</LinearLayout>
```

上述代码采用线性布局的方式，引入了两个 Button，分别设置了 onClick 属性来响应单击事件。

MainActivity.java 代码如下：

```java
public class MainActivity extends Activity {
    @Override
    protected void onCreate(Bundle savedInstanceState) {
        super.onCreate(savedInstanceState);
        // 确定进度的标题栏进度条
        requestWindowFeature(Window.FEATURE_PROGRESS);
        // 不确定进度的标题栏进度条
        requestWindowFeature(Window.FEATURE_INDETERMINATE_PROGRESS);
        setContentView(R.layout.activity_main);
    }

    public void show(View view) {
        setProgressBarVisibility(true);
        setProgress(800);
        setProgressBarIndeterminateVisibility(true);
    }

    public void dismiss(View view) {
        setProgressBarVisibility(false);
        setProgressBarIndeterminateVisibility(false);
    }
}
```

上述代码调用了 Activity 类的 requestWindowFeature 方法传入 Window.FEATURE_PROGRESS 参数来显示带有进度的标题栏；传入 Window.FEATURE_INDETERMINATE_PROGRESS 参数则显示不带进度的标题栏进度条。注意这个方法要在 setContentView 方法之前调用（View 绘制流程）。

调用 setProgressBarVisibility 方法传入布尔变量即可决定标题栏进度条的显示与否，同理 setProgressBarIndeterminateVisibility 用以决定标题栏不确定进度条的显示与否。

运行实例，如图 4.23 所示。

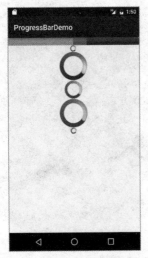

图 4.22　ProgressBar 基础样式　　　　　　　图 4.23　标题进度条一

可以看出连标题栏都没有，更别提进度条了。这是因为主题设置不对。可以为 MainActivity 设置主题，AndroidManifest.xml 代码如下：

```xml
<?xml version="1.0" encoding="utf-8"?>
<manifest xmlns:android="http://schemas.android.com/apk/res/android"
    package="com.example.administrator.progressbartitledemo">
...
    <activity
        android:name="ad.progressbarwindow.MainActivity"
        android:theme="@android:style/Theme.Holo.Light">
...
    </application>

</manifest>
```

再次运行实例，如图 4.24 所示。可以看出，此时进度条在标题栏中显示出来了。查看动态图，请扫描图 4.25 中的二维码。

图 4.24　标题进度条二　　　　　　　　图 4.25　标题进度条二维码

4.4.3 ProgressBar 模拟下载

进度条是如何更新进度的呢？这里模拟了一个下载过程。

主布局代码如下：

```xml
<?xml version="1.0" encoding="utf-8"?>
<LinearLayout xmlns:android="http://schemas.android.com/apk/res/android"
    android:layout_width="match_parent"
    android:layout_height="match_parent"
    android:orientation="vertical">

    <Button
        android:id="@+id/btn_download"
        android:layout_width="wrap_content"
        android:layout_height="wrap_content"
        android:onClick="download"
        android:text=" 下载模拟 " />

    <TextView
        android:layout_width="wrap_content"
        android:layout_height="wrap_content"
        android:text=" 下载进度如下： " />

    <ProgressBar
        android:id="@+id/probar_download"
        style="@android:style/Widget.ProgressBar.Horizontal"
        android:layout_width="match_parent"
        android:layout_height="wrap_content" />
</LinearLayout>
```

上述代码设置了一个 Button，设置了 onClick 属性用于处理单击事件，TextView 用于显示提示信息，ProgressBar 设置了水平进度条的样式。

MainActivity.java 代码如下：

```java
public class MainActivity extends AppCompatActivity {
    private ProgressBar progressBar;

    @Override
    protected void onCreate(Bundle savedInstanceState) {
        super.onCreate(savedInstanceState);
        setContentView(R.layout.activity_main);
        progressBar = (ProgressBar) findViewById(R.id.probar_download);
        progressBar.setMax(100);
        progressBar.setProgress(20);
    }

    public void download(View view) {
        new Thread() {
            @Override
            public void run() {
                for (int i = 0; i < 100; i++) {
```

```
                    try {
                        Thread.sleep(100);
                    } catch (InterruptedException e) {
                        e.printStackTrace();
                    }
                    progressBar.incrementProgressBy(1);
                }
            }
        }.start();
    }
}
```

上述代码在 onCreate 方法中调用 setMax 方法设置进度条的最大进度为 100，调用 setProgresss 方法设置起始进度为 20。对于单击事件这里新开启了一个线程，在 run 方法中通过遍历的方式，每睡眠 100ms 调用一次 ProgressBar 的 incrementProgressBy 方法增加进度，这里传入 1 表示每次增加 1 个进度。最后记得调用 start 方法开启线程。

运行实例并单击"下载模拟"按钮，如图 4.26 所示。查看动态图，请扫描图 4.27 中的二维码。

图 4.26　ProgressBar 模拟　　　　　　图 4.27　ProgressBar 模拟二维码

除了上面的系统默认的进度条颜色外，还可以对进度条颜色进行自定义。在 drawable 文件夹下新建一个 drawable 文件，代码如下：

```xml
<?xml version="1.0" encoding="utf-8"?>
<layer-list xmlns:android="http://schemas.android.com/apk/res/android">
    <item android:id="@android:id/background">
        <shape>
            <corners android:radius="10dp" />
            <solid android:color="#ffffff" />
        </shape>
    </item>
    <!-- 进度条 -->
```

```
        <item android:id="@android:id/progress">
            <clip>
                <shape>
                    <corners android:radius="10dp" />
                    <solid android:color="#3b713d" />
                </shape>
            </clip>
        </item>
</layer-list>
```

上述代码中 layer-list 用来将多个图层堆叠显示，借助这个特性可以做一些特别的效果。layer-list 标签中定义了两个 item，第一个 item 引入了 id 属性并设置其值为"@android:id/background"，表示控件的背景，shape 定义控件形状，有如下几个标签：solid，指定填充的颜色；gradient，颜色渐变；stroke，描边；corners，圆角；padding，间隔。

这里引入了 corners 标签并设置其属性"android:radious"值为 10dp，表示圆角的半径为 10dp；使用了 solid 标签，设置"android:color"为 #ffffff，表示背景颜色为纯白色。

第二个 item 其 id 值为 @android:id/progress 表示控件进度，同样也是引入了 corners 标签和 solid 标签来添加圆角和设定颜色，为了区别，这里为 solid 设置了和控件背景不同的颜色。

设置好这个图片文件后，需要在布局文件中引用这个图片文件，代码如下：

```
<ProgressBar
    android:id="@+id/probar_download"
    style="@android:style/Widget.ProgressBar.Horizontal"
    android:layout_width="match_parent"
    android:layout_height="wrap_content"
    android:progressDrawable="@drawable/progressbar_color" />
```

上述代码设置了 progressDrawable 属性，并设置其值为自定义的 drawable 文件，这时再次运行实例，如图 4.28 所示。

可以看出进度条的颜色和背景都改变了。

4.5 对话框之父——Dialog

对话框是人机交互中的重要控件，在开发中也经常会用到各式各样的对话框。总结一下，对话框主要有以下几种：

- AlertDialog：警告对话框，是最常见的对话框形式，是 Dialog 的直接子类。如果想要实例化 AlertDialog 类，往往需要依靠其内部类 AlertDialog.Builder 类完成。

图 4.28　ProgressBar 改变默认进度条颜色

- 单选和多选对话框：适用于一些需要用户做出选择的场景，在需要选择时弹出，选择结束后关闭，可以节省有限的屏幕空间。

- **ProgressDialog**：进度对话框，在进行网络请求或文件操作等耗时操作时常常会用到。
- 定制对话框，满足个性化需要，对于一些复杂的界面想做成对话框的形式，可以自行定制一个对话框。

总结其常用的方法，如表 4.4 所示。

表 4.4　Dialog 的常用方法

方　　法	说　　明
setTitle	标题设置
setIcon	图标设置
setMessage	提示消息设置
setItems	显示列表设置
setView	自定义显示样式
setSingleChoiceItems	单选框列表
setMultiChoiceItems	复选框列表
setPositiveButton	确定按钮
setNeutralButton	退出按钮
setNegativeButton	取消按钮
create	创建一个对话框
show	显示一个对话框
dismiss	隐藏一个对话框

4.5.1　AlertDialog

在进行删除或应用退出等操作时，为了防止用户误操作，通常要弹出警告对话框进一步提示用户是否进行该操作。这里以提示是否退出应用为例，介绍 AlertDialog 是如何使用的。

不需要布局文件，**MainActivity.java** 代码如下：

```java
public class MainActivity extends Activity {

    @Override
    protected void onCreate(Bundle savedInstanceState) {
        super.onCreate(savedInstanceState);
        setContentView(R.layout.activity_main);
    }

    @Override
    public boolean onKeyDown(int keyCode, KeyEvent event) {
        if (keyCode == KeyEvent.KEYCODE_BACK) {
            showDialog();

        }
        return super.onKeyDown(keyCode, event);
    }

    private void showDialog() {
        Dialog dialog = new AlertDialog.Builder(this)
```

```
                    // 设置标题
                    .setTitle("退出程序？")
                    // 设置提示内容
                    .setMessage("确定退出程序吗？")
                    // 确定按钮
                    .setPositiveButton("确定", new DialogInterface.
                    OnClickListener() {
                        @Override
                        public void onClick(DialogInterface dialog, int
                        which) {
                            finish();
                        }
                    })
                    // 取消按钮
                    .setNegativeButton("取消", new DialogInterface.
                    OnClickListener() {
                        @Override
                        public void onClick(DialogInterface dialog, int
                        which) {
                            Toast.makeText(MainActivity.this, "您点了取消按钮",
                                Toast.LENGTH_SHORT).show();
                        }
                    })
                    // 创建对话框
                    .create();
            // 显示对话框
            dialog.show();
        }
}
```

上述代码中，覆写 Activity 的 onKeyDown 方法用于监听按键事件，通过其参数 keyCode 进行按键的判断，当 keyCode 等于 KeyEvent.KEYCODE_BACK 时认为按下了返回键，调用 showDialog 显示对话框。

创建一个 AlertDialog 主要用到的方法如下：
- setTitle：设置对话框的标题。
- setMessage：设置对话框显示信息。
- setPositiveButton：设置"确定"按钮。需要传入两个参数：第一个是按钮显示的文本；第二个是这个按钮的监听事件。通过匿名内部类的方式实现接口，这里单击"确定"按钮时调用 Activity 的 finish 方法结束这个 Activity。
- setNegativeButton：设置"取消"按钮，同样需要传入两个参数，在监听事件中打印了一个 Toast。

运行实例，如图 4.29 所示。

当然除了这种方式退出 APP 之外，更常用的是连续单击两次退出程序，参考代码如下：

```
public class MainActivity extends Activity {

    private long mEixtTime = 0;
```

```java
    @Override
    protected void onCreate(Bundle savedInstanceState) {
        super.onCreate(savedInstanceState);
        setContentView(R.layout.activity_main);
    }

    @Override
    public boolean onKeyDown(int keyCode, KeyEvent event) {
        if (keyCode == KeyEvent.KEYCODE_BACK) {
            exit();
            return false;
        }
        return super.onKeyDown(keyCode, event);
    }

    public void exit() {
        if ((System.currentTimeMillis() - mEixtTime) > 2000) {
            Toast.makeText(getApplicationContext(), "再按一次退出程序",
                    Toast.LENGTH_SHORT).show();
            mEixtTime = System.currentTimeMillis();
        } else {
            finish();
        }
    }
}
```

其原理是：单击返回时记下时间，再一次单击返回时判断两次单击的时刻差，若时间差小于 2000ms 则调用 finish 方法退出 APP，若时间差大于 2000ms 就提示"再按一次退出程序"。

运行实例，如图 4.30 所示。在 2000ms 内单击两次返回键即可退出程序。查看动态图，请扫描图 4.31 中的二维码。

图 4.29　AlertDialog 基础实例　　图 4.30　双击 back 退出应用实例　　图 4.31　双击 back 退出应用实例二维码

4.5.2 单选和多选对话框

下面介绍如何使用 setMultiChoiceItems 和 setSingleChoiceItems 方法，构造一个可供选择的对话框。

布局文件代码如下：

```xml
<?xml version="1.0" encoding="utf-8"?>
<LinearLayout xmlns:android="http://schemas.android.com/apk/res/android"
    android:layout_width="match_parent"
    android:layout_height="match_parent"
    android:orientation="vertical">
    <Button
        android:layout_width="match_parent"
        android:layout_height="wrap_content"
        android:onClick="singleChoiceItems"
        android:text="单选对话框样式" />
    <Button
        android:layout_width="match_parent"
        android:layout_height="wrap_content"
        android:onClick="multiChoiceItems"
        android:text="多选对话框样式" />
</LinearLayout>
```

上述代码定义了两个按钮并设置了相应的 onClick 属性用于响应 Button 的单击事件。

MainActivity.java 代码如下：

```java
public class MainActivity extends Activity {
    String single[] = {"Java", "C", "C++"};
    String multi[] = {"android", "iOS", "wp"};
    StringBuilder mStringBuilder;
    private String mSingleChoice;

    @Override
    protected void onCreate(Bundle savedInstanceState) {
        super.onCreate(savedInstanceState);
        setContentView(R.layout.activity_main);
    }

    public void singleChoiceItems(View view) {
        Dialog dialog = new AlertDialog.Builder(this)
                .setTitle("单选对话框实例")
                .setPositiveButton("确定", new DialogInterface.
                OnClickListener() {
                    @Override
                    public void onClick(DialogInterface dialog, int 
                    which) {


                        Toast.makeText(MainActivity.this,
                        "选择了" + mSingleChoice, Toast.LENGTH_SHORT).
                        show();
                        dialog.dismiss();
```

```java
                }
            })
            // 设置单选对话框监听
            .setSingleChoiceItems(single, 0, new DialogInterface.
            OnClickListener() {
                @Override
                public void onClick(DialogInterface dialog, int 
                which) {
                    // 根据which决定选择哪一个子项
                    mSingleChoice = single[which];
                }
            }).create();
    dialog.show();
}

public void multiChoiceItems(View view) {
    mStringBuilder = new StringBuilder();
    Dialog dialog = new AlertDialog.Builder(this)
            .setTitle("多选对话框实例")
            // 设置图标
            .setIcon(android.R.drawable.ic_btn_speak_now)
            .setPositiveButton("确定", new DialogInterface.
            OnClickListener() {
                @Override
                public void onClick(DialogInterface dialog, int 
                which) {
                    Toast.makeText(MainActivity.this,
                     "选择了" + mStringBuilder, Toast.LENGTH_SHORT).
                    show();
                }
            })
            .setNegativeButton("取消", new DialogInterface.
            OnClickListener() {
                @Override
                public void onClick(DialogInterface dialog, int 
                which) {
                    // 隐藏对话框
                    dialog.dismiss();
                }
            })
            // 设置多选对话框监听
            .setMultiChoiceItems(multi, null,
                new DialogInterface.OnMultiChoiceClickListener() {
                @Override
                public void onClick(DialogInterface dialog,
                                    int which, boolean isChecked) {
                    // 满足选择条件
                    if (isChecked) {
                        // 根据which决定选择哪一个子项
                        mStringBuilder.append(multi[which] + "、");
                    }
```

```
                }
            }).create();
        dialog.show();
    }
}
```

上述代码自定义了两个方法来创建单选对话框和多选对话框：

- setSingleChoiceItems：单选对话框形式。这里要传入三个参数：第一个参数为数据源；第二个是默认选择项，传入数组下标；第三个是选择事件监听，这里实现 DialogInterface.OnClickListener 接口并覆写了其 onClick 方法，根据其参数值 which 判断哪一项被选中。
- setMultiChoiceItems：多选对话框形式。也要传入三个参数：第一个参数为数据源；第二个是初始选择的下标数组，这里传入 null，默认都不选；第三个是选择事件监听，实现了 DialogInterface.OnMultiChoiceClickListener 接口并覆写了其 onClick 方法，方法中有三个参数，要根据 isChecked 和 which 两个参数判断哪些项被选中。

可以看出，这两个方法参数内容和监听都是较为相似的，学习技术就是要分析出事物的共性，辨别其异性，这样才能举一反三，增进理解，并提高学习速度。

运行实例并单击"单选框样式"按钮，如图 4.32 所示。单击"多选框样式"按钮，如图 4.33 所示。

查看动态图，请扫描图 4.34 中的二维码。

图 4.32　单选对话框实例

图 4.33　多选对话框实例

图 4.34　单选、多选对话框实例二维码

4.5.3　ProgressDialog 进度对话框

在处理一些耗时操作时，考虑到用户体验，需要将实时进度告知用户。除了上面章节中介绍到的 ProgressBar 组件外，还有 ProgressDialog 可以供用户调用，此控件需要的时候即弹出，进度完成后即消失，不占用布局空间。

API 中提供的 ProgressDialog 的常用方法，如表 4.5 所示。

表 4.5　ProgressDialog 的常用方法

方　　法	说　　明
STYLE_HORIZONTAL	常量，水平进度条
STYLE_SPINNER	常量，环形进度条
setMessage(CharSequence message)	设置显示信息
setProgressStyle(int style)	设置进度条样式
onStart	启动进度框
setMax	设置最大进度
setButton(CharSequence text, final OnClickListener listener)	在对话框上设置按钮
setSecondaryProgress(int secondaryProgress)	设置第二进度条进度
incrementProgressBy(int diff)	设置进度条每次增长的进度

下面通过一个实例，对上面的方法进行学习。

主布局文件（activity_main.xml）代码如下：

```xml
<?xml version="1.0" encoding="utf-8"?>
<RelativeLayout xmlns:android="http://schemas.android.com/apk/res/android"
    android:layout_width="match_parent"
    android:layout_height="match_parent">

    <Button
        android:layout_width="match_parent"
        android:layout_height="wrap_content"
        android:onClick="progressDialog"
        android:text=" 模拟网络请求 弹出 ProgressDialog" />
</RelativeLayout>
```

上述代码采用了相对布局，添加了一个 Button 按钮并设置 onClick 属性用于响应单击事件。

MainActivity.java 代码如下：

```java
public class MainActivity extends Activity {
    @Override
    protected void onCreate(Bundle savedInstanceState) {
        super.onCreate(savedInstanceState);
        setContentView(R.layout.activity_main);
    }

    public void progressDialog(View view) {
        // 获得 ProgressDialog 对象
        final ProgressDialog progressDialog = new ProgressDialog(this);
        progressDialog.setTitle(" 正在网络请求 ...");
        // 设置进度对话框样式
        progressDialog.setProgressStyle(ProgressDialog.STYLE_HORIZONTAL);
        // 设置最大进度
        progressDialog.setMax(100);
        // 设置初始进度
        progressDialog.setProgress(10);
```

```java
        // 设置按钮监听
        progressDialog.setButton("隐藏", new DialogInterface.
OnClickListener() {
            @Override
            public void onClick(DialogInterface dialog, int which) {
                progressDialog.dismiss();
            }
        });
        // 启动进度条
        progressDialog.onStart();
        // 新开一个线程模拟网络请求
        new Thread() {
            @Override
            public void run() {
                int i = 10;
                while (i <= 100) {
                    try {
                        // 线程休眠100ms
                        Thread.sleep(100);
                        // 每次增加1
                        progressDialog.incrementProgressBy(1);
                        i++;
                    } catch (InterruptedException e) {
                        e.printStackTrace();
                    }
                }
                // 进度条走完时，调用dismiss方法隐藏进度对话框
                progressDialog.dismiss();
            }
        }.start();
        progressDialog.show();
    }
}
```

上述代码首先创建了一个 ProgressDialog 对象，初始化对象时需要传入上下文对象，然后调用 setTitle 方法为其设置一个标题；调用 setProgressStyle 方法为其设置一个样式，这里设置了水平进度条的样式；调用 setMax 方法设置进度条的最大值；调用 setProgress 方法设置初始进度值；调用 setButton 方法为 ProgressDialog 方法添加一个按钮，需要传入两个参数，第一个参数为按钮显示文本，第二个参数为其添加一个单击监听，覆写了 onClick 方法，方法中调用了 ProgressDialog 的 dismiss 方法，即单击按钮隐藏进度条；调用 onStart 方法启动进度；调用 show 方法显示进度框。

对于模拟进度，这里新开了一个线程，在 run 方法中添加了一个 while 循环，当 i≤100，也就是进度条没有走到头时调用 ProgressDialog 的 incrementProgressBy 方法增加进度。当 i=100 时，结束 while 循环并调用 ProgressDialog 的 dismiss 方法隐藏进度条。最后不要忘记调用 start 方法启动线程。

运行实例如图 4.35 所示。查看动态图，请扫描图 4.36 中的二维码。

图 4.35 ProgressDialog 模拟

图 4.36 ProgressDialog 模拟二维码

4.5.4 定制对话框

系统对话框布局形式是有限的,一般不能满足个性化需要,这里通过一个自定义布局的进度对话框模拟网络请求的过程,并引入了动画方面的相关知识。对这部分知识不熟悉的同学可以先行跳过,后面还会对动画知识进行系统讲解。

主布局文件(activity_main.xml)代码如下:

```xml
<?xml version="1.0" encoding="utf-8"?>
<RelativeLayout xmlns:android="http://schemas.android.com/apk/res/android"
    android:layout_width="match_parent"
    android:layout_height="match_parent">

    <Button
        android:layout_width="match_parent"
        android:layout_height="wrap_content"
        android:onClick="test"
        android:text=" 模拟请求 " />
</RelativeLayout>
```

上述代码在线性布局中添加了一个 Button 并设置其 onClick 属性响应单击事件。

自定义布局(load_layout.xml)对话框布局文件代码如下:

```xml
<?xml version="1.0" encoding="utf-8"?>
<LinearLayout xmlns:android="http://schemas.android.com/apk/res/android"
    android:id="@+id/dialog_view"
    android:layout_width="wrap_content"
    android:layout_height="wrap_content"
    android:background="#ffffffff"
    android:gravity="center"
    android:minHeight="60dp"
    android:minWidth="180dp"
    android:orientation="vertical"
```

```xml
        android:padding="10dp">

        <ImageView
            android:id="@+id/img"
            android:layout_width="100dp"
            android:layout_height="100dp"
            android:src="@drawable/rotate" />

        <TextView
            android:id="@+id/tv_tip"
            android:layout_width="wrap_content"
            android:layout_height="wrap_content"
            android:layout_marginLeft="10dp"
            android:layout_marginTop="20dp"
            android:text=" 数据加载中......" />
</LinearLayout>
```

上述代码中自定义进度对话框布局采用了垂直的线性布局，设置了最小高度和最小宽度的属性。在线性布局中引入了两个控件：ImageView 用于显示进度提示图片；TextView 用于显示进度提示信息。

通过 style 的形式设置了对话框的样式，在 styles.xml 文件中添加如下代码：

```xml
<!-- 自定义 loading dialog -->
<style name="loading_dialog" parent="android:style/Theme.Dialog">
    <!-- 无边框 -->
    <item name="android:windowFrame">@null</item>
    <!-- 无标题 -->
    <item name="android:windowNoTitle">true</item>
    <!-- 窗口浮动 -->
    <item name="android:windowIsFloating">true</item>
</style>
```

style 可以理解成一部分 Android 属性的集合，因为具有通用性，所以被抽离出来放在一起，使用时通过 R.style.* 调用该样式。定义一个样式需要 style 标签进行包裹，parent 属性设置类似 Java 中继承的概念（继承父样式表中的属性），这个是可选的。一个 item 标签包裹一个属性，这里设置了无边框、不显示标题、窗口浮动。

进度条是旋转的，这里引入了旋转动画来模拟进度条，通过 xml 文件的形式定义：

```xml
<?xml version="1.0" encoding="utf-8"?>
<set xmlns:android="http://schemas.android.com/apk/res/android"
    android:shareInterpolator="false">
    <rotate
        android:duration="1500"
        android:fromDegrees="0"
        android:interpolator="@android:anim/accelerate_decelerate_interpolator"
        android:pivotX="50%"
        android:pivotY="50%"
        android:repeatCount="-1"
        android:repeatMode="restart"
```

```
            android:startOffset="-1"
            android:toDegrees="+360" />
</set>
```

文件中的动画属性含义，如表 4.6 所示。

表 4.6 传统动画的属性

属 性	说 明
duration	动画持续时间
interpolator	动画插值器，函数控制动画变化的速率
pivotX	X 方向动画基点，50% 表示水平中心
pivotY	Y 方向动画基点，50% 表示纵向中心
repeatCount	动画重复次数，-1 表示无限次
repeatMode	动画重复模式
startOffset	动画开始延迟时间
toDegrees	旋转角度

MainActivity.java 代码如下：

```java
public class MainActivity extends Activity {
    @Override
    protected void onCreate(Bundle savedInstanceState) {
        super.onCreate(savedInstanceState);
        setContentView(R.layout.activity_main);
    }

    public void test(View view) {
        final Dialog dialog = createloadDialog(MainActivity.this, "稍等，
                正在加载中...");
        new Thread() {
            @Override
            public void run() {
                try {
                    sleep(3000);
                    dialog.dismiss();
                } catch (InterruptedException e) {
                    e.printStackTrace();
                }
            }
        }.start();
        dialog.show();
    }

        // 返回一个 ProgressDialog 对象
    public Dialog createloadDialog(Context context, String msg) {
        LayoutInflater layoutInflater = LayoutInflater.from(context);
        // 由自定义布局获得 View 对象
        View view = layoutInflater.inflate(R.layout.load_layout, null);
        // 加载布局
        LinearLayout linearLayout = (LinearLayout) view.findViewById(R.
            id.dialog_view);
        // 获取 View 对象中的 ImageView
```

```
            ImageView imageView = (ImageView) view.findViewById(R.id.img);
            TextView tipTextView = (TextView) view.findViewById(R.id.tv_tip);
            // 加载动画
            Animation animation = AnimationUtils.loadAnimation(
                    context, R.anim.load_animation);
            // 设置动画
            imageView.startAnimation(animation);
            // 设置加载信息
            tipTextView.setText(msg);
            // 创建自定义样式dialog
            Dialog loadDialog = new Dialog(context, R.style.loading_dialog);
            // 不可以用返回键取消
            loadDialog.setCancelable(false);
            // 设置布局
            loadDialog.setContentView(view, new LinearLayout.LayoutParams(
                    LinearLayout.LayoutParams.MATCH_PARENT,
                    LinearLayout.LayoutParams.MATCH_PARENT));
            return loadDialog;
        }
    }
```

这里说明几点：

- createloadDialog 方法返回一个 ProgressDialog 对象，这里使用了 LayoutInflater 类的 inflate 获得了进度框的 View 对象，并使用 findViewById 方法获取了 View 对象中的控件。加载动画文件时用到了 AnimationUtils 的 loadAnimation 方法，可以得到一个动画对象，开始动画使用了 Animation 类的 startAnimation 方法。
- 对于 Dialog 的实例化，第一个参数是上下文对象，第二个是主题样式，设置了 setCancelable 方法的参数为 false，表示不可以用返回键取消对话框，最后调用 Dialog 的 setContentView 方法，传入对话框布局文件对象和布局方式以渲染对话框布局界面。
- 对于延迟一个方法的执行，这里采用了线程睡眠的方式，让线程睡眠 3s 后调用 Dialog 的 dismiss 方法关闭对话框，模拟请求完成。

运行实例，如图 4.37 所示。查看动态图，请扫描图 4.38 中的二维码。

图 4.37　定制对话框实例

图 4.38　定制对话框实例二维码

第 5 章　Android 控件进阶操作实战

5.1　控之经典——ListView

　　ListView 是最经典的控件之一，虽然现在其江山地位不稳，将要被 RecylerView 取代，但设计理念是很经典的，而且很多程序员还是习惯了 ListView，因此我们还需要对 ListView 进行深入学习。ListView 内容非常多，读者要有足够的耐心进行学习，每一个功能点都有可能应用到项目中。

　　ListView 经常被用在列表显示上，每一个列表项都具有相同的布局，一个 ListView 通常都有三个要素组成：

- ListView 控件。
- 适配器类，用到了设计模式中的适配器模式，它是视图和数据之间的桥梁，负责提供对数据的访问，生成每一个列表项对应的 View。常用的适配器类有 ArrayAdapter、SimpleAdapter 和 SimpleCursorAdapter。
- 数据源。

　　当然最重要、最复杂的部分就是适配器类的编写和设计，在一些复杂的界面，常常需要对适配器类进行相关逻辑处理。

　　ListView 的常用属性如表 5.1 所示。

表 5.1　ListView 的常用属性

属　性	说　明
android:divider	子项分割线
android:dividerHeight	分割线高度
android:listSelector	子项单击效果
android:scrollbars	滑动条

　　ListView 的常用方法如表 5.2 所示。

表 5.2　ListView 的常用方法

方　　法	说　明
addFooterView(View v)	在列表尾部加入一个 View
addHeaderView(View v)	在列表头部加入一个 View
setAdapter(ListAdapter adapter)	设置适配器
setDivider(Drawable divider)	设置子项分隔栏
setDividerHeight(int height)	设置分隔栏高度

5.1.1 ArrayAdapter 适配器

ListView 的数据渲染都需要借助适配器来完成，首先看一下结合最简单的 ArrayAdapter 来实现 ListView。

主布局文件（activity_main.xml）代码如下：

```xml
<?xml version="1.0" encoding="utf-8"?>
<RelativeLayout xmlns:android="http://schemas.android.com/apk/res/android"
    android:layout_width="match_parent"
    android:layout_height="match_parent">
    <ListView
        android:id="@+id/lv"
        android:layout_width="wrap_content"
        android:layout_height="wrap_content"
        android:divider="@android:color/holo_red_dark"
        android:dividerHeight="3dp"
        android:scrollbars="none"/>
</RelativeLayout>
```

上述代码设置了 divider 属性，在 ListView 的子项之间添加分隔栏；设置了 dividerHeight 属性，决定了分隔栏的高度；将 scrollbars 属性的值设置为 none 表示上下拖动时在右侧没有滑动条。

MainActivity.java 代码如下：

```java
public class MainActivity extends Activity {
    private ListView mListView;
    private String mDatas[] = {"Sunday", "Monday", "Tuesday", "Wednesday",
            "Thursday", "Friday", "Saturday", "Sunday", "Monday", "Tuesday",
            "Wednesday","Thursday", "Friday", "Saturday"};// 准备数据源
    private ArrayAdapter<String> mAdapter;

    @Override
    protected void onCreate(Bundle savedInstanceState) {
        super.onCreate(savedInstanceState);
        requestWindowFeature(Window.FEATURE_NO_TITLE);// 隐藏标题栏
        setContentView(R.layout.activity_main);
        mListView = (ListView) findViewById(R.id.lv);
        // 实例化 ArrayAdapter
        mAdapter = new ArrayAdapter<String>(this,
                android.R.layout.simple_list_item_1, mDatas);
        // 设置适配器
        mListView.setAdapter(mAdapter);
    }
}
```

创建一个 ArrayAdapter 对象需要传入三个参数：上下文对象、子项布局 id（用到了 Android 内置的 list 布局）、数据源。上述代码中 ArrayAdapter 的数据源传入的是字符数组，最后调用 ListView 的 setAdapter 方法为 ListView 设置适配器。

运行实例，如图 5.1 所示。

可以看出，每个子项之间存在分隔栏，上下拖动 ListView 时最右边也不会有滑动条出现。

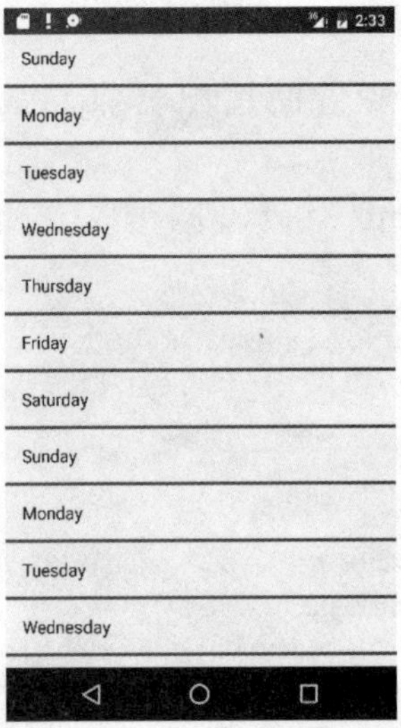

图 5.1　ListView 之 ArrayAdapter

5.1.2　SimpleAdapter 适配器

ArrayAdapter 适用于信息显示比较单一的场景，若显示项中包含多种形式的数据，就不太适用了。下面介绍可以适配多种数据类型的适配器类 SimpleAdapter 的使用方法。

当存在多种数据类型时首先要考虑布局问题，因此首先要设置子项目布局（item_layout.xml）文件，代码如下：

```xml
<?xml version="1.0" encoding="utf-8"?>
<LinearLayout xmlns:android="http://schemas.android.com/apk/res/android"
    android:layout_width="match_parent"
    android:layout_height="match_parent"
    android:orientation="horizontal">

    <ImageView
        android:id="@+id/img"
        android:layout_width="50dp"
        android:layout_height="50dp"
        android:src="@mipmap/ic_launcher" />

    <TextView
        android:id="@+id/tv"
        android:layout_width="wrap_content"
        android:layout_height="50dp"
        android:layout_marginLeft="50dp"
        android:gravity="center"
        android:text="hello"
        android:textSize="28sp" />
</LinearLayout>
```

上述代码采用线性布局，设置其 orientation 属性为 horizontal（水平布局），添加了一个 ImageView 控件用于显示图片，添加了一个 TextView 用于显示文本。

主布局文件（activity_main.xml）代码如下：

```xml
<?xml version="1.0" encoding="utf-8"?>
<RelativeLayout xmlns:android="http://schemas.android.com/apk/res/android"
    android:layout_width="match_parent"
    android:layout_height="match_parent">

    <ListView
        android:id="@+id/listview"
        android:layout_width="match_parent"
        android:layout_height="match_parent" />
</RelativeLayout>
```

上述代码在主布局中添加了一个 ListView 控件并设置了 id 属性，设置宽、高属性的属性值都是 match_parent。

MainActivity.java 代码如下：

```java
public class MainActivity extends Activity {
    private ListView mListView;
    private SimpleAdapter mSimpleAdapter;
    private List<Map<String, Object>> mDatas = new ArrayList<Map<String, Object>>();

    @Override
    protected void onCreate(Bundle savedInstanceState) {
        super.onCreate(savedInstanceState);
        setContentView(R.layout.activity_main);
        mListView = (ListView) findViewById(R.id.listview);
        // 初始化数据集
        initDatas();
        // 实例化 SimpleAdapter
        mSimpleAdapter = new SimpleAdapter(
                this,
                mDatas,
                R.layout.item_layout, new String[]{"img", "name"},
                new int[]{R.id.img, R.id.tv});
        // 设置配置器
        mListView.setAdapter(mSimpleAdapter);
    }

    private void initDatas() {
        Map map1 = new HashMap();
        map1.put("img", R.drawable.fish);
        map1.put("name", "小金鱼");
        Map map2 = new HashMap();
        map2.put("img", R.drawable.horse);
        map2.put("name", "千里马");
        Map map3 = new HashMap();
        map3.put("img", R.drawable.mouse);
        map3.put("name", "米老鼠");
        mDatas.add(map1);
        mDatas.add(map2);
        mDatas.add(map3);
```

```
        }
    }
```

SimpleAdapter 的构造函数如下：

```
SimpleAdapter(Context context, List<? extends Map<String, ?>> data, int
resource, String[] from, int[] to)
```

实例化 SimpleAdapter 时要传入如下参数：
- context：即上下文对象；
- data：一个包裹 Map 集合的 List 数据集，这里传入 datas，即 initDatas 方法初始化的数据集；
- resource：子项布局文件，这里是自定义的 item_layout.xml 文件，传入 R.layout.animal_layout；
- from：一个字符串数组，字符串指的是 Map 中的键值，这里有两个键，即 img 和 name；
- to：一个 int 型数组，表示子项布局中对应控件的 id，这里传入 ImageView 的 id（R.id.img）和 TextView 的 id（R.id.tv）即可。

运行实例，如图 5.2 所示。

图 5.2 ListView 之 SimpleAdapter

5.1.3 BaseAdapter 适配器

上面讲解了 SimpleAdapter 作为 ListView 的适配器，通过源码可以看出 SimpleAdapter 继承自 BaseAdapter，也就是说这个 SimpleAdapter 可以看作是 Android 帮我们实现的适配器，下面研究如何通过继承 BaseAdapter 实现自定义的适配器。

主布局文件（activity_main.xml）代码如下：

```
<?xml version="1.0" encoding="utf-8"?>
<RelativeLayout xmlns:android="http://schemas.android.com/apk/res/android"
    android:layout_width="match_parent"
    android:layout_height="match_parent">

    <ListView
        android:id="@+id/lv"
        android:layout_width="match_parent"
        android:layout_height="match_parent" />
</RelativeLayout>
```

上述代码在相对布局中添加了一个 ListView 控件，设置了 id 属性为 lv，并添加了宽、高属性为 match_parent。

子项布局代码如下：

```
<LinearLayout xmlns:android="http://schemas.android.com/apk/res/android"
    android:layout_width="match_parent"
    android:layout_height="match_parent"
    android:orientation="horizontal">

    <ImageView
```

```xml
        android:id="@+id/img"
        android:layout_width="50dp"
        android:layout_height="50dp"
        android:layout_marginLeft="20dp"
        android:src="@mipmap/ic_launcher" />

    <TextView
        android:id="@+id/tv"
        android:layout_width="wrap_content"
        android:layout_height="50dp"
        android:layout_marginLeft="50dp"
        android:gravity="center"
        android:text="hello"
        android:textSize="28sp" />
</LinearLayout>
```

子布局和 SimpleAdapter 的布局一致。

BaseAdapter 的数据源可能比较复杂，因此这里创建一个 JavaBean 类对数据进行封装：

```java
public class Animal {
    public Animal(String animal, int imgId) {
        this.animal = animal;
        this.imgId = imgId;
    }

    private String animal;
    private int imgId;

    public String getAnimal() {
        return animal;
    }

    public void setAnimal(String animal) {
        this.animal = animal;
    }

    public int getImgId() {
        return imgId;
    }

    public void setImgId(int imgId) {
        this.imgId = imgId;
    }
}
```

这个类中封装了两个属性：String 型的动物名和 int 型的图片 id，添加了一个包含这两个属性的构造方法并设置了相应的 Setter 和 Getter 方法。

下面看一下自定义的适配器类，它继承自 BaseAdapter，代码如下：

```java
public class AnimalAdapter extends BaseAdapter {
    private Context context;
    private List<Animal> datas;
```

```java
// 构造函数需要传入两个必要的参数：上下文对象和数据源
public AnimalAdapter(Context context, List<Animal> datas) {
    this.context = context;
    this.datas = datas;
}
// 返回子项的个数
@Override
public int getCount() {
    return datas.size();
}
// 返回子项对应的对象
@Override
public Object getItem(int position) {
    return datas.get(position);
}
// 返回子项的下标
@Override
public long getItemId(int position) {
    return position;
}
// 返回子项视图
@Override
public View getView(int position, View convertView, ViewGroup parent) {
    Animal animal = (Animal) getItem(position);
    View view;
    ViewHolder viewHolder;
    if (convertView == null) {
        view = LayoutInflater.from(context).inflate(R.layout.item_
        layout, null);
        viewHolder = new ViewHolder();
        viewHolder.animalImage = (ImageView) view.findViewById(R.
         id.img);
        viewHolder.animalName = (TextView) view.findViewById(R.
         id.tv);
        view.setTag(viewHolder);
    } else {
        view = convertView;
        viewHolder = (ViewHolder) view.getTag();
    }
    viewHolder.animalName.setText(animal.getAnimal());
    viewHolder.animalImage.setImageResource(animal.getImgId());
    return view;
}
// 创建 ViewHolder 类
class ViewHolder {
    ImageView animalImage;
    TextView animalName;
}
}
```

自定义的 AnimalAdapter 类继承自 BaseAdapter 类，必须要覆写[①]四个方法，每个方法的具体含义已经在代码中做了注释。此外，为了提高加载效率，这里创建了内部类 ViewHolder，可以避免每次调用 getView 方法时都要通过 findViewById 方法去实例化控件，可以提高运行效率。

MainActivity.java 代码如下：

```java
public class MainActivity extends Activity {
    private ListView mListView;
    private List<Animal> datas = new ArrayList<Animal>();
    private AnimalAdapter mAnimalAdapter;

    @Override
    protected void onCreate(Bundle savedInstanceState) {
        super.onCreate(savedInstanceState);
        // 隐藏标题栏
        requestWindowFeature(Window.FEATURE_NO_TITLE);
        setContentView(R.layout.activity_main);
        // 初始化数据源
        initDatas();
        mListView = (ListView) findViewById(R.id.lv);
        mAnimalAdapter = new AnimalAdapter(this, datas);
        mListView.setAdapter(mAnimalAdapter);
        mListView.setOnItemClickListener(new AdapterView.
        OnItemClickListener() {
            @Override
            public void onItemClick(AdapterView<?> parent, View view,
            int position, long id) {
                Toast.makeText(MainActivity.this,
                        "您单击了" + datas.get(position).getAnimal(),
                        Toast.LENGTH_SHORT).show();
            }
        });
    }

    private void initDatas() {
        Animal animal0 = new Animal("兔八哥", R.drawable.rabbit);
        Animal animal1 = new Animal("眼镜蛇", R.drawable.snack);
        Animal animal2 = new Animal("小金鱼", R.drawable.fish);
        Animal animal3 = new Animal("千里马", R.drawable.horse);
        Animal animal4 = new Animal("米老鼠", R.drawable.mouse);
        Animal animal5 = new Animal("大国宝", R.drawable.panda);
        datas.add(animal0);
        datas.add(animal1);
        datas.add(animal2);
        datas.add(animal3);
        datas.add(animal4);
        datas.add(animal5);
    }
}
```

---

① "覆写"为 Java 里的术语。

上述代码为 ListView 设置了 setOnItemClickListener 方法监听单项单击事件，采用匿名内部类的方式实现了 AdapterView.OnItemClickListener 接口并覆写了其 onItemClick 方法，由参数 positon 通过 List 的 get 方法并传入 Position 来获取 Animal 对象，再通过对象封装的 getAnimal 方法获得对应的动物名，通过 Toast 显示出来。

运行实例，如图 5.3 所示。查看动态图，请扫描图 5.4 中的二维码。

图 5.3　ListView 之 BaseAdapter

图 5.4　ListView 之 BaseAdapter 二维码

5.2　控之经典——ListView 进阶

如何在 ListView 的上添加滑动监听、实现上拉加载功能，是面试中和实际工作中经常会遇到的问题。能否很从容地处理这部分问题，也反映了程序员的基础能力。因此，有必要要对这部分知识点进行研究和学习。

首先看一下 API 文档中滑动监听的定义，其继承结构如下：

public static interface

AbsListView.OnScrollListener

android.widget.AbsListView.OnScrollListener

由继承结构可以看出，OnScrollListener 是一个静态接口，接口中都是未实现需要覆写的方法。OnScrollListener 中有两个需要覆写的方法：

- onScroll(AbsListView view, int firstVisibleItem, int visibleItemCount, int totalItemCount)：正在滑动时不断触发，主要有四个参数，分别是 ListView 对象、当前可以看见的第一个子项（也就是当前屏幕最上方的子项）、可以看到子项的个数、总的子项个数。
- onScrollStateChanged(AbsListView view, int scrollState)：顾名思义，这个是在滑动状态变化的情况下触发，里面有两个参数，分别是 ListView 对象和滑动状态。

那么滑动状态又分为哪些呢？API 文档中也进行了说明，共有三个状态：

- SCROLL_STATE_FLING：手指正在拖着滑动（手指没离开屏幕）。
- SCROLL_STATE_IDLE：滑动停止。
- SCROLL_STATE_TOUCH_SCROLL：手指使劲在屏幕上滑了一下，由于惯性屏幕继续滚动（手指离开屏幕）。

下面通过一个小实例来介绍，如何通过监听滑动变化实现上拉加载的功能。

首先要定义一个底部布局，用于加载时的提示，代码如下：

```xml
<?xml version="1.0" encoding="utf-8"?>
<LinearLayout xmlns:android="http://schemas.android.com/apk/res/android"
    android:id="@+id/ll_footer"
    android:layout_width="match_parent"
    android:layout_height="wrap_content"
    android:orientation="vertical">

    <ProgressBar
        android:id="@+id/progress"
        style="?android:attr/progressBarStyleSmall"
        android:layout_width="wrap_content"
        android:layout_height="wrap_content"
        android:layout_gravity="center" />

    <TextView
        android:id="@+id/tv_wait"
        android:layout_width="wrap_content"
        android:layout_height="wrap_content"
        android:layout_gravity="center"
        android:text="正在加载..."
        android:textSize="10sp" />
</LinearLayout>
```

上述代码中布局采用线性布局，添加了一个 ProgressBar 在 TextView 的上方，设置了 ProgressBar 的 style 属性，其值为"?android:attr/progressBarStyleSmall"，也就是小圆形的进度条形式，设置了 layout_gravity 属性，其值为 center，居中显示。添加了一个 TextView 显示提示信息，同样也设置其 layout_gravity 属性为 center。

为了实现上拉加载的功能，这里自定义了一个控件继承自 ListView，并实现了 OnScrollListener 接口：

```java
public class UpAddListView extends ListView implements AbsListView.OnScrollListener {
    private View footer;
    // 标志位 是否正在加载
    private Boolean isAdding = false;
    private int totalItems, totalItemCount;
    private IUpAddListener iUpAddListener;

    public UpAddListView(Context context) {
        super(context);
        initView(context);
    }

    public UpAddListView(Context context, AttributeSet attrs) {
        super(context, attrs);
        initView(context);
    }
```

```java
public UpAddListView(Context context, AttributeSet attrs, int defStyleAttr) {
    super(context, attrs, defStyleAttr);
    initView(context);
}

@Override
public void onScrollStateChanged(AbsListView view, int scrollState) {
    // 当前可见第一项和当前屏幕可见性个数之和等于总的子项个数 && 滑动停止状态
    if ((totalItemCount == totalItems) && scrollState == SCROLL_STATE_IDLE) {
        if (!isAdding) {
            footer.findViewById(R.id.ll_footer).setVisibility(View.VISIBLE);
            // 回调方法
            iUpAddListener.onAdd();
            isAdding = true;
        }
    }
}

@Override
public void onScroll(AbsListView view, int firstVisibleItem, int visibleItemCount, int totalItemCount) {
    // 当前看到第一个子项 + 可以看到的子项个数
    totalItems = firstVisibleItem + visibleItemCount;
    // 总的子项个数
    this.totalItemCount = totalItemCount;
}

private void initView(Context context) {
    footer = LayoutInflater.from(context).inflate(R.layout.footer_layout, null);
    // 初始状态底部布局不显示
    footer.findViewById(R.id.ll_footer).setVisibility(View.GONE);
    // 添加底部布局
    this.addFooterView(footer);
    // 设置滑动监听
    this.setOnScrollListener(this);
}

public interface IUpAddListener {
    void onAdd();
}

public void setInterface(IUpAddListener iUpAddListener) {
    this.iUpAddListener = iUpAddListener;
}
```

```
    // 加载完毕
    public void addCompleted() {
        isAdding = false;
        footer.findViewById(R.id.ll_footer).setVisibility(View.GONE);
    }
}
```

实现 OnScrollListener 接口要覆写其两个方法：
- onScroll：滚动时会触发。为了在 onScrollStateChanged 中使用这个方法中的参数，我们定义了两个 int 型的全局变量，将这个方法中的参数传递给全局变量以便在 onScrollStateChanged 中使用。
- onScrollStateChanged：滚动状态改变时会触发。这里思考一下，怎么才能判断滚动到最底部了呢？首先滚动到底部，不能再继续滑动了，滑动状态 scrollState 必须是停止状态了，即"scrollState==SCROLL_STATE_IDLE"。滑动停止了就滑动到底了吗？显然不是，还必须要满足一个条件，这个条件就要借助 onScroll 中传递的三个参数了，当满足当前屏幕中第一个可见项 + 当前屏幕可以看到的子项个数 = 总体子项个数时，就说明滚动到底了。

这里使用了回调方法用于数据传递，定义了一个内部接口，里面有一个 onAdd 的抽象方法。此外，设置了 setInterface 方法，进行接口注册。

addCompleted 方法，在加载完成后调用，将加载标志位设置成 false 并将底部栏隐藏。

下面将自定义控件引入到主布局（activity_main.xml）文件：

```xml
<?xml version="1.0" encoding="utf-8"?>
<RelativeLayout xmlns:android="http://schemas.android.com/apk/res/android"
    android:layout_width="match_parent"
    android:layout_height="match_parent">

    <ad.listviewonscrolllistenerdemo.UpAddListView
        android:id="@+id/lv"
        android:layout_width="match_parent"
        android:layout_height="match_parent" />
</RelativeLayout>
```

引入自定义控件需要在标签中加入完整的包.类名，这里是 ad.listviewonscrolllistenerdemo.UpAddListView，设置了宽、高属性为 match_parent，占据整个界面。

MainActivity.java 代码如下：

```java
public class MainActivity extends Activity implements UpAddListView.IUpAddListener {
    private UpAddListView mUpAddListView;
    private List datas = new ArrayList<String>();
    private ArrayAdapter mArrayAdapter;

    @Override
    protected void onCreate(Bundle savedInstanceState) {
        super.onCreate(savedInstanceState);
        setContentView(R.layout.activity_main);
        initViews();
```

```java
        initDatas();
    }

    private void initViews() {
        mUpAddListView = (UpAddListView) findViewById(R.id.lv);
        mUpAddListView.setInterface(this);
        mArrayAdapter = new ArrayAdapter(this,
                android.R.layout.simple_list_item_1, datas);
        mUpAddListView.setAdapter(mArrayAdapter);
    }

    private void initDatas() {
        for (int i = 0; i < 10; i++) {
            datas.add("测试数据" + i);
        }
    }

    private void addMoreDatas() {
        for (int i = 0; i < 2; i++) {
            datas.add("新数据" + i);
        }
    }

    @Override
    public void onAdd() {
        // 为了便于观察,并模仿请求操作时间,这里采用延迟执行的方法
        Handler handler = new Handler();
        handler.postDelayed(new Runnable() {
            @Override
            public void run() {
                // 添加更多数据
                addMoreDatas();
                // 通知刷新
                mArrayAdapter.notifyDataSetChanged();
                // 完成加载
                mUpAddListView.addCompleted();
            }
        }, 2000);
    }
}
```

为了方便,这里采用了 ArrayAdapter 作为适配器类,初始时添加了 10 条测试数据,每次上拉加载时调用 addMoreDatas 方法添加两条新数据。Activity 实现了在 UpAddListView 中定义的 IUpAddListener 接口,覆写了其 onAdd 方法,此处为了模拟数据加载的时间,采用了延迟 Handler 类的 postDelayed 方法,延迟 2000ms 再进行加载和加载完成的操作。加载完成后记得调用 notifyDataSetChanged 方法,刷新 ListView 显示。

运行实例,如图 5.5 所示。滑动到底部然后上拉,如图 5.6 所示。

查看动态图,请扫描图 5.7 中的二维码。

图 5.5　ListView 上拉加载一　　图 5.6　ListView 上拉加载二　　图 5.7　ListView 上拉加载二维码

5.3　控之经典——GridView

上面的章节中，介绍了 ListView 的方法，本节将对 GridView（网格视图）的属性和方法进行讲解，与 ListView 一般用于列表项的展示相比，GridView 是按照行列的方式来显示内容的，一般用于显示图片。图文等内容，例如实现九宫格图，用 GridView 是首选，也是最简单的。首先我们来看一下 GridView 有哪些常用属性和相关方法，如表 5.3 所示。

表 5.3　ListView 的常用属性和相关方法

属　　性	相　关　方　法	说　　明
android:columnWidth	setColumnWidth(int)	设置列的宽度
android:horizontalSpacing	setHorizontalSpacing(int)	定义列之间水平间距
android:numColumns	setNumColumns(int)	设置列数
android:stretchMode	setStretchMode(int)	缩放模式
android:verticalSpacing	setVerticalSpacing(int)	定义行之间默认垂直间距

下面通过一个简单实例看一下如何使用 GridView 控件。

主布局文件代码如下：

```xml
<?xml version="1.0" encoding="utf-8"?>
<RelativeLayout xmlns:android="http://schemas.android.com/apk/res/android"
    android:layout_width="match_parent"
    android:layout_height="match_parent">

    <GridView
        android:id="@+id/gv"
        android:layout_width="match_parent"
        android:layout_height="match_parent"
        android:horizontalSpacing="5dp"
        android:numColumns="3"
        android:stretchMode="columnWidth"
```

```
        android:verticalSpacing="5dp" />
</RelativeLayout>
```

在 RelativeLayout 布局中引入了一个 GridView 控件，设置了宽、高属性为 match_parent，控件占据整个界面；设置了 numColumns 属性，显示 3 列；设置了 stretchMode 属性为 columnWidth，表示图片的缩放与列宽的大小一致。

子项布局文件代码如下：

```
<?xml version="1.0" encoding="utf-8"?>
<LinearLayout xmlns:android="http://schemas.android.com/apk/res/android"
    android:layout_width="match_parent"
    android:layout_height="match_parent"
    android:orientation="vertical">

    <ImageView
        android:id="@+id/iv"
        android:layout_width="100dp"
        android:layout_height="100dp" />

    <TextView
        android:id="@+id/tv"
        android:layout_width="100dp"
        android:layout_height="wrap_content"
        android:gravity="center"
        android:textSize="20sp" />
</LinearLayout>
```

子项布局文件中包括一个 ImageView 控件和一个 TextView 控件，采用线性布局并设置 orientation 属性为 vertical。

为了数据操作方便，同样这里也设置了一个 JavaBean 类对数据进行封装，代码如下：

```
public class Animal {
  // 和上一节 JavaBean 类一样
}
```

其中主要包含 animal（即动物名）和 imgId（即图片 ID）这两个属性，并设置了相应的 setter 和 getter 方法。

这里自定义了 GridView 的适配器类，这个类继承自 BaseAdapter，代码如下：

```
public class GridAdapter extends BaseAdapter {
    private Context context;
    private List<Animal> datas;

    public GridAdapter(Context context, List<Animal> datas) {
        this.context = context;
        this.datas = datas;
    }
    // 返回子项的个数
    @Override
    public int getCount() {
```

```
        return datas.size();
    }
    // 返回子项对应的对象
    @Override
    public Object getItem(int position) {
        return datas.get(position);
    }
    // 返回子项的下标
    @Override
    public long getItemId(int position) {
        return position;
    }
    // 返回子项视图
    @Override
    public View getView(int position, View convertView, ViewGroup parent) {
        Animal animal = (Animal) getItem(position);
        View view;
        ViewHolder viewHolder;
        if (convertView == null) {
            view = LayoutInflater.from(context).inflate(R.layout.item_
              layout, null);
            viewHolder = new ViewHolder();
            viewHolder.animalImage = (ImageView) view.findViewById(R.
              id.iv);
            viewHolder.animalName = (TextView) view.findViewById(R.
              id.tv);
            view.setTag(viewHolder);
        } else {
            view = convertView;
            viewHolder = (ViewHolder) view.getTag();
        }
        viewHolder.animalName.setText(animal.getAnimal());
        viewHolder.animalImage.setImageResource(animal.getImgId());
        return view;
    }
    // 创建 ViewHolder 类
    class ViewHolder {
        ImageView animalImage;
        TextView animalName;
    }
}
```

细心的读者可以看到，这里的适配器类除了名字和 ListView 的适配器类不同之外，其余都相同，GridView 和 ListView 都是继承自 AbsListView，用法也有很多相似之处。学习要有举一反三的能力，找出控件使用时的共性，可以提高学习速度并能加深理解。这里就不再重复解释上面的代码，不清楚的同学可以翻看前面的 ListView 部分。

MainActivity.java 代码如下：

```java
public class MainActivity extends AppCompatActivity {
    private GridView mGridView;
    private GridAdapter mGridAdapter;
    private List<Animal> mDatas = new ArrayList<Animal>();

    @Override
    protected void onCreate(Bundle savedInstanceState) {
        super.onCreate(savedInstanceState);
        setContentView(R.layout.activity_main);
        mGridView = (GridView) findViewById(R.id.gv);
        // 初始化数据
        initDatas();
        // 实例化适配器
        mGridAdapter = new GridAdapter(this, mDatas);
        // 设置适配器
        mGridView.setAdapter(mGridAdapter);
        mGridView.setOnItemClickListener(new AdapterView.
        OnItemClickListener() {// 设置子项单击监听
            @Override
            public void onItemClick(AdapterView<?> parent, View view,
            int position, long id) {
                String name = mDatas.get(position).getAnimal();
                ImageView imageView = new ImageView(MainActivity.this);
                imageView.setScaleType(ImageView.ScaleType.CENTER);
                imageView.setLayoutParams(new LinearLayout.
                LayoutParams(
                        ViewGroup.LayoutParams.WRAP_CONTENT,
                        ViewGroup.LayoutParams.WRAP_CONTENT));
                imageView.setImageResource(mDatas.get(position).
                getImgId());
                Dialog dialog = new AlertDialog.Builder(MainActivity.
                this)
                        .setIcon(android.R.drawable.ic_btn_speak_now)
                        .setTitle("您选择的动物是: " + name)
                        .setView(imageView)
                        .setNegativeButton("取消", new DialogInterface.
                        OnClickListener() {
                            @Override
                            public void onClick(DialogInterface dialog,
                            int which) {
                            }
                        }).create();
                dialog.show();
            }
        });
    }

    private void initDatas() {
        Animal animal0 = new Animal("兔八哥", R.drawable.rabbit);
```

```
        Animal animal1 = new Animal("眼镜蛇", R.drawable.snack);
        Animal animal2 = new Animal("小金鱼", R.drawable.fish);
        Animal animal3 = new Animal("千里马", R.drawable.horse);
        Animal animal4 = new Animal("米老鼠", R.drawable.mouse);
        Animal animal5 = new Animal("大国宝", R.drawable.panda);
        Animal animal6 = new Animal("千里马", R.drawable.horse);
        Animal animal7 = new Animal("米老鼠", R.drawable.mouse);
        Animal animal8 = new Animal("大国宝", R.drawable.panda);
        mDatas.add(animal0);
        mDatas.add(animal1);
        mDatas.add(animal2);
        mDatas.add(animal3);
        mDatas.add(animal4);
        mDatas.add(animal5);
        mDatas.add(animal6);
        mDatas.add(animal7);
        mDatas.add(animal8);
    }
}
```

总的来讲，实现 GridView 需要三个步骤：
- 准备数据源（initDatas()）；
- 新建适配器（GridAdapter extends BaseAdapter）；
- 加载适配器（setAdapter()）。

这里我们还实现了 GridView 的子项单击监听（setOnItemClickListener），单击某个子项时，弹出 Dialog 对话框，通过 List 的 get 方法获取单击子项对应的 Animal 对象，然后再通过 Animal 类的 getAnimal 方法获取对应的动物名。同理，根据上面的方法获得图片资源对象，调用 Dialog 的 setView 方法，将资源图片对象传入，可以把对应子项的图片显示在 Dialog 对话框中。

运行实例，结果如图 5.8 所示，单击任意子项（这里单击眼镜蛇），结果如图 5.9 所示。查看动态图，请扫描图 5.10 中的二维码。

图 5.8　GridView 实例　　　图 5.9　GridView 单击效果　　　图 5.10　GridView 单击效果二维码

5.4 控之经典——GridView 进阶

5.4.1 GridView 动态图删除子项

用过 UC 浏览器的人相信都不会对图 5.11 所示的功能陌生。

长按图标时会在左上角显示出一个删除的图片，单击这个图片就可以删除与之对应的子项，下面介绍如何借助 GridView 来实现这个功能。

主布局文件代码如下：

图 5.11 UC 浏览器单击删除 Tab

```xml
<?xml version="1.0" encoding="utf-8"?>
<RelativeLayout xmlns:android="http://schemas.android.com/apk/res/android"
    android:layout_width="match_parent"
    android:layout_height="match_parent">

    <GridView
        android:id="@+id/gv"
        android:layout_width="match_parent"
        android:layout_height="match_parent"
        android:horizontalSpacing="5dp"
        android:numColumns="3"
        android:stretchMode="columnWidth"
        android:verticalSpacing="5dp" />
</RelativeLayout>
```

其功能和上一小节的布局文件一致，这里就不再介绍。

子项布局文件代码如下：

```xml
<?xml version="1.0" encoding="utf-8"?>
<FrameLayout xmlns:android="http://schemas.android.com/apk/res/android"
    android:layout_width="match_parent"
    android:layout_height="match_parent">

    <LinearLayout
        android:layout_width="fill_parent"
        android:layout_height="wrap_content"
        android:layout_marginRight="4dip"
        android:layout_marginTop="4dip"
        android:gravity="center"
        android:orientation="vertical">

        <ImageView
            android:id="@+id/iv"
            android:layout_width="60dip"
            android:layout_height="55dip" />

        <TextView
            android:id="@+id/tv"
            android:layout_width="70dip"
```

```xml
        android:layout_height="wrap_content"
        android:layout_marginTop="10dip"
        android:gravity="center"
        android:textColor="@android:color/black"
        android:textSize="15sp"
        android:textStyle="bold" />
    </LinearLayout>

    <ImageView
        android:id="@+id/delete_markView"
        android:layout_width="20dip"
        android:layout_height="20dip"
        android:layout_gravity="right|top"
        android:adjustViewBounds="true"
        android:src="@drawable/delete"
        android:visibility="gone" />
</FrameLayout>
```

上述代码中子项布局采用 FrameLayout 帧布局模式，设置删除图片的 layout_gravity 属性为 right|top，即布局在右上角并设置其 visibility 属性为 gone，即初始时不显示该图片。

JavaBean 类 Animal 和上一节一样，这里就不再进行介绍。

下面看一下自定义的适配器类：

```java
public class GridAdapter extends BaseAdapter {
    private Context context;
    private List<Animal> datas;
    final int position = 0;
    private boolean mIsShowDelete;

    public GridAdapter(Context context, List<Animal> datas) {
        this.context = context;
        this.datas = datas;
    }
    // 返回子项的个数
    @Override
    public int getCount() {
        return datas.size();
    }
    // 返回子项对应的对象
    @Override
    public Object getItem(int position) {
        return datas.get(position);
    }
    // 返回子项的下标
    @Override
    public long getItemId(int position) {
        return position;
    }
    // 返回子项视图
    @Override
```

```java
public View getView(final int position, View convertView, final
    ViewGroup parent) {
    Animal animal = (Animal) getItem(position);
    View view;
    ViewHolder viewHolder;
    if (convertView == null) {
        view = LayoutInflater.from(context).inflate(R.layout.item_
            layout, null);
        viewHolder = new ViewHolder();
        viewHolder.animalImage = (ImageView) view.findViewById(R.
            id.iv);
        viewHolder.animalName = (TextView) view.findViewById(R.
            id.tv);
        viewHolder.deleteImage = (ImageView) view.findViewById(R.
            id.delete_markView);
        view.setTag(viewHolder);
    } else {
        view = convertView;
        viewHolder = (ViewHolder) view.getTag();
    }
    viewHolder.animalName.setText(animal.getAnimal());
    viewHolder.animalImage.setImageResource(animal.getImgId());
    viewHolder.deleteImage.setVisibility(mIsShowDelete ? View.
        VISIBLE : View.GONE);
    if (mIsShowDelete) {
        viewHolder.deleteImage.setOnClickListener(new View.
            OnClickListener() {
            @Override
            public void onClick(View v) {
                datas.remove(position);
                setmIsShowDelete(false);
            }
        });
    }
    return view;
}
// 创建 ViewHolder 类
class ViewHolder {
    ImageView animalImage, deleteImage;
    TextView animalName;
}

public void setmIsShowDelete(boolean mIsShowDelete) {
    this.mIsShowDelete = mIsShowDelete;
    notifyDataSetChanged();
}
}
```

这里设置了一个全局变量 mIsShowDelete 作为 DeleteImage 是否显示的标志位,并添加了一个 setIsShowDelete 方法用于改变 mIsShowDelete 的值,在这个方法中还调用了 notifyDataSetChanged 方法刷新 GridView 显示。

这里还为 DeleteImage 添加了一个单击事件的监听，覆写了 onClick 方法，在这个方法中调用 List 集合的 remove 方法去除这个子项的数据并调用 setmIsShowDelete 传入 false 隐藏"删除图片"。

MainActivity.java 代码如下：

```java
public class MainActivity extends Activity {
    private GridView mGridView;
    private GridAdapter mGridAdapter;
    private boolean isShowDelete;
    private List<Animal> datas = new ArrayList<Animal>();

    @Override
    protected void onCreate(Bundle savedInstanceState) {
        super.onCreate(savedInstanceState);
        setContentView(R.layout.activity_main);
        mGridView = (GridView) findViewById(R.id.gv);
        // 初始化数据
        initDatas();
        // 实例化适配器类
        mGridAdapter = new GridAdapter(this, datas);
        // 设置适配器类
        mGridView.setAdapter(mGridAdapter);
        // 设置长按事件监听
        mGridView.setOnItemLongClickListener(newAdapterView.
        OnItemLongClickListener() {
            @Override
            public boolean onItemLongClick(AdapterView<?> parent,
                                  View view, int position, long id) {
                if (isShowDelete) {
                    // 删除图片显示时长按隐藏
                    isShowDelete = false;
                    mGridAdapter.setmIsShowDelete(isShowDelete);
                } else {
                    // 删除图片隐藏时长按显示
                    isShowDelete = true;
                    mGridAdapter.setmIsShowDelete(isShowDelete);
                }
                return false;
            }
        });
    }

    private void initDatas() {
        Animal animal0 = new Animal("兔八哥", R.drawable.rabbit);
        Animal animal1 = new Animal("眼镜蛇", R.drawable.snack);
        Animal animal2 = new Animal("小金鱼", R.drawable.fish);
        Animal animal3 = new Animal("千里马", R.drawable.horse);
        Animal animal4 = new Animal("米老鼠", R.drawable.mouse);
        Animal animal5 = new Animal("大国宝", R.drawable.panda);
        datas.add(animal0);
        datas.add(animal1);
```

```
            datas.add(animal2);
            datas.add(animal3);
            datas.add(animal4);
            datas.add(animal5);
        }
    }
```

这里设置了 GridView 的长按事件监听，当"删除图片"显示时，将 isShowDelete 标志位设置成 false，然后调用 GridAdapter 适配器类的 setmIsShowDelete 方法隐藏右上角的"删除图片"。反之，当"删除图片"隐藏时，则调用 setmIsShowDelete 方法传入 true 参数显示右上角的"删除图片"。

运行实例并长按任一子项，如图 5.12 所示。单击右上角的"删除图片⊗"则删除当前子项，右上角的"删除图片"也随之隐藏。查看动态图，请扫描图 5.13 中的二维码。

图 5.12　GridView 单击删除子项

图 5.13　GridView 单击删除子项二维码

5.4.2　GridView 动态图增加子项

下面研究如何实现动态增加子项，在上述基础上动态增加子项的功能。布局文件不做调整，因此，这里就不再介绍代码。

在 GridAdapter 适配器类中动态增加子项的功能，参考代码如下：

```
public class GridAdapter extends BaseAdapter {
    // 和上一节一样，省略部分相同代码
    // 返回子项视图
    @Override
    public View getView(final int position, View convertView, final ViewGroup parent) {
        View view;
        ViewHolder viewHolder;
        if (convertView == null) {
            view = LayoutInflater.from(context).inflate(R.layout.item_layout, null);
            viewHolder = new ViewHolder();
            viewHolder.animalImage = (ImageView) view.findViewById(R.id.iv);
```

```java
            viewHolder.animalName = (TextView) view.findViewById(R.id.tv);
            viewHolder.deleteImage = (ImageView) view.findViewById(R.id.delete_
            markView);
            // 设置 tag
            view.setTag(viewHolder);
        } else {
            view = convertView;
            // 由 tag 获取对象
            viewHolder = (ViewHolder) view.getTag();
        }
        if (position < datas.size()) {
            Animal animal = (Animal) getItem(position);
            viewHolder.animalName.setText(animal.getAnimal());
            viewHolder.animalImage.setImageResource(animal.getImgId());
            // 根据标志位 isShowDelete 决定是否显示删除图片按钮
            viewHolder.deleteImage.setVisibility(isShowDelete ? View.VISIBLE:
            View.GONE);
            if (isShowDelete) {
                viewHolder.deleteImage.setOnClickListener(new View.
                OnClickListener() {
                    @Override
                    public void onClick(View v) {
                        datas.remove(position);
                        setIsShowDelete(false);
                    }
                });
            }
        } else {
            viewHolder.animalName.setText(" 单击添加 ");
            viewHolder.animalImage.setImageResource(R.drawable.add);
            viewHolder.deleteImage.setVisibility(View.GONE);
        }
        return view;
    }
// 和上一节一样，省略部分相同代码
}
```

注意，因为这里多了最后一个子项用来作为"添加项"，所以在 getCount 方法的返回中要返回 datas.size()+1。

在 getView 方法中添加了判断，在 position<datas.size 时，加载 datas 里面的数据，而在 position=datas.size() 的地方加载"添加项"。

MainActivity 中也做了一些调整，代码如下：

```java
public class MainActivity extends Activity {
    // 和上一节一样，省略部分相同代码
    mGridView.setOnItemClickListener(new AdapterView.
    OnItemClickListener() {
        @Override
        public void onItemClick(AdapterView<?> parent, View view,
                                int position, long id) {
            // 单击了最后一张 "+" 图片
```

```
                    if (position == parent.getChildCount() - 1) {
                        addDatas();
                    }
                }
            });
            mGridView.setOnItemLongClickListener(new AdapterView.
                    OnItemLongClickListener() {
                // 和上一节一样，省略部分相同代码
            });
        }

        private void addDatas() {
            Animal animalAdd = new Animal(" 大国宝 ", R.drawable.panda);
            datas.add(animalAdd);
            mGridAdapter.notifyDataSetChanged();
        }

        private void initDatas() {
            // 和上一节一样，省略部分相同代码
        }
    }
```

这里添加了子项单击事件监听（OnItemClickListener），判断当"position==parent.getChildCount-1"时，即单击最后一个"添加项"时调用 addDatas 方法，添加一条记录到 datas 里面。注意，添加完成数据后，要调用 notifyDataSetChanged 方法刷新列表。

运行项目实例并单击最后一个"单击添加"图片，如图 5.14 所示，单击最后一个子项（"单击添加"），将会新插入一个子项。查看动态图，请扫描图 5.15 中的二维码。

图 5.14　GridView 单击增加子项　　　　　图 5.15　GridView 单击增加子项二维码

5.5　新控件——RecyclerView 控件

Android 5.0 引入了一个全新的列表控件 RecyclerView，说它新，是相对于其他控件而言。它更为灵活，同时也拥有比 ListView 和 GridView 控件较多的优点，例如子项 View 的创建、View 的回收以及重用等机制。

为了使用 RecyclerView 控件，需要创建一个 Adapter 和一个 LayoutManager 类。Adapter 继承自 RecyclerView.Adapetr 类，主要用来将数据和布局子项进行绑定。LayoutManager，布局管理器，设置每一项 View 在 RecyclerView 中的位置布局以及控件子项 View 的显示或者隐藏。当 View 重用或者回收时，LayoutManger 都会向 Adapter 请求新的数据进行替换原来数据的内容。这种回收重用的机制可以提高性能，避免创建很多的 View 或者是频繁地调用 findViewById 方法。

RecyclerView 提供了三种内置的 LayoutManager：
- LinearLayoutManager：线性布局，横向或者纵向滑动列表。
- GridLayoutManager：网格布局。
- StaggeredGridLayoutManager：流式布局（瀑布流效果）。

当然除了上面的三种内部布局之外，还可以继承 RecyclerView.LayoutManager 来实现一个自定义的 LayoutManager。

下面通过一个简单实例对 RecyclerView 有一个简单的理解。RecyclerView 控件需要引入 RecyclerView 兼容包，选中项目并右击，在弹出的快捷菜单中选择 Open Module Settings，如图 5.16 所示。

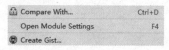

图 5.16　Open Module Settings

切换到 Dependencies 标签，单击右上角的"＋"并选择第一项 Library dependency 选项，如图 5.17 所示。

图 5.17　添加 Library dependency

在弹出的界面中输入 recyclerview，然后单击右边的搜索按钮，如图 5.18 所示。

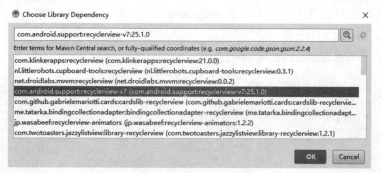

图 5.18　搜索 Library

选中兼容包，然后单击 OK 按钮，如图 5.19 所示。

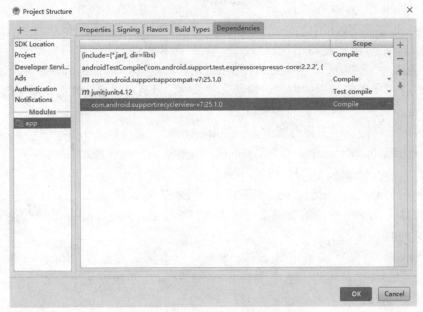

图 5.19 添加 Library dependency 成功

再次选中兼容包，然后单击 OK 按钮，等待 Gradle 编译完成就可以使用 RecyclerView 了。

5.5.1 RecyclerView 线性布局

主布局文件（activity_main.xml）中引入一个 RecyclerView，代码如下：

```xml
<?xml version="1.0" encoding="utf-8"?>
<RelativeLayout xmlns:android="http://schemas.android.com/apk/res/android"
    android:layout_width="match_parent"
    android:layout_height="match_parent">

    <android.support.v7.widget.RecyclerView
        android:id="@+id/recyclerview"
        android:layout_width="match_parent"
        android:layout_height="match_parent"
        android:scrollbars="vertical" />
</RelativeLayout>
```

上述代码在相对布局中添加了一个 RecyclerView 控件，引入时要输入完整的"包.类"名，可以看出 RecyclerView 在 V7 包中。

为 RecyclerView 添加一个子项布局文件（item.xml），代码如下：

```xml
<?xml version="1.0" encoding="utf-8"?>
<RelativeLayout xmlns:android="http://schemas.android.com/apk/res/android"
    android:layout_width="match_parent"
    android:layout_height="50dp">

    <TextView
```

```
        android:id="@+id/id_num"
        android:layout_width="match_parent"
        android:layout_height="50dp"
        android:gravity="center"
        android:text="1" />
</RelativeLayout>
```

和 ListView 相似，RecyclerView 同样需要创建一个适配器，代码如下：

```
public class MyAdapter extends RecyclerView.Adapter<MyAdapter.MyViewHolder> {
    private Context mContext;
    private List<String> mDatas;

    public MyAdapter(Context context, List<String> datas) {
        mContext = context;
        mDatas = datas;
    }

    @Override
    public MyViewHolder onCreateViewHolder(ViewGroup parent, int viewType) {
        MyViewHolder holder = new MyViewHolder(LayoutInflater.from(
                mContext).inflate(R.layout.item, parent,
                false));
        return holder;
    }

    @Override
    public void onBindViewHolder(MyViewHolder holder, int position) {
        holder.tv.setText(mDatas.get(position));
    }

    @Override
    public int getItemCount() {
        return mDatas.size();
    }

    class MyViewHolder extends RecyclerView.ViewHolder {
        TextView tv;

        public MyViewHolder(View view) {
            super(view);
            tv = (TextView) view.findViewById(R.id.id_num);
        }
    }
}
```

适配器类继承自 RecyclerView.Adapter，这里有几个方法需要解释一下：
- 构造方法 MyAdapter：传入了 Context 对象和数据集合，这点和 ListView 一样。
- OnCreateViewHolder 方法：这是必须要覆写的方法，返回一个 ViewHolder 对象。创建这个内部类需要传入一个 View 对象，View 对象的获得同样是使用

LayoutInflator 类的相关方法。
- onBindViewHolder 方法：绑定控件数据。
- getItemCount 方法：返回数据项个数。

MainActivity.java 代码如下：

```java
public class MainActivity extends Activity {
    private RecyclerView mRecyclerView;
    private List<String> mDatas;

    @Override
    protected void onCreate(Bundle savedInstanceState) {
        super.onCreate(savedInstanceState);
        setContentView(R.layout.activity_main);
        initData();
        mRecyclerView = (RecyclerView) findViewById(R.id.recyclerview);
        mRecyclerView.setLayoutManager(new LinearLayoutManager(this));
        mRecyclerView.setAdapter( new MyAdapter(this, mDatas));
    }

    protected void initData() {
        mDatas = new ArrayList<String>();
        for (int i = 1; i < 20; i++) {
            mDatas.add("" + i);
        }
    }
}
```

上述代码在 onCreate 方法中通过 findViewById 方法得到布局中添加的 RecyclerView 对象，调用 RecyclerView 类的 setLayoutManager 方法设置布局类型，这里传入的是 Android 提供的 LinearLayoutManager 对象（线性布局）。调用 setAdapter 方法为 RecyclerView 添加自定义的适配器。

运行实例，如图 5.20 所示。

图 5.20　RecyclerView 线性布局

可以看出子项自上而下显示在屏幕中,可以向下滑动查看下面的内容,和ListView的效果基本一致。

5.5.2 RecyclerView 网格布局

修改 MainActivity.java 代码如下:

```java
// 省略部分相同代码
@Override
protected void onCreate(Bundle savedInstanceState) {
    // 省略部分相同代码
    mRecyclerView.setLayoutManager(new GridLayoutManager(this,4));
    mRecyclerView.setAdapter( new MyAdapter(this, mDatas));
}
// 省略部分相同代码
}
```

这时 setLayoutManager 方法传入了一个 GridLayoutManager 对象,这个对象需要传入两个参数:上下文对象和列数。再次运行实例,如图 5.21 所示。

图 5.21　RecyclerView 网格布局

5.5.3 RecyclerView 瀑布流布局

上面讲解了两种基本的用法,下面讲解瀑布流布局的用法。

为了提高演示效果,这里修改了子项布局的代码,添加了 ImageView 控件用来显示图片,代码如下:

```xml
<?xml version="1.0" encoding="utf-8"?>
<LinearLayout xmlns:android="http://schemas.android.com/apk/res/android"
    android:layout_width="wrap_content"
    android:layout_height="wrap_content"
    android:orientation="vertical">

    <TextView
        android:id="@+id/id_num"
        android:layout_width="wrap_content"
```

```xml
        android:layout_height="wrap_content"
        android:gravity="center"
        android:text="1" />

    <ImageView
        android:layout_width="wrap_content"
        android:layout_height="wrap_content"
        android:src="@drawable/pic" />

</LinearLayout>
```

瀑布流布局需要随机调整子项的高，修改自定义的适配器代码如下：

```java
public class MyAdapter extends RecyclerView.Adapter<MyAdapter.MyViewHolder> {
    private Context mContext;
    private List<String> mDatas;
    private List<Integer> mHights;

    public MyAdapter(Context context, List<String> datas) {
        mContext = context;
        mDatas = datas;
        mHights = new ArrayList<>();
        initHeights();
    }
    // 省略部分相同代码

    @Override
    public void onBindViewHolder(MyViewHolder holder, int position) {
        ViewGroup.LayoutParams layoutparams = holder.itemView.
        getLayoutParams();
        layoutparams.height = mHights.get(position);
        holder.itemView.setLayoutParams(layoutparams);
        holder.tv.setText(mDatas.get(position));
    }
    // 省略部分相同代码
    private void initHeights(){
        for (int i = 0; i < mDatas.size(); i++) {
            mHights.add((int) (50 + Math.random() * 300));
        }
    }
}
```

在构造方法中调用了 initHeights 方法初始化了一个高度的 List 集合，在 onBindViewHolder 方法中为子项设置了随机的高，调用 View 的 getLayoutParams 方法得到布局参数对象 layoutparams，将高度集合 List 中的值赋给这个对象的高。

修改 MainActivity.java 代码如下：

```java
    // 省略部分相同代码
    @Override
    protected void onCreate(Bundle savedInstanceState) {
        super.onCreate(savedInstanceState);
```

```
                // 省略部分相同代码
mRecyclerView.setLayoutManager(new StaggeredGridLayoutManager(4,
        StaggeredGridLayoutManager.VERTICAL));
mRecyclerView.setAdapter( new MyAdapter(this, mDatas));
                // 省略部分相同代码
```

上述代码修改 setLayoutManager 方法中的参数为 StaggeredGridLayoutManager 对象，创建这个对象需要传入两个参数：第一个参数为每行列数，第二个参数为布局的方向，这里传入 StaggeredGridLayoutManager.VERTICAL 常量（垂直布局）。运行实例，结果如图 5.22 所示。

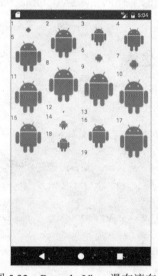

图 5.22　RecyclerView 瀑布流布局

5.6　多页面切换器——ViewPager 控件

一般 APP 都是由多个页面组成，页面的切换是开发中比较重要的部分，Android 提供了封装好的页面切换控件 ViewPager 供开发者使用，其继承结构如下：

public class

ViewPager

extends ViewGroup

java.lang.Object

　　□[①]　　android.view.View

　　　　□　　android.view.ViewGroup

　　　　　　□　　android.support.v4.view.ViewPager

ViewPager 继承自 ViewGroup，可以看出来它是一个容器类，类前包名是 android.support.v4，这是一个兼容包。注意，在布局文件中引入该控件时，标签要写全，即：< android.support.v4.view.ViewPager >。API 文档中对 ViewPager 进行了描述，总结如下：

- ViewPager 类直接继承自 ViewGroup 类，作为一个容器类，可以向其中添加 View 类。

[①]　方框"□"代表继承，继承是 Java 三大特性之一。

- 数据源和显示之间需要一个适配器类 PagerAdapter 进行适配。
- ViewPager 经常和 Fragemnet 一起使用，并且提供专门的适配器类 FragmentPagerAdapter 和 FragmentStatePagerAdapter 类供开发者调用。

实现 PageAdapter 必须覆写四个方法：

- public Object instantiateItem(ViewGroup container, int position)：初始化一个子项。
- public void destroyItem(ViewGroup container, int position,Object object)：销毁一个子项。
- public int getCount()：返回子项的个数。
- public boolean isViewFromObject(View arg0, Object arg1)：返回一个布尔型变量，判断子项是否来自 Object。

5.6.1 ViewPager 的基本用法

下面通过一个实例看一下 ViewPager 的基本用法。

主布局文件（activity_main.xml）代码如下：

```xml
<?xml version="1.0" encoding="utf-8"?>
<RelativeLayout xmlns:android="http://schemas.android.com/apk/res/android"
    android:layout_width="match_parent"
    android:layout_height="match_parent">

    <android.support.v4.view.ViewPager
        android:id="@+id/viewPager"
        android:layout_width="match_parent"
        android:layout_height="match_parent" />
</RelativeLayout>
```

这里引入了一个 ViewPager 控件，通过包.类名的方式引入，设置了宽高属性为 match_parent。ViewPager 是布局容器类，这里添加了三个子布局用来切换页面。

子布局一（layout1.xml）代码如下：

```xml
<?xml version="1.0" encoding="utf-8"?>
<LinearLayout xmlns:android="http://schemas.android.com/apk/res/android"
    android:layout_width="match_parent"
    android:layout_height="match_parent">

    <TextView
        android:layout_width="match_parent"
        android:layout_height="match_parent"
        android:gravity="center"
        android:text="页面1"
        android:textSize="30sp" />
</LinearLayout>
```

上述子布局中仅添加了一个 TextView 用于区别不同的页面，其余两个子布局同样也是只包含一个 TextView，不同的仅是 text 的属性值不同，这里就不再介绍代码。

ViewPager 同样需要适配器类，代码如下：

```java
public class MyViewPagerAdapter extends PagerAdapter {
    private List<View> datas;

    public MyViewPagerAdapter(List<View> datas) {
        this.datas = datas;
    }
    // 返回页卡数量
    @Override
    public int getCount() {
        return datas.size();
    }
    // 判断View是否来自Object
    @Override
    public boolean isViewFromObject(View view, Object object) {
        return view == object;
    }

    // 初始化一个页卡
    @Override
    public Object instantiateItem(ViewGroup container, int position) {
        container.addView(datas.get(position));
        return datas.get(position);
    }

    // 销毁一个页卡
    @Override
    public void destroyItem(ViewGroup container, int position, Object object) {
        container.removeView(datas.get(position));
    }
}
```

自定义适配器类 MyViewPagerAdapter 继承自 PagerAdapter，创建了构造函数，用于在初始化时传入 datas 数据集。PagerAdapter 是抽象类，继承这个类需要覆写它的四个抽象方法，这四个方法的说明请参考代码中的注释。

MainActivity.java 代码如下：

```java
public class MainActivity extends Activity {
    private ViewPager mViewPager;
    private List<View> mDatas;
    private MyViewPagerAdapter myViewPagerAdapter;

    @Override
    protected void onCreate(Bundle savedInstanceState) {
        super.onCreate(savedInstanceState);
        setContentView(R.layout.activity_main);
        mViewPager = (ViewPager) findViewById(R.id.viewPager);
        // 初始化数据集
        initDatas();
        myViewPagerAdapter = new MyViewPagerAdapter(mDatas);
        // 设置适配器
        mViewPager.setAdapter(myViewPagerAdapter);
```

```
    }

    private void initDatas() {
        mDatas = new ArrayList<>();
        View view1 = LayoutInflater.from(this).inflate(R.layout.layout1, null);
        View view2 = LayoutInflater.from(this).inflate(R.layout.layout2, null);
        View view3 = LayoutInflater.from(this).inflate(R.layout.layout3, null);
        mDatas.add(view1);
        mDatas.add(view2);
        mDatas.add(view3);
    }
}
```

总结以下，ViewPager 的实现可以分为三个步骤：

- 准备数据源（initDatas），这里是调用 LayoutInflater 的静态方法 from 并传入上下文对象获得一个 LayoutInflater 对象，和前面获取 LayoutInflater 对象的方式稍有不同，但查看源码可以看出，它们其实调用的方法是一致的，是 Android 封装好的方法。参考如下源码：

```
public static LayoutInflater from(Context context) {
    LayoutInflater LayoutInflater = (LayoutInflater) context.
            getSystemService(Context.LAYOUT_INFLATER_SERVICE);
    if (LayoutInflater == null) {
        throw new AssertionError("LayoutInflater not found.");
    }
    return LayoutInflater;
}
```

可以看出，from 是一个静态方法，这个方法内部也是通过 getSystemService 方法并传入 Context.LAYOUT_INFLATER_SERVICE 常量来获取 LayoutInflater 对象，然后返回这个对象。

- 准备适配器类并初始化（MyViewPagerAdapter），传入布局数据源。
- 设置适配器，调用 ViewPager 的 setAdapter 方法传入初始化的自定义适配器。

运行实例后并向右滑动即可切换到第二个页面，如图 5.23 所示。查看动态图，请扫描图 5.24 中的二维码。

图 5.23　ViewPager 基本用法第二个页面

图 5.24　ViewPager 基本用法二维码

可以看出，左右滑动屏幕就可以切换不同的 View 了。

5.6.2 ViewPager 导航条

上面的是通过页面中的内容来区别不同的页面，其实 ViewPager 还提供了导航条来区别不同页面和切换页面。下面介绍如何添加顶部或底部导航。Android 提供了两种方式供开发者选择，即 PagerTitleStrip 和 PagerTabStrip，下面分别介绍。

1. PagerTitleStrip

API 中这么定义：PagerTitleStrip 是一个非交互的页面指示器，一般指示 ViewPager 中的前一页、当前页和下一页三个页面。可以通过 PagerTitleStrip 标签添加到 xml 布局中。我们可以设置 layout_gravity 属性为 TOP 或者 BOTTOM 来决定在页面顶部或者底部显示，添加 PagerTitleStrip 要在适配器中覆写 getPageTitle 方法。

上面是抽象的理论描述，下面通过实例来看一下如何在 ViewPager 中添加 PagerTitleStrip 控件。

主布局文件代码如下：

```xml
<?xml version="1.0" encoding="utf-8"?>
<RelativeLayout xmlns:android="http://schemas.android.com/apk/res/android"
    android:layout_width="match_parent"
    android:layout_height="match_parent">

    <android.support.v4.view.ViewPager
        android:id="@+id/viewPager"
        android:layout_width="match_parent"
        android:layout_height="match_parent">

        <android.support.v4.view.PagerTitleStrip
            android:id="@+id/pagerTitleStrip"
            android:layout_width="match_parent"
            android:layout_height="wrap_content">
        </android.support.v4.view.PagerTitleStrip>
    </android.support.v4.view.ViewPager>
</RelativeLayout>
```

PagerTitleStrip 标签也要设置全路径并放在 ViewPager 标签内，默认没有添加 layout_gravity 属性，标签显示在页面顶部，若想设置在底部，添加这一属性设置其值为 BOTTOM 即可。

修改适配器类如下：

```java
public class MyViewPagerAdapter extends PagerAdapter {
    private List<View> datas;
    private List<String> titles;

    public MyViewPagerAdapter(List<View> datas, List<String> titles) {
        this.datas = datas;
        this.titles = titles;
    }
```

```
    // 省略部分相同代码
    @Override
    public CharSequence getPageTitle(int position) {
        return titles.get(position);
    }
}
```

为了方便观察，对上一个实例增加或修改的代码部分进行了加粗。首先修改了构造方法，多传入了一个标题的数据集；然后覆写了一个 getPagerTitle 的方法，这个方法可以根据 position 参数返回对应的 title。

MainActivity.java 代码如下：

```
public class MainActivity extends Activity {
    private ViewPager mViewPager;
    private PagerTitleStrip mPagerTitleStrip;
    private List<View> mDatas;
    private List<String> mTitles;
    private MyViewPagerAdapter myViewPagerAdapter;

    @Override
    protected void onCreate(Bundle savedInstanceState) {
        super.onCreate(savedInstanceState);
        setContentView(R.layout.activity_main);
        mViewPager = (ViewPager) findViewById(R.id.viewPager);
        mPagerTitleStrip = (PagerTitleStrip) findViewById(R.id.pagerTitleStrip);
        initDatas();
        myViewPagerAdapter = new MyViewPagerAdapter(mDatas, mTitles);
        mViewPager.setAdapter(myViewPagerAdapter);
    }

    private void initDatas() {
        mDatas = new ArrayList<>();
        mTitles = new ArrayList<>();
        View view1 = LayoutInflater.from(this).inflate(R.layout.layout1, null);
        View view2 = LayoutInflater.from(this).inflate(R.layout.layout2, null);
        View view3 = LayoutInflater.from(this).inflate(R.layout.layout3, null);
        mDatas.add(view1);
        mDatas.add(view2);
        mDatas.add(view3);
        mTitles.add("第一页");
        mTitles.add("第二页");
        mTitles.add("第三页");
    }
}
```

与上一个实例相比，这里添加了一个标题的数据集 titles，初始化 MyViewPagerAdapter 时传入了两个参数：页面布局数据集（mDatas）和标题数据集（mTitles），标题数据集数据将传到自定义的适配器中。

运行实例并向右滑动屏幕，如图 5.25 所示。可以看出页面切换到第二个页面，同时顶部的页面标签也切换到了"第二页"。查看动态图,请扫描图 5.26 中的二维码。

图 5.25　ViewPager 之 PagerTitleStrip 用法　　　图 5.26　ViewPager 之 PagerTitleStrip 用法二维码

单击顶部的标题栏,不会进行页面切换,正如 API 文档中所描述的,non-interactive indicator 只能作为一个页面指示器,不具有交互作用。下面介绍具有交互效果的 PagerTabStrip。

2. PagerTabStrip

API 中这么描述 PagerTabStrip:

```
PagerTabStrip is an interactive indicator of the current, next, and
previous pages of a ViewPager. It is intended to be used as a child view of a
ViewPager widget in your XML layout. Add it as a child of a ViewPager in your
layout file and set its android:layout_gravity to TOP or BOTTOM to pin it to
the top or bottom of the ViewPager. The title from each page is supplied by
the method getPageTitle(int) in the adapter supplied to the ViewPager.
    For a non-interactive indicator, see PagerTitleStrip.
```

大致含义如下:PagerTabStrip 是一个关于当前页、下一页和上一页可交互的页面指示器,作为一个子项布局在 ViewPager 控件内部。同时,也可以通过设置 layout_gravity 属性为 TOP 或 BOTTOM 来决定显示在页面顶部或底部。每个页面标题是通过适配器类中覆写 getPageTitle 方法提供给 ViewPager 的。最后一句也点明了,若要使用一个非交互指示器,可以参考 PagerTitleStrip。

从 API 文档可以看出,两个方式使用方法一样,因此,这里只要在布局文件中更换一下标签,代码如下:

```xml
<?xml version="1.0" encoding="utf-8"?>
<RelativeLayout xmlns:android="http://schemas.android.com/apk/res/android"
    android:layout_width="match_parent"
    android:layout_height="match_parent">
    <android.support.v4.view.ViewPager
        android:id="@+id/viewPager"
        android:layout_width="match_parent"
        android:layout_height="match_parent">
        <android.support.v4.view.PagerTabStrip
            android:id="@+id/pagerTabStrip"
```

```
            android:layout_width="match_parent"
            android:layout_height="wrap_content">
        </android.support.v4.view.PagerTabStrip>
    </android.support.v4.view.ViewPager>
</RelativeLayout>
```

上述代码将标签换成 android.support.v4.view.PagerTabStrip。

MainActivity.java 中，将 PagerTitleStrip 换成 PagerTabStrip 即可，其余代码不变：

```
private PagerTabStrip pagerTabStrip= (PagerTabStrip)findViewById(R.id.pagerTabStrip);
```

运行实例并单击顶部 Tab "第二页"，结果如图 5.27 所示。

图 5.27　ViewPager 之 PagerTabStrip 用法第二个页面

单击顶部指示页，可以进行页面切换，还有动画效果。除此之外，相对 PagerTitleStrip 而言，PagerTabStrip 当前页的下面还多了一个小横标。以上功能基本实现了，下面来研究如何让外观变得更漂亮。Android 提供了一些方法用于改变指示栏的样式。常用方法如表 5.4 所示。

表 5.4　PagerTabStrip 的常用方法

方　　法	说　　明
setBackgroundColor(int color)	设置背景颜色
setBackgroundResource(int resId)	设置背景图片
setDrawFullUnderline(boolean drawFull)	设置是否显示分隔栏
setTabIndicatorColor(int color)	设置指示器颜色
setTextColor(int color)	设置指示器文字颜色

在 MainActivity.java 的 onCreate 方法中加入如下代码：

```
mPagerTabStrip = (PagerTabStrip) findViewById(R.id.pagerTabStrip);
// 取消标题栏子 View 之间的分割线
mPagerTabStrip.setDrawFullUnderline(false);
// 改变指示器颜色为白色
mPagerTabStrip.setTabIndicatorColor(Color.YELLOW);
// 该变字体颜色为白色
mPagerTabStrip.setTextColor(Color.GREEN);
```

```
// 设置字体大小
mPagerTabStrip.setTextSize(1,24);
// 设置标题栏背景图片
mPagerTabStrip.setBackgroundResource(
        android.R.drawable.screen_background_light_transparent);
```

再次运行实例并向右滑动切换界面,如图 5.28 所示。查看动态图,请扫描图 5.29 中的二维码。

图 5.28　PagerTabStrip 自定义标题栏　　　　图 5.29　PagerTabStrip 自定义标题栏二维码

可以看出,除了通过左右滑动切换页面之外,还可以单击顶部标题来切换页面。

第 6 章　Android 系统组件操作实战

在进行 Android 应用程序开发时，开发者接触最多的是 Android 应用程序框架层，在整个应用程序框架中有几个重要的组件，总结如下：

- Activity：Android 中最常用的组件，在这个组件中放置各种控件，用于显示信息或用户交互，一个 Activity 可以认为是一个交互窗口。Activity 的启动由 Intent 触发，启动方式可以分为显式启动和隐式启动两种，显式 Intent 可以明确地指向一个 Activity 组件，而隐式则指向一个或多个目标 Activity 组件。Activity 组件是可以停止的，在开发中，常使用 Activity 的 finish 方法结束一个 Activity 的运行。
- Intent：上面讲到了 Activity 是一个界面，界面之间要进行切换，这里就要用到 Intent 组件。Intent 在页面跳转时还可以携带部分数据信息，包括 String 型、int 型、Bitmap，甚至一个对象。
- Service：Activity 是一个界面，Service 可以理解为一个没有交互界面的 Activity，它们一般运行在后台，不需要和用户进行交互。可以通过两种方式启动：第一种方式，启动者和服务绑定在一起，生命周期一致，启动者一旦退出，服务也将终止，也就是所谓的"同生共死"方式（bindService）。第二种方式，启动者和服务之间没有关联，即使启动者退出了，服务仍然在后台运行（startService）。
- BroadcastReceiver：一种消息性组件，用于在不同的组件甚至是不同的应用之间传递消息。BroadcastReceiver 工作在系统内部，无法被用户感知。BroadcastReceiver 广播有两种注册方式：静态注册和动态注册。静态注册是在 AndroidManifest.xml 文件中注册，这种广播在应用安装时会被系统解析，此种形式的广播应用不需要启动就可以接收到相应的广播。动态注册广播则需要通过 Context.registerReceiver 来实现，并且在不需要的时候通过 Context.unRegisterReceiver 来解除广播，此种形式的广播必须在相应应用启动的情况下才能接收广播，因为应用不启动就无法注册广播。在实际开发中，可以通过 Context 的一系列 send 方法来发送广播，感兴趣的广播接收者进行接收即可，其发送和接收的过程匹配是通过广播接收者 <intent-filter> 来描述的。

本章将结合具体实例对以上系统组件进行研究和学习。

6.1　Activity 生命周期

什么是生命周期？和生物体一样，Activity 也有生命周期，从"出生"到"活跃"再到"死亡"，这都是一系列有序的过程。

和程序员创建和调用的普通方法不同，生命周期的方法是由系统直接回调，这些回调的时机都是有规律的。详细了解这些规律有助于更好地控制程序，这也是为什么必须要学习和研究 Activity 生命周期的原因。

API 文档中提供了一张 Activity 的生命周期图，如图 6.1 所示。

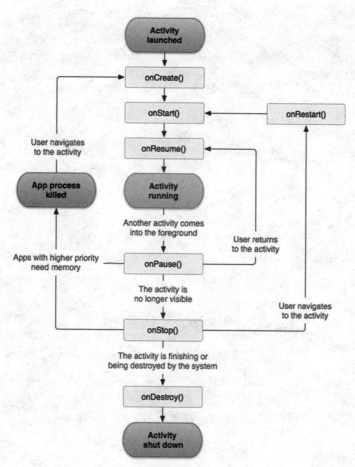

图 6.1　Activity 生命周期

从图 6.1 中可以看出，Activity 类中定义了七个回调方法：

- onCreate：这个方法基本每个 Activity 都会覆写，在 Activity 第一次被载入时调用，主要用于控件或数据的初始化操作。
- onStart：活动由不可见变为可见时调用。
- onResume：在准备好和用户交互时调用，此时 Activity 位于栈顶。
- onPause：当前 Activity 失去焦点，但不是全部不可见时调用，最常见的情况就是对话框弹出，而 Activity 可见时才调用。
- onStop：活动完全不可见时调用，此时 Activity 不再处于栈顶。
- onDestroy：活动被销毁时调用。
- onRestart：活动由不可见变为可见，即由停止状态转变成活动状态时调用。

下面通过一个实例来介绍这些生命周期方法被回调的时机。

新建一个项目，默认生成的主布局文件（activity_main.xml）代码如下：

```xml
<?xml version="1.0" encoding="utf-8"?>
<RelativeLayout    xmlns:android="http://schemas.android.com/apk/res/android"
    xmlns:tools="http://schemas.android.com/tools"
    android:layout_width="match_parent"
    android:layout_height="match_parent"
    android:paddingBottom="@dimen/activity_vertical_margin"
```

```xml
        android:paddingLeft="@dimen/activity_horizontal_margin"
        android:paddingRight="@dimen/activity_horizontal_margin"
        android:paddingTop="@dimen/activity_vertical_margin"
        tools:context="com.example.administrator.activitydemo.MainActivity">

        <TextView
            android:layout_width="wrap_content"
            android:layout_height="wrap_content"
            android:text="Hello World!" />
</RelativeLayout>
```

MainActivity.java 代码如下:

```java
public class MainActivity extends AppCompatActivity {
    private String TAG = "MainActivity";

    @Override
    protected void onCreate(Bundle savedInstanceState) {
        super.onCreate(savedInstanceState);
        setContentView(R.layout.activity_main);
        Log.d(TAG, "onCreate");
    }

    @Override
    protected void onStop() {
        super.onStop();
        Log.d(TAG, "onStop");
    }

    @Override
    protected void onDestroy() {
        super.onDestroy();
        Log.d(TAG, "onDestroy");
    }

    @Override
    protected void onPause() {
        super.onPause();
        Log.d(TAG, "onPause");
    }

    @Override
    protected void onResume() {
        super.onResume();
        Log.d(TAG, "onResume");
    }

    @Override
    protected void onStart() {
        super.onStart();
        Log.d(TAG, "onStart");
    }
```

```
    @Override
    protected void onRestart() {
        super.onRestart();
        Log.d(TAG, "onRestart");
    }
}
```

上述代码覆写了 Activity 的七个生命周期方法，在每个方法中都打印了 Log，通过 Log 信息的打印与否和打印时间来确定是否调用回调和回调调用的顺序。

启动项目，Log 信息如图 6.2 所示。

```
01-15 10:45:53.294 10348-10348/ad.activitylife D/MainActivity: onCreate
01-15 10:45:53.295 10348-10348/ad.activitylife D/MainActivity: onStart
01-15 10:45:53.295 10348-10348/ad.activitylife D/MainActivity: onResume
```

图 6.2 Activity 生命周期之启动

可以看出，启动 Activity 至 Activity 显示到前台，会回调三个生命周期的方法，调用顺序为 onCreate → onStart → onResume。

单击 Home 键，Log 信息如图 6.3 所示。

```
01-15 10:48:13.709 10348-10348/ad.activitylife D/MainActivity: onPause
01-15 10:48:13.966 10348-10348/ad.activitylife D/MainActivity: onStop
```

图 6.3 Activity 生命周期之单击 Home 键

可以看出，Activity 前台至 Activity 隐藏到后台，会调用两个生命周期的方法，调用顺序为 onPause → onStop。

在 recent 里单击这个项目，Log 信息如图 6.4 所示。

```
01-15 10:48:43.583 10348-10348/ad.activitylife D/MainActivity: onRestart
01-15 10:48:43.583 10348-10348/ad.activitylife D/MainActivity: onStart
01-15 10:48:43.583 10348-10348/ad.activitylife D/MainActivity: onResume
```

图 6.4 Activity 生命周期之再次进入

可以看出，Activity 由后台又显示到了前台显示，因为 Activity 已经启动过了（不会再调用 onCreate 了），所以会调用 onRestart → onStart → onResume 方法。

单击返回键退出 Activity，Log 信息如图 6.5 所示。

```
01-15 10:49:25.425 10348-10348/ad.activitylife D/MainActivity: onPause
01-15 10:49:26.269 10348-10348/ad.activitylife D/MainActivity: onStop
01-15 10:49:26.269 10348-10348/ad.activitylife D/MainActivity: onDestroy
```

图 6.5 Activity 生命周期之退出

可以看出，退出 Activity 会依次调用 onPause、onStop 和 onDestroy 方法，这个实例完全验证了生命周期图。

一个 Activity 的生命周期流程无论如何都会完全走完吗？猜想是得不出结论的，通过一个实例看一下，修改 MainActivity 代码如下：

```
public class MainActivity extends AppCompatActivity {
    private String TAG = "MainActivity";
    private TextView mTextView;

    @Override
    protected void onCreate(Bundle savedInstanceState) {
```

```
        super.onCreate(savedInstanceState);
        setContentView(R.layout.activity_main);
        Log.d(TAG, "onCreate");
        mTextView.setText("test");
    }
```

这里定义了一个 TextView，但并没有实例化这个 TextView，因此这个 TextView 必定为 null，这时运行程序必定会出现空指针异常，Log 信息如图 6.6 所示。

```
01-15 10:51:25.494 12309-12309/ad.activitylife D/MainActivity: onCreate
01-15 10:51:25.495 12309-12309/ad.activitylife E/AndroidRuntime: FATAL EXCEPTION: main
```

图 6.6　Activity 生命周期之 crash

可以看出仅回调了 onCreate 方法，程序就 crash 了，后面的生命周期方法也没有回调，因此，只有在正常情况下才会按照流程回调生命周期方法，出现异常错误时生命周期可能会中断，开发中应注意。

6.2　指向器——Intent

一般来讲，一个应用程序都会包含多个 Activity，这些 Activity 怎么进行跳转呢？前面的实例中或多或少也接触了这样一个组件——Intent，借助它的 startActivity 方法就可以任性地在这些 Activity 中跳来跳去了。Intent 除了可以切换 Activity 之外，还能做些什么呢？下面通过一个实例了解一下它的进阶用法。

主布局文件（activity_main.xml）代码如下：

```xml
<?xml version="1.0" encoding="utf-8"?>
<LinearLayout xmlns:android="http://schemas.android.com/apk/res/android"
    android:layout_width="match_parent"
    android:layout_height="match_parent">

    <Button
        android:layout_width="match_parent"
        android:layout_height="wrap_content"
        android:onClick="jump"
        android:text="跳转到另一个Activity" />
</LinearLayout>
```

主布局文件中只是添加了一个 Button 按钮，设置了这个 Button 的 onClick 属性，其值为 jump。

AnotherActivity 的布局文件（activity_another.xml）代码如下：

```xml
<?xml version="1.0" encoding="utf-8"?>
<LinearLayout xmlns:android="http://schemas.android.com/apk/res/android"
    android:layout_width="match_parent"
    android:layout_height="match_parent">

    <TextView
        android:id="@+id/tv"
        android:layout_width="match_parent"
```

```xml
        android:layout_height="wrap_content"
        android:text="MainActivity 捎来消息"
        android:textSize="18sp" />
</LinearLayout>
```

AnotherActivity 的布局文件中添加了一个 TextView 控件，用来显示 MainActivity 传过来的信息。

MainActivity.java 代码如下：

```java
public class MainActivity extends Activity {
    @Override
    protected void onCreate(Bundle savedInstanceState) {
        super.onCreate(savedInstanceState);
        setContentView(R.layout.activity_main);
    }

    public void jump(View view) {
        Intent intent = new Intent();
        intent.setClass(this, AnotherActivity.class);
        intent.putExtra("info", "这是 MainActivity 传递的信息");
        startActivity(intent);
    }
}
```

上述代码在单击事件的响应方法 jump 中添加了一个 Intent 对象，调用 Intent 的 setClass 方法指定跳转的根 Activity 和目标 Activity，即根 Activity 为 MainActivity，目标 Activity 为 AnotherActivity。调用了 Intent 的 putExtra 方法用于在 Intent 跳转中传递信息，这个方法需要传入两个参数，第一个参数可以当做 key，第二个参数即为 value。最后调用 Activity 的 startActivity 方法传入 Intent 对象 intent 启动 Intent。

AnotherActivity.java 代码如下：

```java
public class AnotherActivity extends Activity {
    private TextView mTextView;

    @Override
    protected void onCreate(Bundle savedInstanceState) {
        super.onCreate(savedInstanceState);
        setContentView(R.layout.activity_another);
        mTextView = (TextView) findViewById(R.id.tv);
        Intent intent = getIntent();
        String info = intent.getStringExtra("info");
        mTextView.setText("MainActivity 捎来消息：" + info);
    }
}
```

上述代码调用了 findViewById 方法传入布局文件中设置的 id 获得 TextView 对象，通过 getIntent 方法获取 Intent 对象，通过 Intent 的 getStringExtra 方法传入对应的 key 即可获取通过 Intent 跳转传过来的信息，然后通过 TextView 的 setText 方法显示出来。

AndroidManifest.xml 代码如下：

```xml
<activity android:name="ad.intentstring.AnotherActivity" />
```

注意，当项目中有多个 Activity 时，每一个 Activity 都需要在 AndroidManifest.xml 中配置，这一点初学者经常会忘记，若没有配置这个 AnotherActivity，运行项目单击跳转的按钮，将会出现 crash，Log 信息如图 6.7 所示。

```
FATAL EXCEPTION: main
Process: ad.intentstring, PID: 25852
java.lang.IllegalStateException: Could not execute method for android:onClick
    at android.view.View$DeclaredOnClickListener.onClick(View.java:4725)
    at android.view.View.performClick(View.java:5637)
    at android.view.View$PerformClick.run(View.java:22429)
    at android.os.Handler.handleCallback(Handler.java:751)
    at android.os.Handler.dispatchMessage(Handler.java:95)
    at android.os.Looper.loop(Looper.java:154)
    at android.app.ActivityThread.main(ActivityThread.java:6119) <1 internal calls>
    at com.android.internal.os.ZygoteInit$MethodAndArgsCaller.run(ZygoteInit.java:886)
    at com.android.internal.os.ZygoteInit.main(ZygoteInit.java:776)
Caused by: java.lang.reflect.InvocationTargetException <1 internal calls>
    at android.view.View$DeclaredOnClickListener.onClick(View.java:4720)
    at android.view.View.performClick(View.java:5637)
    at android.view.View$PerformClick.run(View.java:22429)
    at android.os.Handler.handleCallback(Handler.java:751)
    at android.os.Handler.dispatchMessage(Handler.java:95)
    at android.os.Looper.loop(Looper.java:154)
    at android.app.ActivityThread.main(ActivityThread.java:6119)
    at java.lang.reflect.Method.invoke(Native Method) <2 more...>
Caused by: android.content.ActivityNotFoundException: Unable to find explicit activity
```

图 6.7　ActivityNotFoudException 异常

Log 信息很明确地说明异常为 ActivityNotFoundException，也就是说没有找到 AnotherActivity，这时在 AndroidManifest.xml 配置即可。

运行项目并单击按钮跳转到 MainActivity，如图 6.8 所示，可以看出 MainActivity 传递过来的 String 型信息已经被显示出来了。查看动态图，请扫描图 6.9 中的二维码。

图 6.8　Intent 跳转实例　　　　　　图 6.9　Intent 跳转实例二维码

上面介绍了源 Activity 向目标 Activity 传值的方法，若想要再回传值到源 Activity，可以借助 startActivityForResult 方法来启动 Intent，用这个方法跳转之后，源 Activity 中还需要覆写 onActivityResult 这个方法，方便接收回传的信息。修改上面的实例进行讲解（主布局文件代码没变，不再介绍）。

AnotherActivity 布局文件（activity_another.xml）代码如下：

```xml
<?xml version="1.0" encoding="utf-8"?>
<LinearLayout xmlns:android="http://schemas.android.com/apk/res/android"
    android:layout_width="match_parent"
    android:layout_height="match_parent"
    android:orientation="vertical">

    <TextView
        android:id="@+id/tv"
        android:layout_width="match_parent"
        android:layout_height="wrap_content"
        android:text="MainActivity 捎来消息"
        android:textSize="18sp" />

    <Button
        android:layout_width="match_parent"
        android:layout_height="wrap_content"
        android:onClick="back"
        android:text=" 回传信息 " />
</LinearLayout>
```

上述代码添加了一个 Button 按钮，并设置了 onClick 属性其值为 back，修改 MainActivity.java 代码如下：

```java
public class MainActivity extends Activity {
    public  static final  int REQUEST_CODE=0;
    @Override
    protected void onCreate(Bundle savedInstanceState) {
        super.onCreate(savedInstanceState);
        setContentView(R.layout.activity_main);
    }
    public void jump(View view){
        Intent intent=new Intent();
        intent.setClass(this, AnotherActivity.class);
        intent.putExtra("info", "这是 MainActivity 传递的信息");
        startActivityForResult(intent, REQUEST_CODE);
    }
    @Override
    protected void onActivityResult(int requestCode, int resultCode,
    Intent data) {
        super.onActivityResult(requestCode, resultCode, data);
        if(requestCode==REQUEST_CODE){
            if(resultCode==RESULT_OK){
                Toast.makeText(MainActivity.this,
                    data.getStringExtra("infoBack"),Toast.LENGTH_
                    LONG).show();
            }
        }
    }
}
```

上述代码调用了 startActivityForResult 方法进行跳转，传入了两个参数，第一个是 Intent 对象，第二个是自定义的请求码，任意一个整数即可。覆写了 onActivityResult 方法，该方法有三个参数，第一个是请求码，用于判断返回的数据是不是我们请求的数据，与请求时传入的请求码一致时进行数据获取；第二个参数是结果码，结果码有三种，RESULT_OK 表示回传成功；第三个参数是 Intent 对象，回传的数据封装在这个对象里。数据获取时首先进行双重判断，满足请求码和结果码时，调用 getStringExtra 方法，传入对应的 key 即可获取回传信息。

修改 AnotherActivity.java 代码如下：

```java
public class AnotherActivity extends Activity {
    // 省略部分相同代码
    public void back(View view) {
        getIntent().putExtra("infoBack", "这是回传的信息");
        setResult(RESULT_OK, getIntent());
        finish();
    }
}
```

上述代码调用 getIntent 方法获取 Intent 对象，并调用 Intent 的 putExtra 方法传入 key 和 value 值，调用 setResult 方法回传数据，需要传入结果码和 Intent 对象两个参数，调用 Activity 的 finish 方法关闭 AnotherActivity。

运行实例并单击"跳转到另一个 ACTIVITY"这个按钮跳转到 AnotherActivity，如图 6.10 所示。单击"回传信息"，如图 6.11 所示。

关闭 AnotherActivity，这时 Toast 在 MainActivity 显示出回传的信息。查看动态图，请扫描图 6.12 中的二维码。

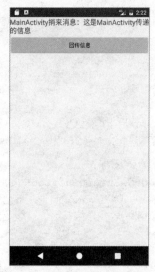
图 6.10　Intent 跳转回传实例一

图 6.11　Intent 跳转回传实例二

图 6.12　Intent 跳转回传实例二维码

6.3　指向器——Intent 隐式启动方式

前面的实例中实例化 Intent 都是直接将类名作为参数传入的，因此这种也称为显式 Intent。除了这种显式方式之外，Android 还提供了隐式的方式启动 Intent。隐式 Intent 要在 AndroidManifest.xml 中注册 intent-filter 属性，这个属性在 Activity 中注册时表明了这个 Activity 具备响应某种操作的能力。下面通过实例来看一下隐式 Intent 的使用。

新建项目，主布局文件（activity_main.xml）代码如下：

```xml
<?xml version="1.0" encoding="utf-8"?>
<LinearLayout xmlns:android="http://schemas.android.com/apk/res/android"
    android:layout_width="match_parent"
    android:layout_height="match_parent"
    android:orientation="vertical">

    <Button
        android:id="@+id/btn"
        android:layout_width="match_parent"
        android:layout_height="wrap_content"
        android:text="隐式 Intent" />
</LinearLayout>
```

目标 Activity（AnotherActivity.java）的布局文件代码如下：

```xml
<?xml version="1.0" encoding="utf-8"?>
<LinearLayout xmlns:android="http://schemas.android.com/apk/res/android"
    android:layout_width="match_parent"
    android:layout_height="match_parent"
    android:orientation="vertical">

    <TextView
        android:layout_width="match_parent"
        android:layout_height="wrap_content"
        android:gravity="center_horizontal"
        android:padding="5dp"
        android:text="这是 AnotherActivity" />

</LinearLayout>
```

MainActivity.java 代码如下：

```java
public class MainActivity extends AppCompatActivity implements View.OnClickListener {
    private Button mButton;

    @Override
    protected void onCreate(Bundle savedInstanceState) {
        super.onCreate(savedInstanceState);
        setContentView(R.layout.activity_main);
        mButton = (Button) findViewById(R.id.btn);
        mButton.setOnClickListener(this);
    }

    @Override
    public void onClick(View v) {
        switch (v.getId()) {
            case R.id.btn:
                Intent intent = new Intent();
```

```
                intent.setAction("ad.intentimplicit.AnotherActivity");
                startActivity(intent);
                break;
            }
        }
    }
```

上述代码通过 new 的方式创建了一个新的 Intent 对象，调用了 Intent 的 setAction 方法，传入了一个 String 型的参数。这时，所有注册这个 Action 的 Activity 都可以响应到这个操作，最后同样也是调用 startActivity 方法启动 Intent。

上面也讲到了要响应这个 Action 必须要在 AndroidManifest.xml 的 intent-filter 标签中注册同样的 Action 字符串，代码如下：

```
        //省略部分相同代码
        <activity android:name="ad.intentimplicit.AnotherActivity">
            <intent-filter>
                <action android:name="ad.intentimplicit.AnotherActivity" />
            </intent-filter>
        </activity>
       //省略部分相同代码
```

上述代码在 AnotherActivity 标签中添加了一个 intent-filter 标签，在这个标签中创建了一个 Action 标签，设置其 name 属性的值和 setAction 方法中传入的参数是一致的，这样才能响应这个 Intent。为了保证 Action 中 name 值的唯一性，这里一般采用“包 . 类”的方式进行命名。

运行这个实例，会出现如图 6.13 所示的 crash：

```
FATAL EXCEPTION: main
Process: ad.intentimplicit, PID: 3098
android.content.ActivityNotFoundException: No Activity found to handle Intent { act=ad.intentimplicit.AnotherActivity }
    at android.app.Instrumentation.checkStartActivityResult(Instrumentation.java:1809)
    at android.app.Instrumentation.execStartActivity(Instrumentation.java:1523)
    at android.app.Activity.startActivityForResult(Activity.java:4225)
    at android.support.v4.app.BaseFragmentActivityJB.startActivityForResult(BaseFragmentActivityJB.java:50)
    at android.support.v4.app.FragmentActivity.startActivityForResult(FragmentActivity.java:79)
    at android.app.Activity.startActivityForResult(Activity.java:4183)
    at android.support.v4.app.FragmentActivity.startActivityForResult(FragmentActivity.java:859)
    at android.app.Activity.startActivity(Activity.java:4522)
    at android.app.Activity.startActivity(Activity.java:4490)
    at ad.intentimplicit.MainActivity.onClick(MainActivity.java:28)
    at android.view.View.performClick(View.java:5637)
    at android.view.View$PerformClick.run(View.java:22429)
    at android.os.Handler.handleCallback(Handler.java:751)
    at android.os.Handler.dispatchMessage(Handler.java:95)
    at android.os.Looper.loop(Looper.java:154)
    at android.app.ActivityThread.main(ActivityThread.java:6119) <1 internal calls>
    at com.android.internal.os.ZygoteInit$MethodAndArgsCaller.run(ZygoteInit.java:886)
    at com.android.internal.os.ZygoteInit.main(ZygoteInit.java:776)
```

图 6.13　没有添加 catagory 导致的 crash

其原因是每一个 intent-filter 必须要设置一个 category。

修改 AndroidManifest.xml 代码如下：

```
<activity android:name=".AnotherActivity">
    <intent-filter>
        <action android:name="com.example.administrator.intentaction.
        AnotherActivity" />
        <category android:name="android.intent.category.DEFAULT"/>
    </intent-filter>
</activity>
```

再次运行实例并单击 Button，结果如图 6.14 所示。

可以看出，通过隐式方式也正常地跳到了 AnotherActivity。若再次创建一个 ThirdActivity，并在 AndroidManifest.xml 中为其设置和 AnotherActivity 一样的 Action 时会有什么现象呢？可以通过下列实例验证。

新建一个 ThirdActivity：

```java
public class ThirdActivity extends Activity {
    @Override
    protected void onCreate(Bundle savedInstanceState) {
        super.onCreate(savedInstanceState);
        setContentView(R.layout.third_layout);
    }
}
```

AndroidManifest.xml 代码如下：

```xml
            // 省略部分相同代码
        <activity android:name="ad.intentimplicit.AnotherActivity">
            <intent-filter>
                <action
                    android:name="ad.intentimplicit.AnotherActivity" />
                <category
                    android:name="android.intent.category.DEFAULT" />
            </intent-filter>
        </activity>

        <activity android:name="ad.intentimplicit.ThirdActivity">
            <intent-filter>
                <action
                    android:name="ad.intentimplicit.AnotherActivity" />
                <category
                    android:name="android.intent.category.DEFAULT" />
            </intent-filter>
        </activity>
    </application>
</manifest>
```

再次运行实例，如图 6.15 所示。

图 6.14　隐式 Intent 启动

图 6.15　隐式 Intent 启动同 Action 实例

单击"隐式 INTENT"按钮会跳转到 Activity 选择的界面，可以看出两个 Activity 都响应了，用户可以自行选择要跳转的 Activity。

6.4 Mini 型 Activity——Fragment

自从 iPad 发布以来，Android 平板也像雨后春笋般涌现出来，各种型号、各种尺寸的平板设备一下子涌进了移动市场，不同屏幕之间 APP 的适配也让各大开发厂商颇费脑筋。若想要获得较好的适配，必须开发平板和手机两套 APP，这样会用大量的人力去开发和维护。好在 Android 自 3.0 版本引入了 Fragment 的概念，它可以帮助我们更灵活地控制 APP 在平板或手机上的布局和显示，降低开发成本和开发时间，使得应用开发更灵活、维护更方便。

Fragment 是什么？它是一种可以嵌在 Activity 中的程序片段，除了遵循 Activity 的生命周期之外，它还有自己的生命周期，有自己的布局文件，可以把它理解成一个迷你版的 Activity。

怎么引入 Fragment 到 Activity 中呢？Android 提供了两种方式：一种称为静态方式；一种称为动态方式。下面分别通过实例来更好地理解 Fragment 的用法。

6.4.1 静态方式

静态方式创建两个 Fragment 类，然后在布局文件中引用这两个 Fragment。下面首先创建这两个 Fragment。

Fragment1 布局文件（fragment1_layout.xml）代码如下：

```xml
<?xml version="1.0" encoding="utf-8"?>
<RelativeLayout xmlns:android="http://schemas.android.com/apk/res/android"
    android:layout_width="match_parent"
    android:layout_height="match_parent">

    <TextView
        android:layout_width="wrap_content"
        android:layout_height="wrap_content"
        android:text="Fragment1" />
</RelativeLayout>
```

上述代码采用相对布局的布局方式，在布局中添加了一个 TextView，设置其 text 属性值为 Fragment1，用来区别不同的 Fragment。

Fragment2 布局文件（fragment2_layout.xml）代码如下：

```
        // 省略部分相同代码
        android:text="Fragment2" />
</RelativeLayout>
```

上述代码在 Fragment2 的布局中添加了一个 TextView 控件并设置其 text 属性为 Fragment2。

Fragment1.java 代码如下：

```java
public class Fragment1 extends Fragment {
    @Override
    public View onCreateView(LayoutInflater inflater,
                    ViewGroup container, Bundle savedInstanceState) {
        View view = inflater.inflate(R.layout.fragment1_layout, container, false);
        return view;
    }
}
```

上述代码在Fragment中载入布局文件，通过覆写onCreateView方法，根据里面参数inflater对象调用其inflate方法，由布局文件获取View对象，将View对象返回。inflate方法需要传入三个参数：自定义的布局文件、父布局对象、否添加到父布局（传入false）。添加Fragment要导包，共有两种：android.support.v4.app.Fragment 和 android.app.Fragment。需要注意的是这两种包是不兼容的，也就是说要么全用前一种Fragment。要么全用后一种Fragment，混用可能会出现如下错误提示：

```
Caused by: android.app.Fragment$InstantiationException: Trying to instantiate a class com.example.administrator.fragmentadd.Fragment1 that is not a Fragment
```

Fragment2.java 代码如下：

```java
public class Fragment2 extends Fragment {
    @Override
    public View onCreateView(LayoutInflater inflater, ViewGroup container,
    Bundle savedInstanceState) {
        View view = inflater.inflate(R.layout.fragment2_layout, container,
        false);
        return view;
    }
}
```

它和Fragment1方式一致。

在主布局文件中引用上面创建的两个Fragment，代码如下：

```xml
<?xml version="1.0" encoding="utf-8"?>
<LinearLayout xmlns:android="http://schemas.android.com/apk/res/android"
    android:layout_width="match_parent"
    android:layout_height="match_parent"
    android:orientation="vertical">

    <TextView
        android:layout_width="wrap_content"
        android:layout_height="wrap_content"
        android:text="Hello World!" />

    <fragment
        android:id="@+id/fragment1"
        android:name="ad.fragmentstatic.Fragment1"
        android:layout_width="wrap_content"
        android:layout_height="wrap_content" />
```

```
    <fragment
        android:id="@+id/fragment2"
        android:name="ad.fragmentstatic.Fragment2"
        android:layout_width="wrap_content"
        android:layout_height="wrap_content" />
</LinearLayout>
```

载入 Fragment 就像载入一个控件，通过 fragment 标签载入，其 name 属性的值为上面自定义的 Fragment，是"包.类名"的方式。

MainActivity.java 代码为创建项目时默认生成的代码，具体如下：

```
public class MainActivity extends AppCompatActivity {

    @Override
    protected void onCreate(Bundle savedInstanceState) {
        super.onCreate(savedInstanceState);
        setContentView(R.layout.activity_main);
    }
}
```

运行实例，如图 6.16 所示。

图 6.16 Fragment 静态载入

可以看出，已经将 Fragment1 和 Fragment2 引入到 MainActivity 中。

6.4.2 动态方式

上面通过在布局文件中载入 Fragment，下面通过一个实例学习在代码中动态添加 Fragment 的方法，动态方法将更加灵活。学习实例之前需要了解一下动态添加 Fragment 要用到的类和方法。

常用类有三个：

- android.app.Fragment：初始化 Fragment。
- android.app.FragmentManager：用于在 Activity 中操作 Fragment，其对象通过

Activity 的 getFragmentManager 方法获取。
- android.app.FragmentTransaction：保证 Fragment 操作的原子性。

常用方法如表 6.1 所示。

表 6.1 Fragment 的常用方法

方 法	说 明
getFragmentManager	获取 FragmentManager 对象
benginTransatcion	FragmentManager 类的方法，用于获取 FragmentTransaction 对象
add	FragmentTransaction 类的方法，用于向 Activity 中添加一个 Fragment
remove	从 Activity 中移除一个 Fragment
replace	替换 Fragment
hide	隐藏 Fragment

修改主布局文件代码如下：

```xml
<?xml version="1.0" encoding="utf-8"?>
<LinearLayout xmlns:android="http://schemas.android.com/apk/res/android"
    android:layout_width="match_parent"
    android:layout_height="match_parent"
    android:orientation="vertical">

    <TextView
        android:layout_width="match_parent"
        android:layout_height="wrap_content"
        android:layout_margin="10dp"
        android:gravity="center"
        android:text="Fragment 动态加载实例 "
        android:textSize="18sp" />

    <Button
        android:layout_width="match_parent"
        android:layout_height="wrap_content"
        android:gravity="center"
        android:onClick="addFragment"
        android:text=" 动态载入 Fragment"
        android:textSize="18sp" />

    <fragment
        android:id="@+id/fragment1"
        android:name="ad.fragmentdynamic.Fragment1"
        android:layout_width="wrap_content"
        android:layout_height="wrap_content" />

    <FrameLayout
        android:id="@+id/ll_fragment"
        android:layout_width="wrap_content"
        android:layout_height="wrap_content"></FrameLayout>
</LinearLayout>
```

上述代码中添加了一个 Button 按钮，其单击事件用来动态载入 Fragment2，Fragment1 还是通过静态的方式载入，取消了 Fragment2 静态载入部分代码，添加了一个 FrameLayout 布局用来承载显示动态载入的 Fragment2。

修改 MainActivity.java 代码如下：

```java
public class MainActivity extends Activity {
    @Override
    protected void onCreate(Bundle savedInstanceState) {
        super.onCreate(savedInstanceState);
        setContentView(R.layout.activity_main);
    }

    public void addFragment(View view) {
        Fragment2 fragment2 = new Fragment2();
        // 获取 FragmentManager 对象
        FragmentManager fragmentManager = getFragmentManager();
        // 通过 FragmentManager 的 beginTransaction 方法获取 FragmentTransaction 对象
        FragmentTransaction fragmentTransaction = fragmentManager.beginTransaction();
        // 调用 FragmentTransaction 的 replace 方法传入占位 id 和 Fragment 对象
        // 动态添加 Fragment
        fragmentTransaction.add(R.id.ll_fragment, fragment2);
        // 调用 commit 方法提交操作
        fragmentTransaction.commit();
    }
}
```

addFragment 也就是按钮"动态载入 Fragment"的单击响应，这个方法中首先通过 new 的方式创建了一个 Fragment2 的对象，然后通过 getFragmentManager 方法获取一个 FragmentManager 对象，调用这个 FragmentManger 的 beginTransaction 方法获取一个 FragmentTransaction 对象，其关键点是调用 FragmentTransaction 的 add 方法，这个方法传入两个参数，第一个参数是动态载入 Fragment 的占位控件 id，第二个参数为要动态载入的 Fragment 对象，最后调用 FragmentTransaction 的 commit 方法提交操作，这部分操作类似数据库中的"事务"。

运行实例，如图 6.17 所示。单击"动态载入 FRAGMENT"按钮，即可动态载入 Fragment2，如图 6.18 所示。

图 6.17　Fragment 动态载入一

图 6.18　Fragment 动态载入二

下面通过一个实例看一下如何使用 replace、remove 和 hide 方法。

在主布局文件中添加三个按钮用来处理这三个方法，代码如下：

```xml
<Button
    android:layout_width="match_parent"
    android:layout_height="wrap_content"
    android:gravity="center"
    android:onClick="replace"
    android:text="Fragment3 replace Fragment2"
    android:textSize="18sp" />

<Button
    android:layout_width="match_parent"
    android:layout_height="wrap_content"
    android:gravity="center"
    android:onClick="remove"
    android:text="Fragment3 remove"
    android:textSize="18sp" />

<Button
    android:layout_width="match_parent"
    android:layout_height="wrap_content"
    android:gravity="center"
    android:onClick="hide"
    android:text="Fragment2 hide"
    android:textSize="18sp" />
```

上述代码中，每个 Button 都设置了对应的 onClick 和 text 属性，又添加了一个 Fragment3，代码和 Fragment2 及 Fragment1 一致。

修改 MainActivity.java 代码如下：

```java
public class MainActivity extends Activity {
    private Fragment3 fragment3;
    private Fragment2 fragment2;
    private FragmentManager fragmentManager;

    @Override
    protected void onCreate(Bundle savedInstanceState) {
    // 省略部分相同代码
    }

    public void addFragment(View view) {
    // 省略部分相同代码
    }

    public void replace(View view) {
        fragment3 = new Fragment3();
        FragmentTransaction fragmentTransaction =
                fragmentManager.beginTransaction();
        fragmentTransaction.replace(R.id.ll_fragment, fragment3);
        fragmentTransaction.commit();
    }
```

```
public void remove(View view) {
    FragmentTransaction fragmentTransaction =
            fragmentManager.beginTransaction();
    fragmentTransaction.remove(fragment3);
    fragmentTransaction.commit();
}

public void hide(View view) {
    FragmentTransaction fragmentTransaction =
            fragmentManager.beginTransaction();
    fragmentTransaction.hide(fragment2);
    fragmentTransaction.commit();
}
```

上述代码中，Button 的响应方法 replace 中调用了 FragmentTransaction 的 replace 方法，这个方法同样需要传入两个参数，第一个参数为要替换的宿主 id，第二个参数是用于替换的 Fragment；remove 方法只需要传入一个参数，即要移除的 Fragment 对象；hide 方法传入需要隐藏的 Fragment 对象。

查看动态图，请扫描图 6.19 中的二维码（执行顺序：单击"动态载入 FRAGMENT"→单击"FRAGMENT3 REPLACE FRAGMENT2"→单击"FRAGMENT3 REMOVE"→单击"动态载入 FRAGMENT"→单击"FRAGMENT2 HIDE"）。

图 6.19　Fragment 动态载入二维码

6.5　Mini 型 Activity——Fragment 生命周期

同 Activity 一样，Fragment 也有自己的生命周期，其生命周期受宿主 Activity 的影响，和 Activity 的生命周期也有很多类似之处，对比 Activity 的生命周期和 Fragment 的生命周期，Fragment 比 Activity 多了几个额外的回调方法，如图 6.20 所示。下面对这几个额外的方法进行介绍：

- onAttach：当 Fragment 和宿主 Activity 发生关联时调用。
- onCreateView：创建 Fragment 视图时调用。
- onActivityCreated：宿主 Activity 的 onCreate 方法回调时调用。
- onDestroyView：和 onCreateView 方法相对应，当该 Fragment 的视图被移除时调用。
- onDetach：与 onAttach 方法对应，当该 Fragment 与宿主 Activity 取消关联时调用。

我们在两个 Fragment 中覆写所有的生命周期方法，然后通过 Log 打印的方式，学习 Fragment 的生命周期。

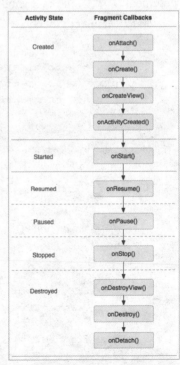

图 6.20　Fragment 生命周期

Fragment1 布局文件（fragment1_layout.xml）代码如下：

```xml
<?xml version="1.0" encoding="utf-8"?>
<RelativeLayout xmlns:android="http://schemas.android.com/apk/res/android"
    android:layout_width="match_parent"
    android:layout_height="match_parent">

    <TextView
        android:layout_width="match_parent"
        android:layout_height="match_parent"
        android:gravity="center"
        android:text="Fragment1"
        android:textSize="20sp" />
</RelativeLayout>
```

上述代码添加一个 TextView 控件，设置其 text 属性值为 Fragment1。

Fragment2 布局文件（fragment2_layout.xml）代码如下：

```xml
        // 省略部分相同代码
        android:text="Fragment2"
        android:textSize="20sp" />
</RelativeLayout>
```

上述代码添加一个 TextView 控件，设置其 text 属性值为 Fragment2。

主布局文件代码如下：

```xml
<?xml version="1.0" encoding="utf-8"?>
<RelativeLayout xmlns:android="http://schemas.android.com/apk/res/android"
    android:layout_width="match_parent"
    android:layout_height="match_parent">

    <LinearLayout
        android:id="@+id/ll_btns"
        android:layout_width="match_parent"
        android:layout_height="wrap_content"
        android:layout_alignParentTop="true"
        android:orientation="horizontal">

        <Button
            android:id="@+id/btn_add"
            android:layout_width="match_parent"
            android:layout_height="wrap_content"
            android:layout_weight="1"
            android:gravity="center"
            android:onClick="btnAdd"
            android:padding="10dp"
            android:text=" 添加 Fragment"
            android:textSize="18sp" />

        <Button
            android:id="@+id/replace"
```

```xml
            android:layout_width="match_parent"
            android:layout_height="wrap_content"
            android:layout_weight="1"
            android:gravity="center"
            android:onClick="btnReplace"
            android:padding="10dp"
            android:text=" 替换 Fragment"
            android:textSize="18sp" />
    </LinearLayout>

    <LinearLayout
        android:id="@+id/ll"
        android:layout_width="match_parent"
        android:layout_height="match_parent"
        android:layout_below="@+id/ll_btns"
        android:orientation="vertical"/>
</RelativeLayout>
```

上述代码中最外层布局采用相对布局，里面添加了两个线性布局，其中上面的线性布局的 orientation 属性设置为 horizontal（水平布局），在这个线性布局中添加了两个 Button 控件并为每个 Button 控件添加了 onClick 属性，响应单击事件；下面的线性布局作为 Fragment 的容器布局。

Fragment1.java 代码如下：

```java
public class Fragment1 extends Fragment {
    @Override
    public View onCreateView(LayoutInflater inflater,
                             ViewGroup container, Bundle
                                savedInstanceState) {
        Log.d("Fragment1", " onCreateView ");
        View view = inflater.inflate(R.layout.fragment1_layout, container,
        false);
        return view;
    }

    @Override
    public void onCreate(Bundle savedInstanceState) {
        super.onCreate(savedInstanceState);
        Log.d("Fragment1", "onCreate");
    }

    @Override
    public void onStart() {
        super.onStart();
        Log.d("Fragment1", "onStart");
    }

    @Override
    public void onResume() {
        super.onResume();
        Log.d("Fragment1", "onResume");
```

```java
    }

    @Override
    public void onPause() {
        super.onPause();
        Log.d("Fragment1", "onPause");
    }

    @Override
    public void onStop() {
        super.onStop();
        Log.d("Fragment1", "onStop");
    }

    @Override
    public void onDestroyView() {
        super.onDestroyView();
        Log.d("Fragment1", "onDestroyView");
    }

    @Override
    public void onDestroy() {
        super.onDestroy();
        Log.d("Fragment1", " onDestroy ");
    }

    @Override
    public void onDetach() {
        super.onDetach();
        Log.d("Fragment1", "onDetach");
    }

    @Override
    public void onAttach(Activity activity) {
        super.onAttach(activity);
        Log.d("Fragment1", "onAttach");
    }

    @Override
    public void onActivityCreated(Bundle savedInstanceState) {
        super.onActivityCreated(savedInstanceState);
        Log.d("Fragment1", "onActivityCreated");
    }
}
```

上述代码覆写了生命周期中的所有方法（共 11 个）并分别在这些方法中添加了 Log.d 方法，对应打印出日志信息，根据打印顺序可以知道这些方法是否被调用及调用顺序。

Fragment2 代码和 Fragment1 代码类似，也是覆写了所有生命周期中的方法，并设置了对应的日志打印方法，这里不再介绍。

MainActivity.java 代码如下：

```java
public class MainActivity extends Activity {
    @Override
    protected void onCreate(Bundle savedInstanceState) {
        super.onCreate(savedInstanceState);
        setContentView(R.layout.activity_main);
    }

    public void btnAdd(View view) {
        Fragment1 fragment1 = new Fragment1();
        // 获取 FragmentManager 对象
        FragmentManager fragmentManager = getFragmentManager();
        // 通过 FragmentManager 的 beginTransaction 方法获取 FragmentTransaction 对象
        FragmentTransaction fragmentTransaction = fragmentManager.beginTransaction();
        // 调用 FragmentTransaction 的 replace 方法传入占位 id 和 Fragment 对象
        fragmentTransaction.add(R.id.ll, fragment1);
        // 将 Fragment 放入栈中，返回键不会直接退出，而是返回到 Fragment 添加前的页面
        fragmentTransaction.addToBackStack(null);
        // 调用 commit 方法提交操作
        fragmentTransaction.commit();
    }

    public void btnReplace(View view) {
        Fragment2 fragment2 = new Fragment2();
        // 获取 FragmentManager 对象
        FragmentManager fragmentManager = getFragmentManager();
        // 通过 FragmentManager 的 beginTransaction 方法获取 FragmentTransaction 对象
        FragmentTransaction fragmentTransaction = fragmentManager.beginTransaction();
        // 调用 FragmentTransaction 的 replace 方法传入占位 id 和 Fragment 对象
        fragmentTransaction.replace(R.id.ll, fragment2);
        // 将 Fragment 放入栈中，返回键不会直接退出，而是返回到 Fragment 添加前的页面
        fragmentTransaction.addToBackStack(null);
        // 调用 commit 方法提交操作
        fragmentTransaction.commit();
    }
}
```

上述代码中，btnAdd 和 btnReplace 方法用来响应两个 Button 的单击事件，这里需要注意的是，这两个方法中都添加了 addToBackStack 方法，需要传入一个名字（String 型），这里传入 null 即可。这个方法有什么作用？在没添加这个方法之前，单击返回键会直接退出程序，添加了这个方法，单击返回键时会退回到 Fragment 未添加之前的状态，再次单击返回键才会退出程序，也就是说这个方法可以把添加的 Fragment 放入栈中，返回键将变成出栈操作。默认创建工程时，MainActivity 不是继承自 Activity 而是继承自 AppCompatActivity，而导致这个方法无效，这时需要修改一下继承自 Activity，再次运行测试即可。

上面解释了 addToBackStack 的用法，下面将对生命周期的调用顺序进行研究，首先

运行项目，如图 6.21 所示。可以看出，图中没有添加任何 Fragment，下面区域为空，这时单击"添加 FRAGMENT"按钮，添加 Fragment1 到下面的占位中，如图 6.22 所示。

图 6.21　Fragment 生命周期实例　　　　图 6.22　Fragment 生命周期实例添加

这时观察日志输出，如图 6.23 所示。

```
01-18 12:58:46.833 4064-4064/ad.fragmentlifecycle D/Fragment1: onAttach
01-18 12:58:46.833 4064-4064/ad.fragmentlifecycle D/Fragment1: onCreate
01-18 12:58:46.833 4064-4064/ad.fragmentlifecycle D/Fragment1:  onCreateView
01-18 12:58:46.835 4064-4064/ad.fragmentlifecycle D/Fragment1: onActivityCreated
01-18 12:58:46.835 4064-4064/ad.fragmentlifecycle D/Fragment1: onStart
01-18 12:58:46.835 4064-4064/ad.fragmentlifecycle D/Fragment1: onResume
```

图 6.23　Fragment 生命周期之添加

可以看出，Fragment1 的 onAttach、onCreate、onCreateView、onActivityCreated、onStart 和 onResume 方法依次被调用了，也就是说一个 Fragment 从创建到显示依次会调用上面的方法。单击"替换 FRAGMENT"按钮，如图 6.24 所示。

可以看出，Fragment1 变成了 Fragment2。观察日志输出，如图 6.25 所示。

可以看出，Fragment2 先执行了 onAttach 和 onCreate 方法，然后 Fragment1 依次执行了 onPause、onStop 和 onDestoryView 方法。这里需要注意，并没有调用 Fragment1 的 onDetach 方法；最后 Fragment2 依次执行了 onCreateView、onActivityCreated、onStart 和 onResume 方法，和上面的 Fragment1 从创建到显示时调用的方法一致。

单击返回键返回到 Fragment1，观察日志信息，如图 6.26 所示。

图 6.24　Fragment 生命周期实例替换

可以看出，Fragment2 依次执行了 onPause、onStop、onDestroyView、onDestroy、onDetach 方法，Fragment1 回到屏幕显示，依次执行 onCreateView、onActivityCreated、onStart 和 onResume 方法（因为上面没有调用 Fragment1 的 onDetach 方法，Fragment1 没

有解除绑定，因此这里也没有调用 Fragment1 的 onAttach 方法）。

```
01-18 13:02:18.904 4064-4064/ad.fragmentlifecycle D/Fragment2: onAttach
01-18 13:02:18.904 4064-4064/ad.fragmentlifecycle D/Fragment2: onCreate
01-18 13:02:18.904 4064-4064/ad.fragmentlifecycle D/Fragment1: onPause
01-18 13:02:18.904 4064-4064/ad.fragmentlifecycle D/Fragment1: onStop
01-18 13:02:18.904 4064-4064/ad.fragmentlifecycle D/Fragment1: onDestroyView
01-18 13:02:18.904 4064-4064/ad.fragmentlifecycle D/Fragment2:  onCreateView
01-18 13:02:18.904 4064-4064/ad.fragmentlifecycle D/Fragment2: onActivityCreated
01-18 13:02:18.905 4064-4064/ad.fragmentlifecycle D/Fragment2: onStart
01-18 13:02:18.905 4064-4064/ad.fragmentlifecycle D/Fragment2: onResume
```

图 6.25　Fragment 生命周期之替换

```
01-18 13:05:07.793 4064-4064/ad.fragmentlifecycle D/Fragment2: onPause
01-18 13:05:07.793 4064-4064/ad.fragmentlifecycle D/Fragment2: onStop
01-18 13:05:07.793 4064-4064/ad.fragmentlifecycle D/Fragment2: onDestroyView
01-18 13:05:07.793 4064-4064/ad.fragmentlifecycle D/Fragment2: onDestroy
01-18 13:05:07.793 4064-4064/ad.fragmentlifecycle D/Fragment2: onDetach
01-18 13:05:07.793 4064-4064/ad.fragmentlifecycle D/Fragment1:  onCreateView
01-18 13:05:07.794 4064-4064/ad.fragmentlifecycle D/Fragment1: onActivityCreated
01-18 13:05:07.794 4064-4064/ad.fragmentlifecycle D/Fragment1: onStart
01-18 13:05:07.794 4064-4064/ad.fragmentlifecycle D/Fragment1: onResume
```

图 6.26　Fragment 生命周期之返回

单击回到桌面（Fragment 由前台转换成后台），如图 6.27 所示。

```
01-18 13:05:57.441 4064-4064/ad.fragmentlifecycle D/Fragment1: onPause
01-18 13:05:57.748 4064-4064/ad.fragmentlifecycle D/Fragment1: onStop
```

图 6.27　Fragment 生命周期之 Home

根据打印日志，可以看出依次执行了 onPause 和 onStop 方法，然后通过多任务栏再次返回到项目（Fragment 由后台再次转换成前台），Log 信息如图 6.28 所示。

```
01-18 13:06:28.852 4064-4064/ad.fragmentlifecycle D/Fragment1: onStart
01-18 13:06:28.852 4064-4064/ad.fragmentlifecycle D/Fragment1: onResume
```

图 6.28　Fragment 生命周期之再次进入

可以看出依次执行了 onStart 和 onResume。再次单击返回键，移除 Fragment1，回到程序初始运行时的状态，观察 Log 信息，如图 6.29 所示。

```
01-18 13:07:02.577 4064-4064/ad.fragmentlifecycle D/Fragment1: onPause
01-18 13:07:02.577 4064-4064/ad.fragmentlifecycle D/Fragment1: onStop
01-18 13:07:02.577 4064-4064/ad.fragmentlifecycle D/Fragment1: onDestroyView
01-18 13:07:02.578 4064-4064/ad.fragmentlifecycle D/Fragment1:  onDestroy
01-18 13:07:02.578 4064-4064/ad.fragmentlifecycle D/Fragment1: onDetach
```

图 6.29　Fragment 生命周期之移除

可以看出依次执行了 onPause、onStop、onDestroyView、onDestroy 和 onDetach 方法，这样生命周期完整地走过一遍。可以发现，生命周期方法都是成对的，例如 onAttach 和 onDetach、onCreateView 和 onDestroyView。读者可以自行测试其他场景，更好地理解生命周期的调用过程。

6.6　FragmentPagerAdapter&FragmentStatePagerAdapter

前面的 ViewPager 部分讲到了 ViewPager 结合布局文件来实现界面切换的功能，此方式有明显不足的地方,所有页面的逻辑代码都只能写在 MainActivity 中,这样势必会造成 MainActivity 过于臃肿,不利于代码阅读和维护。

下面结合 FragmentPagerAdapter 和 FragmentStatePagerAdapter 来实现同样的功能,并且不同页面拥有不同的 Fragment,这样不同页面的操作逻辑就可以在相应的 Fragment 中进行处理,代码可维护性大大提高。

6.6.1 FragmentPagerAdapter 实现页面切换

FragmentPagerAdapter 类的继承结构如下:
public abstract class
FragmentPagerAdapter
extends PagerAdapter
java.lang.Object
　　□　android.support.v4.view.PagerAdapter
　　　　□　android.support.v4.app.FragmentPagerAdapter

由继承结构可以看出,FragmentPagerAdapter 继承自 PagerAdapter,子页面由 Fragment 组成,实现 FragmentPagerAdapter 时必须要覆写的方法是 getItem 和 getCount 方法。

下面通过一个实例进行实现,分三个步骤实现:
- 准备 Fragment 的数据集。
- 编写适配器类。
- 初始化数据集,设置适配器。

首先创建每个子页面的布局,代码如下:

```xml
<?xml version="1.0" encoding="utf-8"?>
<LinearLayout xmlns:android="http://schemas.android.com/apk/res/android"
    android:layout_width="match_parent"
    android:layout_height="match_parent">

    <TextView
        android:layout_width="match_parent"
        android:layout_height="match_parent"
        android:gravity="center"
        android:text="页面1"
        android:textSize="30sp" />
</LinearLayout>
```

上述代码仅添加一个 TextView,设置 text 不同值用来区别不同的页面。其余两个子布局文件和上面的子布局文件都是相似的,仅 text 属性值不同,这里不再介绍。

主布局文件代码如下:

```xml
<?xml version="1.0" encoding="utf-8"?>
<RelativeLayout xmlns:android="http://schemas.android.com/apk/res/android"
    android:layout_width="match_parent"
    android:layout_height="match_parent">
```

```xml
<android.support.v4.view.ViewPager
    android:id="@+id/viewPager"
    android:layout_width="match_parent"
    android:layout_height="match_parent">

    <android.support.v4.view.PagerTabStrip
        android:id="@+id/pagerTabStrip"
        android:layout_width="match_parent"
        android:layout_height="wrap_content">
    </android.support.v4.view.PagerTabStrip>
</android.support.v4.view.ViewPager>
</RelativeLayout>
```

上述代码中添加了一个 ViewPager 控件并为其添加了 PagerTabStrip 指示栏。

ViewPager 自定义适配器 MyFragmentViewPagerAdapter 继承自 FragmentPagerAdapter，代码如下：

```java
public class MyFragmentViewPagerAdapter extends FragmentPagerAdapter {
    private List<Fragment> datas;
    private List<String> titles;

    public MyFragmentViewPagerAdapter(FragmentManager fm,
                        List<Fragment> datas, List<String> titles) {
        super(fm);
        this.titles = titles;
        this.datas = datas;
    }

    @Override
    public Fragment getItem(int position) {
        return datas.get(position);
    }

    @Override
    public int getCount() {
        return datas.size();
    }

    @Override
    public CharSequence getPageTitle(int position) {
        return titles.get(position);
    }
}
```

这里构建构造方法时传入了 FragmentManager 参数并传入了 Fragment 类的数据集合 String 型的标题集。必须要覆写的方法只有 getItem（获取子项）和 getCount（获取子项个数）两个，为了显示标题栏，这里覆写了 getPageTitle 方法。可以看出这个适配器并没有实现页面销毁的方法，所有的页面都保存在内存当中，因此，当页面较大、较多时，内存压力可能会比较大。

每一个页面都是一个 Fragment，上面定义了三个子页面的布局，因此也要有对应的三个 Fragment，三个 Fragment 只是引用的布局不同，因此只介绍一个 Fragment 的代码。

MyFragment1.java 代码如下：

```java
public class MyFragment1 extends Fragment {
    @Override
    public View onCreateView(LayoutInflater inflater,
                    ViewGroup container, Bundle savedInstanceState) {
        return inflater.inflate(R.layout.view1, null);
    }
}
```

MainActivity.java 代码如下：

```java
public class MainActivity extends FragmentActivity {
    private ViewPager mViewPager;
    private PagerTabStrip mPagerTabStrip;
    // 数据源
    private List<Fragment> mDatas;
    private List<String> mTitles;
    private MyFragmentViewPagerAdapter myFragmentViewPagerAdapter;

    @Override
    protected void onCreate(Bundle savedInstanceState) {
        super.onCreate(savedInstanceState);
        setContentView(R.layout.activity_main);
        mViewPager = (ViewPager) findViewById(R.id.viewPager);
        mPagerTabStrip = (PagerTabStrip) findViewById(R.id.pagerTabStrip);
        // 取消标题栏和子 View 之间的分割线
        mPagerTabStrip.setDrawFullUnderline(false);
        mPagerTabStrip.setTabIndicatorColor(Color.WHITE);
        mPagerTabStrip.setTextColor(Color.WHITE);
        mPagerTabStrip.setBackgroundResource(android.R.drawable.alert_dark_frame);
        initDatas();
        myFragmentViewPagerAdapter = new MyFragmentViewPagerAdapter(
                getSupportFragmentManager(), mDatas, mTitles);
        mViewPager.setAdapter(myFragmentViewPagerAdapter);
    }

    private void initDatas() {
        mDatas = new ArrayList<>();
        mTitles = new ArrayList<>();
        mDatas.add(new MyFragment1());
        mDatas.add(new MyFragment2());
        mDatas.add(new MyFragment3());
        mTitles.add("第一页");
        mTitles.add("第二页");
        mTitles.add("第三页");
    }
}
```

初始化数据源时，传入的是 Fragment 对象，初始化适配器类 MyFragmentViewPager

Adapter 时要传入 FragmentManager 对象，这里使用 getSupportFragmentManager 方法获取，不过，要注意这时 MainActivity 要继承自 FragmentActivity，才好调用这个方法。最后要注意的是 FragmentPagerAdapter 是 v4 兼容包中的抽象类，为了保证兼容性，推荐所有的都导入 v4 包中，即：

```java
import android.support.v4.app.Fragment;
import android.support.v4.app.FragmentManager;
import android.support.v4.app.FragmentPagerAdapter;
```

运行实例，然后向左滑动或单击顶部的"第二页"即可切换到第二个页面，如图 6.30 所示。

图 6.30　FragmentPagerAdapter 页面

6.6.2　FragmentStatePagerAdapter 实现页面切换

FragmentStatePagerAdapter 和 FragmentPagerAdapter 类似，同样也是继承自 PagerAdapter，不同的是，此适配器更适用于大量页面的情形，因为不被显示的页面会被回收，可以大大降低内存的使用率。在使用 FragmentStatePagerAdapter 作为适配器时，其余都不用改动，只要覆写 instantiateItem（初始化子页面）和 destroyItem（销毁子页面）两个方法即可。

代码如下：

```java
public class MyFragmentViewPagerAdapter extends FragmentStatePagerAdapter {
    private List<Fragment> datas;
    private List<String> titles;

    public MyFragmentViewPagerAdapter(FragmentManager fm,
                      List<Fragment> datas, List<String> titles) {
        super(fm);
        this.titles = titles;
        this.datas = datas;
    }

    @Override
    public Fragment getItem(int position) {
```

```
        return datas.get(position);
    }

    @Override
    public int getCount() {
        return datas.size();
    }

    @Override
    public CharSequence getPageTitle(int position) {
        return titles.get(position);
    }

    @Override
    public Object instantiateItem(ViewGroup container, int position) {
        return super.instantiateItem(container, position);
    }
    @Override
    public void destroyItem(ViewGroup container, int position, Object object) {
        super.destroyItem(container, position, object);
    }
}
```

此种适配器方式可以销毁不可见的页面（即不在标题栏中显示的页面，标题栏中一般存在三个页面），回收内存，在实际开发中推荐使用此类。

为了验证其回收页面的能力，在第一个 Fragment 中覆写了 onDestroyView 方法，代码如下：

```
public class MyFragment1 extends Fragment {
    @Override
    public View onCreateView(LayoutInflater inflater,
                            ViewGroup container, Bundle
                            savedInstanceState) {
        return inflater.inflate(R.layout.view1,null);
    }

    @Override
    public void onDestroyView() {
        super.onDestroyView();
        Log.d("yayun", "onDestroyView: ");
    }
}
```

上述代码在 onDestroyView 方法中添加了 Log，若这个方法被回调则会打印日志。

修改 MainActivity.java 代码如下：

```
public class MainActivity extends FragmentActivity
        implements ViewPager.OnPageChangeListener{
    private ViewPager mViewPager;
    private PagerTabStrip mPagerTabStrip;
    private List<Fragment> mDatas;
```

```java
    private List<String> mTitles;
    private MyFragmentViewPagerAdapter myFragmentViewPagerAdapter;
    @Override
    protected void onCreate(Bundle savedInstanceState) {
        super.onCreate(savedInstanceState);
        //省略部分相同代码
        initDatas();
        myFragmentViewPagerAdapter=new MyFragmentViewPagerAdapter(
                getSupportFragmentManager(),mDatas,mTitles);
        mViewPager.setAdapter(myFragmentViewPagerAdapter);
        mViewPager.setOnPageChangeListener(this);
    }
    private void initDatas() {
        //省略部分相同代码
    }

    @Override
    public void onPageScrolled(int position, float positionOffset, int positionOffsetPixels) {

    }

    @Override
    public void onPageSelected(int position) {
        Toast.makeText(MainActivity.this," 当前是第: "+
                (position+1)+" 页 ",Toast.LENGTH_SHORT).show();
    }

    @Override
    public void onPageScrollStateChanged(int state) {

    }
}
```

实现 onPageChangeListener 接口,需要覆写三个方法:

- void onPageScrolled(int position, float positionOffset, int positionOffsetPixels):页面滚动时触发。
- void onPageSelected(int position):页面被选中时触发。
- void onPageScrollStateChanged(int state):页面滚动状态切换时触发。

在页面选择触发的方法里通过 position 参数获得当前页面信息,然后由 Toast 输出信息。如图 6.31 所示。

切换到第三个页面时,观察 Log 信息,如图 6.32 所示。

图 6.31 FragmentStatePagerAdapter 实例

```
01-18 14:59:02.549 31463-31463/ad.fragmentpageradapterdemo D/yayun: onDestroyView:
```

图 6.32 FragmentStatePagerAdapter 页面回收功能

可以看出，第一个页面的 onDestroyView 方法被回调了，也就是说第一个页面被销毁了，也就验证了这个类的页面回收能力。

6.7 Android 广播接收器之 BroadcastReceiver

"广播"这个词十分恰当，Android 中的广播和生活中的广播十分相似，都是面向"全局"发送，选择"感兴趣的对象"进行收听。

Android 中既可以主动发送自定义广播，也可以收听系统广播；对于广播接收，可以接收系统广播，同样也可以接收自定义的广播。接收广播需要注册广播，注册广播有两种形式：一种是在 AndroidManifest.xml 文件中用 receiver 标签注册，在这个标签中添加 intent-filter 标签，添加过滤器；另一种是在代码中调用 registerReceiver 方法动态注册，考虑到性能问题，最后不要忘记调用 unregisterReceiver 方法取消注册。对于主动发送的自定义广播，可以把要发送的信息和用于过滤的信息（Action/Category）包装到 Intent 中，然后通过调用 senBroadcast、sendOrderBroadcast 方法，将这个 Intent 对象以广播的形式发送出去。

6.7.1 静态注册 BroadcastReceiver

静态注册的优点是不管应用程序有没有启动或在后台运行，都可以收到注册的广播。下面通过一个实例看一下如何通过静态注册来监听网络状态改变的广播。

首先在 AndroidManifest.xml 文件中注册监听，代码如下：

```xml
<receiver android:name=".NetStatusChangeReceiver">
    <intent-filter>
        <action android:name="android.net.conn.CONNECTIVITY_CHANGE"
         />
    </intent-filter>
</receiver>
```

静态注册即在 AndroidManifest.xml 中注册，receiver 标签和 activity 标签同级放在 application 标签内部，为 receiver（接收器）添加一个 name 属性，这个属性值为自定义的广播接收器类。为了确定接收哪个广播，这里添加了一个 intent-filter，在 intent-filter 中添加想要监听的广播就行了。网络变化时 Android 系统会发出一条值为 android.net.conn.CONNECTIVITY_CHANGE，这是一个字符串常量，这里添加 action 的 name 值为这个字符串常量，就可以监听网络变化的监听了。同时，监听网络变化的广播需要网络状态权限，这里要配置。

至于如何监听网络变化的监听，就需要添加一个自定义的广播接收器，在 onReceive 方法中打印 Toast 信息。

自定义的广播接收器类代码如下：

```java
public class NetStatusChangeReceiver extends BroadcastReceiver {
    @Override
    public void onReceive(Context context, Intent intent) {
```

```
            Toast.makeText(context, "网络状态改变", Toast.LENGTH_SHORT).show();
    }
}
```

BroadcastReceiver 是一个抽象类，继承这个抽象类必须覆写其抽象方法 onReceive 方法，这个方法中有两个参数：上下文对象和一个 Intent 对象。Intent 对象中包含了广播中的信息，这里接收到广播后通过 Toast 显示。

运行实例后，选择"设置"→"无线和网络"→"移动数据网络开关"，如图 6.33 所示。

可以看出，关闭或打开"移动数据网络"开关，都会弹出 Toast 信息。

6.7.2 动态注册 BroadcastReceiver

静态注册应用退出后仍会接收广播，这样管理起来不方便还会占用系统资源，在实际开发中不推荐使用。而动态注册更为灵活，在需要时调用 registerReceiver 方法注册广播，在不需要接收该广播时调用 unRegisterReceiver 方法取消注册。下面通过一个实例看一下如何动态添加电池信息的监听。

图 6.33 静态注册广播实例

主布局文件代码如下：

```xml
<?xml version="1.0" encoding="utf-8"?>
<LinearLayout xmlns:android="http://schemas.android.com/apk/res/android"
    android:layout_width="match_parent"
    android:layout_height="match_parent"
    android:orientation="vertical">

    <Button
        android:id="@+id/mybtn"
        android:layout_width="match_parent"
        android:layout_height="wrap_content"
        android:text=" 获取电池电量" />
</LinearLayout>
```

上述代码添加了一个 Button 控件，单击这个 Button 的时候动态注册监听。

自定义的广播接收器类（BatteryInfoBroadcastReceiver）代码如下：

```java
public class BatteryInfoBroadcastReceiver extends BroadcastReceiver {

    @Override
    public void onReceive(Context context, Intent intent) {
        if (Intent.ACTION_BATTERY_CHANGED.equals(intent.getAction())) {
            int level = intent.getIntExtra("level", 0);
            int scale = intent.getIntExtra("scale", 0);
            int voltage = intent.getIntExtra("voltage", 0);
            int temperature = intent.getIntExtra("temperature", 0);
            String technology = intent.getStringExtra("technology");
```

```
            Dialog dialog = new AlertDialog.Builder(context)
                    .setTitle("电池电量")
                    .setMessage(
                            "电池电量为:"
                                    + String.valueOf(level * 100 / scale) + "%\n"
                                    + "电池电压为:"
                                    + String.valueOf((float) voltage / 1000) + "v"
                                    + "\n电池类型为:"
                                    + technology + "\n"
                                    + "电池温度为:"
                                    + String.valueOf((float) temperature / 10) +
                            "°C")
                    .setNegativeButton("关闭",
                            new DialogInterface.OnClickListener() {

                                public void onClick(DialogInterface arg0,
                                        int arg1) {

                                }
                            }).create();
            dialog.show();
        }
    }
}
```

在 onReceive 方法中通过 intent.getAction 获得广播传递过来的 Action 值，将这个值与 Intent 的静态常量（ACTION_BATTERY_CHANGED）进行等值比较，认为若这两个值相同时，这时的广播为电量变化的广播。这个静态常量的值如下：

```
public static final String ACTION_BATTERY_CHANGED = "android.intent.action.BATTERY_CHANGED";
```

onReceive 方法中有一个参数 Intent，这个 Intent 对象包含了广播传送过来的信息，通过 Intent 的 getExtra 方法获得这些信息，包括电池电量（level）、电池总量（scale）、电池电压（voltage）、电池温度（temperature）、电池类型（technology）。最后通过一个对话框将这些信息显示出来。

MainActivity.java 代码如下：

```
public class MainActivity extends Activity {
    private Button mButton = null;
    private BatteryInfoBroadcastReceiver mReceiver = null;

    public void onCreate(Bundle savedInstanceState) {
        super.onCreate(savedInstanceState);
        super.setContentView(R.layout.activity_main);
        mButton = (Button) super.findViewById(R.id.mybtn);
        mButton.setOnClickListener(new OnClickListenerImpl());
    }

    private class OnClickListenerImpl implements View.OnClickListener {

        public void onClick(View v) {
```

```
            mReceiver = new BatteryInfoBroadcastReceiver();
            IntentFilter filter = new IntentFilter(
                    Intent.ACTION_BATTERY_CHANGED);
            MainActivity.this.registerReceiver(mReceiver, filter);
        }
    }

    @Override
    protected void onDestroy() {
        super.onDestroy();
        MainActivity.this.unregisterReceiver(mReceiver);
    }
}
```

通过 findViewById 方法由布局文件中定义的 id 属性得到 Button 对象，为这个 Button 对象添加了单击事件监听，在这个单击事件监听中通过 new 的方式创建了这个自定义广播接收器类对象。调用 registerReceiver 方法需要传入两个参数，第一个是自定义的广播接收器类对象，第二个是 IntentFilter 对象，这个 IntentFilter 对象中的 action 值决定了接收怎样的广播，这里传入的是 Intent.ACTION_BATTERY_CHANGED 常量，与广播接收器中用于广播判断的值是一致的。

在 onDestroy 方法中（应用退出时）调用了 unregisterReceiver 方法取消监听。运行实例，如图 6.34 所示。

单击"获取电池电量"按钮会弹出电池信息的对话框，这里是获取的模拟器电池信息，这些信息不全，读者可以用真机进行测试。

图 6.34　动态注册广播监测电池信息

6.7.3　广播接收器 BroadcastReceiver 实用实例

随着微信等即时通信工具的流行，现如今短信的使用场合越来越少，最重要的功能莫过于接收验证码了，注册 APP 或找回密码时经常需要接收验证码，下面的这个实例结合广播实现验证码自动填入的功能。

第一步：监听短信内容
主布局文件代码如下：

```
<?xml version="1.0" encoding="utf-8"?>
<RelativeLayout xmlns:android="http://schemas.android.com/apk/res/android"
    android:layout_width="match_parent"
    android:layout_height="match_parent">

    <TextView
        android:id="@+id/textView"
        android:layout_width="wrap_content"
        android:layout_height="wrap_content"
        android:padding="5dp"
```

```
            android:text="test"
            android:textSize="24sp" />
</RelativeLayout>
```

主布局文件采用相对布局方式，添加了一个 TextView 控件用来显示监听的短信内容。自定义了一个广播接收器用于接收短信广播，代码如下：

```java
public class SMSBroadcastReceiver extends BroadcastReceiver {
    private static MessageListener mMessageListener;

    public SMSBroadcastReceiver() {
        super();
    }

    @Override
    public void onReceive(Context context, Intent intent) {
        Object[] pdus = (Object[]) intent.getExtras().get("pdus");
        for (Object pdu : pdus) {
            SmsMessage smsMessage = SmsMessage.createFromPdu((byte[]) pdu);
            String content = smsMessage.getMessageBody();
            mMessageListener.OnReceived(content);
        }
    }

    // 回调接口
    public interface MessageListener {
        void OnReceived(String message);
    }

    public void setOnReceivedMessageListener(MessageListener
    messageListener) {
        this.mMessageListener = messageListener;
    }
}
```

上述代码在 onReceive 方法中通过参数 intent 得到一个 Object 对象数组，遍历这个对象数组，在遍历的 for 循环中，调用 SmsMessage 的静态方法 createFromPdu 方法获得 SmsMessage 对象，然后调用 getMessageBody 方法获得短信内容。

这里需要将接收器 SMSBroadcastReceiver 接收到的短信内容传到 Activity 中去显示，用到了接口回调的方式进行组件间的信息传递，在 SMSBroadcastReceiver 中添加了一个内部接口 MessageListener，接口中定义了一个抽象方法 OnReceived。接口回调需要实现注册，因此这里添加了一个 setOnReceivedMessageListener 方法用于监听注册。

MainActivity.java 代码如下：

```java
public class MainActivity extends Activity {
    private TextView mTextView;
    private SMSBroadcastReceiver mSMSBroadcastReceiver;

    @Override
    protected void onCreate(Bundle savedInstanceState) {
```

```java
        super.onCreate(savedInstanceState);
        setContentView(R.layout.activity_main);
        init();
    }

    private void init() {
        mTextView = (TextView) findViewById(R.id.textView);
        mSMSBroadcastReceiver = new SMSBroadcastReceiver();
        mSMSBroadcastReceiver.setOnReceivedMessageListener(new
        SMSBroadcastReceiver.MessageListener() {
            public void OnReceived(String message) {
                mTextView.setText(message);
            }
        });
    }
}
```

这里采用了匿名内部类的方式实现了接口回调,和 OnClickListener 接口的使用方式基本一致,在覆写的方法 OnReceived 中调用 TextView 的 setText 方法将接收到的短信内容显示在 TextView 中。

最后接收短信需要权限,并且自定义的广播接收器类要在 AndroidManifest.xml 中注册,代码如下:

```xml
<receiver android:name=".SMSBroadcastReceiver">
    <intent-filter>
        <action android:name="android.provider.Telephony.SMS_
        RECEIVED" />
    </intent-filter>
</receiver>
```

这里添加了两个权限:android.permission.RECEIVE_SMS,接收短信的权限;android.permission.READ_SMS,读取短信的权限。

运行实例,然后用另一部手机发送一条信息到本机,如图 6.35 所示。

图 6.35 广播实例接收短信

接收到的短信内容显示在 TextView 中。

第二步:截取短信中的验证码

截取短信中的内容原理比较简单,这里以 6 位数字为例,只需要从短信内容中截取连续 6 位的数字组合即可。这个截取算法代码如下:

```java
public String getDynamicPassword(String str) {
    // 6是验证码的位数,一般为6位
    Pattern continuousNumberPattern = Pattern.compile("(?<![0-9])([0-9]{"
        + 6 + "})(?![0-9])");
    Matcher m = continuousNumberPattern.matcher(str);
```

```
        String dynamicPassword = "";
        while (m.find()) {
            dynamicPassword = m.group();
        }

        return dynamicPassword;
}
```

这里用到了 Java 的 Pattern 和 Matcher 类，这两个类经常用于字符串的匹配，使用方式也比较固定，首先调用 Pattern 的静态方法 compile 方法得到一个 Pattern 对象，compile 需要传入用于匹配的正则表达式；然后调用 Pattern 类的 matcher 方法传入要匹配的字符串得到 Matcher 对象；最后添加一个 while 循环，在这个循环中调用 Matcher 类的 group 方法即可截取匹配正则表达式的字符串。

修改 activity_layout.xml 代码如下：

```xml
<?xml version="1.0" encoding="utf-8"?>
<RelativeLayout xmlns:android="http://schemas.android.com/apk/res/android"
    android:layout_width="match_parent"
    android:layout_height="match_parent">

    <TextView
        android:id="@+id/textView"
        android:layout_width="wrap_content"
        android:layout_height="wrap_content"
        android:padding="5dp"
        android:text=" 验证码： "
        android:textSize="24sp" />

    <EditText
        android:id="@+id/edit"
        android:layout_width="wrap_content"
        android:layout_height="wrap_content"
        android:layout_toRightOf="@+id/textView" />
</RelativeLayout>
```

主布局文件中添加了一个 EditText 用来显示验证码。

修改 MainActivity.java 代码如下：

```java
public class MainActivity extends Activity {
    private EditText mCode;
    private SMSBroadcastReceiver mSMSBroadcastReceiver;

    @Override
    protected void onCreate(Bundle savedInstanceState) {
        super.onCreate(savedInstanceState);
        setContentView(R.layout.activity_main);
        init();
    }

    private void init() {
        mCode = (EditText) findViewById(R.id.edit);
```

```
        mSMSBroadcastReceiver = new SMSBroadcastReceiver();
        mSMSBroadcastReceiver.setOnReceivedMessageListener(new
SMSBroadcastReceiver.MessageListener() {
            public void OnReceived(String message) {
                mCode.setText(getDynamicPassword(message));
            }
        });
    }

    public String getDynamicPassword(String str) {
        // 6是验证码的位数,一般为 6 位
        Pattern continuousNumberPattern = Pattern.compile("(?<![0-9])
([0-9]{" + 6 + "})(?![0-9])");
        Matcher m = continuousNumberPattern.matcher(str);
        String dynamicPassword = "";
        while (m.find()) {
            dynamicPassword = m.group();
        }

        return dynamicPassword;
    }
}
```

上述代码通过 findViewById 方法得到 EditText 对象,在回调的方法 OnReceived 方法中调用 EditText 的 setText 方法将截取后的验证码显示在 EditText 中。

在真机上运行这个程序并用另一部手机发送验证码到这部手机上,如图 6.36 所示。查看应用界面,如图 6.37 所示。

图 6.36　广播实例发送短信

图 6.37　广播实例截取短信验证码

可以看出,验证码准确地截取到了应用中。

6.8　Android 自定义广播 Broadcast

接收系统广播的场景在开发中使用最多,上一节已经对接收和处理系统广播进行了介绍,本节将介绍如何发送和接收自定义的广播。Android 提供了两个方法供开发者调用来发送广播,即 sendBroadcast(普通广播)和 sendOrderedBroadcast(有序广播),下面通过实例来学习如何发送和接收普通广播及有序广播。

6.8.1 普通广播发送和接收实例

普通广播具有如下特点：
- 所有的监听该广播的接收器都可以接收到该广播。
- 同级别的接收器接收到广播的顺序是随机的。
- 广播接收器不能截断广播的传播。

为了接收自定义的广播，这里新建了一个广播接收器类，代码如下：

```java
public class MyBroadcastReceiver extends BroadcastReceiver {
    @Override
    public void onReceive(Context context, Intent intent) {
        if (intent.getAction().equals("com.yayun.broadcast")) {
            String info = intent.getStringExtra("info");
            Toast.makeText(context, "接收到了自定义的广播:"
                    + info, Toast.LENGTH_SHORT).show();
        }
    }
}
```

上述代码在 onReceive 方法中首先调用 Intent 的 getAction 方法获得 Intent 的 Action 值，然后将这个值与 com.yayun.broadcast 进行判断，认为这两个值相同时才是接收到了自定义的广播，然后调用 Intent 的 getStringExtra 方法得到广播传递过来的信息，最后使用 Toast 将信息打印出来。

主布局文件代码如下：

```xml
<?xml version="1.0" encoding="utf-8"?>
<RelativeLayout xmlns:android="http://schemas.android.com/apk/res/android"
    android:layout_width="match_parent"
    android:layout_height="match_parent">

    <Button
        android:onClick="send"
        android:layout_width="match_parent"
        android:layout_height="wrap_content"
        android:gravity="center"
        android:text="发送自定义广播"
        android:textSize="24dp" />
</RelativeLayout>
```

这里添加了一个 Button，单击这个 Button 时触发广播的发送。

MainActivity.java 代码如下：

```java
public class MainActivity extends AppCompatActivity {

    @Override
    protected void onCreate(Bundle savedInstanceState) {
        super.onCreate(savedInstanceState);
        setContentView(R.layout.activity_main);
    }

    public void send(View view) {
```

```
            Intent intent = new Intent("com.yayun.broadcast");
            intent.putExtra("info", "这是广播信息");
            sendBroadcast(intent);
        }
    }
```

上述代码在 Button 的单击监听事件中，创建了一个 Intent 对象，创建这个 Intent 对象时传入一个字符串作为构造函数的参数，这个字符串就是这个 Intent 的 Action 值，用于广播的过滤。然后调用 putExtra 方法在这个 Intent 中包裹一条字符串信息，这条信息将会传递到广播接收器中，最后调用 sendBroadcast 方法发送广播。

不要忘记在 AndroidManifest.xml 中注册广播接收器类，代码如下：

```
<receiver android:name=".MyBroadcastReceiver">
    <intent-filter>
        <action android:name="com.yayun.broadcast"></action>
    </intent-filter>
</receiver>
```

在 intent-filter 标签中，添加 action 标签，并设置其 name 属性值为 com.yayun.broadcast，方便过滤自定义的广播。

运行实例，如图 6.38 所示。

单击"发送自定义广播"按钮，Toast 信息随即显示出来了，并显示了广播中传递的字符串。

6.8.2 有序广播发送和接收实例

有序广播具有如下特性：
- 同级别的接收器先后顺序是随机的。
- 级别高的接收器先收到广播，级别低的接收器后收到广播。
- 可以截断广播的继续传播，级别高的接收器在接收到广播后决定是否需要截断。
- 同级别动态注册高于静态注册。

图 6.38　发送自定义广播

修改 MainActivity.java 代码如下：

```
public class MainActivity extends AppCompatActivity {
    //省略部分相同代码
    public void send(View view) {
        Intent intent = new Intent("com.yayun.broadcast");
        intent.putExtra("info", "这是广播信息");
        sendOrderedBroadcast(intent,null);
    }
}
```

这里将原来的 sendBroadcast 方法改成了 sendOrderedBroadcast 方法，这个方法需要传入两个参数，第一个仍然为 Intent 对象，第二个参数是一个定义权限的字符串，为了方便，这里传入 null 即可。

为了验证有序广播，还必须在上面的实例中再新添加一个新的广播接收器，代码如下：

```java
public class MyBroadcastReceiver2 extends BroadcastReceiver {
    @Override
    public void onReceive(Context context, Intent intent) {
        if (intent.getAction().equals("com.yayun.broadcast")) {
            String info = intent.getStringExtra("info");
            Toast.makeText(context, "接收到了自定义的广播2:"
                    + info, Toast.LENGTH_SHORT).show();
        }
    }
}
```

上述代码添加了一个新的广播接收器 MyBroadcastReceiver2，继承自 Broadcast-Receiver，覆写了 onReceive 方法，在这个方法中修改 Toast 中显示的信息用以区别。

修改 AndroidManifest.xml 代码如下：

```xml
<receiver android:name=".MyBroadcastReceiver">
    <intent-filter android:priority="10">
        <action android:name="com.yayun.broadcast"></action>
    </intent-filter>
</receiver>
<receiver android:name=".MyBroadcastReceiver2">
    <intent-filter android:priority="20">
        <action android:name="com.yayun.broadcast"></action>
    </intent-filter>
</receiver>
```

这里在 AndroidManifest.xml 中添加了另一个广播接收器 MyBroadcastReceiver2，并为两个广播接收器添加了 priority 属性，设置了不同的值，值越大，优先级将越高。可以运行实例证明一下。可以看出，MyBroadcastReceiver2 先接收到了广播，如图 6.39 所示。而后 MyBroadcastReceiver 才接收到广播，如图 6.40 所示。

查看动态图，请扫描图 6.41 中的二维码。

图 6.39　有序广播实例一　　　图 6.40　有序广播实例二　　　图 6.41　有序广播实例二维码

有序广播有一个特性——截断广播，下面以修改实例来看一下如何截断广播，代码如下：

```
public class MyBroadcastReceiver2 extends BroadcastReceiver {
    @Override
    public void onReceive(Context context, Intent intent) {
        // 省略部分相同代码
        abortBroadcast();
    }
}
```

上述代码在 MyBroadcastReceiver2 中的 onReceive 方法中添加了方法 abortBroadcast 方法截断广播。运行实例，如图 6.42 所示。可以看出 Toast 仅显示了一条信息，也就是说广播被第一个接收到的接收器截断了。查看动态图，请扫描图 6.43 中的二维码。

图 6.42　有序广播之截断实例　　　　图 6.43　有序广播之截断实例二维码

下面来验证有序广播的最后一个特性——动态注册优先级高于静态注册。
修改 MainActivity.java 代码如下：

```
public class MainActivity extends AppCompatActivity {
    private MyBroadcastReceiver myBroadcastReceiver;

    @Override
    protected void onCreate(Bundle savedInstanceState) {
        super.onCreate(savedInstanceState);
        setContentView(R.layout.activity_main);
        myBroadcastReceiver = new MyBroadcastReceiver();
        IntentFilter intentFilter = new IntentFilter("com.yayun.broadcast");
        registerReceiver(myBroadcastReceiver, intentFilter);
    }

    public void send(View view) {
        // 省略部分相同代码
    }

    @Override
```

```
    protected void onDestroy() {
        super.onDestroy();
        unregisterReceiver(myBroadcastReceiver);
    }
}
```

这里在 onCreate 方法中调用了 registerReceiver 方法动态注册了 MyBroadcastReceiver 广播接收器，在实例化 IntentFilter 时添加了参数 com.yayun.broadcast，这个参数即 IntentFilter 的 Action，用于过滤。

由于动态注册了 MyBroadcastReceiver 广播接收器，这时 AndroidMainifest.xml 中就不需要再静态注册（删除广播接收器的相关代码），修改代码如下：

```
<receiver android:name=".MyBroadcastReceiver2">
<intent-filter >
    <action android:name="com.yayun.broadcast"></action>
</intent-filter>
</receiver>
```

这时再次运行，如图 6.44 所示。可以看出，MyBroadcastReceiver 的 onReceive 方法先接收到了广播，而后 MyBroadcastReceiver2 的 onReceive 方法才接收到广播，如图 6.45 所示。

图 6.44　有序广播之动态注册和静态注册优先级一　　图 6.45　有序广播之动态注册和静态注册优先级二

也就是说，动态注册的 MyBroadcastReceiver 先于静态注册的 MyBroadcastReceiver2 先接收到了广播信息。

6.9　Android Service——startService 和 bindService

Service 是 Android 四大组件之一，在 Android 开发中有很重要的作用。它有如下特点：用户无法与它直接进行交互、没有显示的界面。常见的应用方式有：音乐播放器、天气 APP 定时刷新等。官方对 Service 有如下定义：

```
A Service is an application component representing either an
application's desire to perform a longer-running operation while not
interacting with the user or to supply functionality for other applications
to use. Each service class must have a corresponding <service> declaration
in its package's AndroidManifest.xml. Services can be started with Context.
startService() and Context.bindService().
```

大意如下：Service 是一个没有界面、不和用户进行交互、在后台执行耗时操作的系统组件，每一个 Service 都必须使用 <service> 标签在 AndroidManifest.xml 中进行注册，开发者开始使用 startService 和 bindService 来开启一个服务。

对于 startService 启动的 Service，即使启动这个 Service 的应用组件已经被销毁了，这个 Service 仍然会在后台运行；而对于 bindService 启动的 Service，这是以一种绑定的方式启动 Service，也就是说若启动这个 Service 的应用组件被销毁时，其绑定的 Service 也就被销毁了。

和 Activity 一样，Service 也拥有生命周期方法，可以参考官方 API 中提供的图 6.46。

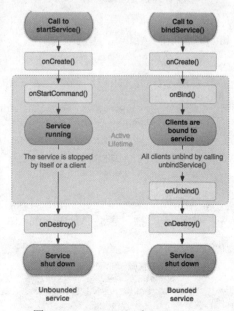

图 6.46　startService 和 bindService

使用 Service 的步骤如下：

- 继承 Service 类（MyService），实现抽象方法。
- 在 AndroidManifest.xml 中进行注册，如：

```
<!-- service 配置开始 -->
<service android:name="MyService"></service>
<!-- service 配置结束 -->
```

- 通过 startService 或 bindService 启动自定义的 Service 类。

6.9.1　startService 启动服务

下面通过实例看一下如何使用 startService 启动一个 Service。

主布局文件代码如下：

```
<?xml version="1.0" encoding="utf-8"?>
<LinearLayout xmlns:android="http://schemas.android.com/apk/res/android"
    android:layout_width="match_parent"
    android:layout_height="match_parent"
    android:orientation="vertical">

    <Button
        android:layout_width="match_parent"
        android:layout_height="wrap_content"
        android:onClick="startService"
        android:text="startService" />

    <Button
```

```xml
        android:layout_width="match_parent"
        android:layout_height="wrap_content"
        android:onClick="stopService"
        android:text="stopService" />

</LinearLayout>
```

上述代码添加了两个 Button 并添加了 onClick 属性来监听单击事件。

自定义一个 Service 类，代码如下：

```java
public class MyServiceToStart extends Service {
    private static final String TAG = "YAYUN_START";

    @Override
    public void onCreate() {
        super.onCreate();
        Log.d(TAG, "onCreate: ");
    }

    @Override
    public int onStartCommand(Intent intent, int flags, int startId) {
        Log.d(TAG, "onStartCommand: ");
        return super.onStartCommand(intent, flags, startId);
    }

    @Nullable
    @Override
    public IBinder onBind(Intent intent) {
        return null;
    }

    @Override
    public void onDestroy() {
        super.onDestroy();
        Log.d(TAG, "onDestroy: ");
    }
}
```

这里自定义的 MyServiceToStart 继承自 Service，Service 类是抽象类，必须覆写其抽象方法 onBind。为了分析其生命周期，这里还覆写了 onCreate、onStartCommand 和 onDestroy 方法。

MainActivity.java 代码如下：

```java
public class MainActivity extends AppCompatActivity {
    private Intent mIntentStart;

    @Override
    protected void onCreate(Bundle savedInstanceState) {
        super.onCreate(savedInstanceState);
        setContentView(R.layout.activity_main);
        mIntentStart = new Intent(this, MyServiceToStart.class);
```

```
    }

    public void startService(View view) {
        startService(mIntentStart);
    }

    public void stopService(View view) {
        stopService(mIntentStart);
    }
}
```

这里创建了一个 Intent 对象来启动 Service，创建这个 Intent 对象时需要传入两个参数：启动 Service 的上下文对象和自定义的 Service 对象。在上面的 Button（STARTSERVICE）的单击事件监听中调用 startService 传入这个新建的 Intent 对象启动 Service，在下面 Button（STOPSERVICE）的单击事件监听中调用 stopService 传入上面新建的 Intent 对象停止 Service。

AndroidManifest.xml 中必须要注册这个自定义的 Service，代码如下：

```
<service android:name=".MyServiceToStart" />
```

添加一个 service 标签在 application 标签中，name 属性值为自定义 Service（"包.类"名）。运行实例，单击 STARTSERVICE 按钮并查看 Log，如图 6.47 所示。

```
01-21 03:24:56.436 5563-5563/ad.servicestartservice D/YAYUN_START: onCreate:
01-21 03:24:56.436 5563-5563/ad.servicestartservice D/YAYUN_START: onStartCommand:
```

图 6.47　STARTSERVICE 之 Log 信息

可以看出，Service 的 onCreate 和 onStartCommand 方法被回调，再次单击 STARTSERVICE 按钮查看 Log，如图 6.48 所示。

```
01-21 03:24:56.436 5563-5563/ad.servicestartservice D/YAYUN_START: onCreate:
01-21 03:24:56.436 5563-5563/ad.servicestartservice D/YAYUN_START: onStartCommand:
01-21 03:25:24.832 5563-5563/ad.servicestartservice D/YAYUN_START: onStartCommand:
```

图 6.48　再次 STARTSERVICE 之 Log 信息

可以看出，不管单击多少次 STARTSERVICE 按钮，onCreate 方法都只会调用一次，而每单击一次 STARTSERVICE 按钮，onStartCommand 都会被回调。

单击 STOPSERVICE 按钮并查看 Log，如图 6.49 所示。

```
01-21 03:25:52.484 5563-5563/ad.servicestartservice D/YAYUN_START: onDestroy:
```

图 6.49　STOPSERVICE 之 Log 信息

可以看出，自定义 Service 的 onDestroy 方法会被回调。

上面已讲过，通过 startService 启动的 Service 在 Activity 销毁时并不会销毁 Service，可以通过 Log 信息的打印进行验证。单击 STARTSERVICE 按钮启动 Service，然后单击返回键，Activity 被销毁了，但是启动的 Service 的 onDestroy 方法并没有回调，也就是说 Service 并没有被销毁。再次启动 Activity 并单击 STARTSERVICE 按钮，Log 信息只会打印 onStartCommand，再一次证明了该 Service 并没有被销毁。

6.9.2 bindService 启动服务

下面通过一个实例看一下 bindService 的用法及生命周期方法的调用。

在上面的主布局文件（activity_main.xml）中再次添加两个 Button，代码如下：

```xml
<Button
    android:layout_width="match_parent"
    android:layout_height="wrap_content"
    android:onClick="bindService"
    android:text="bindService" />

<Button
    android:layout_width="match_parent"
    android:layout_height="wrap_content"
    android:onClick="unBindService"
    android:text="unBindService" />
```

上述代码在原来两个 Button 下方又添加了两个 Button 来触发 bindService 和 unBindService。

新建一个自定义的 Service——MyServiceToBind 来绑定 Service，代码如下：

```java
public class MyServiceToBind extends Service {
    private static final String TAG = "YAYUN_BIND";

    @Override
    public void onCreate() {
        super.onCreate();
        Log.d(TAG, "onCreate: ");
    }

    @Nullable
    @Override
    public IBinder onBind(Intent intent) {
        Log.d(TAG, "onBind: ");
        return null;
    }

    @Override
    public boolean onUnbind(Intent intent) {
        Log.d(TAG, "onUnbind: ");
        return super.onUnbind(intent);
    }

    @Override
    public void onDestroy() {
        super.onDestroy();
        Log.d(TAG, "onDestroy: ");
    }
}
```

MyServiceToBind 同样继承自 Service，除了覆写 onBind 方法之外，还覆写了 onCreate、onUnBind 和 onDestroy 方法并在这些方法中加入了 Log。

修改 MainActivity.java 代码如下：

```java
public class MainActivity extends AppCompatActivity {
    private static final String TAG = "YAYUN_MAIN_ACTIVITY";
    private Intent mIntentStart, mIntentBind;
    private boolean mIsUnBind = true;

    @Override
    protected void onCreate(Bundle savedInstanceState) {
        super.onCreate(savedInstanceState);
        setContentView(R.layout.activity_main);
        mIntentStart = new Intent(this, MyServiceToStart.class);
        mIntentBind = new Intent(this, MyServiceToBind.class);
    }

    private ServiceConnection serviceConnection = new ServiceConnection()
    {

        @Override
        public void onServiceConnected(ComponentName name, IBinder service) {
            Log.d(TAG, "onServiceConnected: ");
        }

        @Override
        public void onServiceDisconnected(ComponentName name) {
            Log.d(TAG, "onServiceDisconnected: ");
        }
    };
    public void startService(View view) {
        startService(mIntentStart);
    }

    public void stopService(View view) {
        stopService(mIntentStart);
    }

    public void bindService(View view) {
        mIsUnBind = false;
        bindService(mIntentBind, serviceConnection, BIND_AUTO_CREATE);
    }

    public void unBindService(View view) {
        if (mIsUnBind) return;
        unbindService(serviceConnection);
        mIsUnBind = true;
    }
}
```

这里在 bindService 方法中调用了 bindService 绑定 Service。调用 bindService 方法需要传入三个参数，Intent 对象、ServiceConnection 对象和标志位。其中，ServiceConnection 是个接口，实现这个接口必须覆写其两个抽象方法，这两个方法分别在 Service 连接

（onServiceConnected）和 Service 断开（onServiceDisconnected）时被回调。对于标志位，这里传入了 BIND_AUTO_CREATE 常量表示 Activity 和 Service 绑定时自动创建服务。为了防止解绑服务时出现问题，这里添加了一个标志位 mIsUnBind，保证只有绑定了 Service 才能解绑，并且在绑定 Service 后保证只能解绑一次。

最后不要忘记在 AndroidManifest.xml 中注册上面的 Service，代码如下：

```
<service android:name=".MyServiceToStart"/>
<service android:name=".MyServiceToBind"/>
```

运行实例，单击 BINDSERVICE 按钮查看 Log，如图 6.50 所示。

```
01-21 08:47:22.152 21213-21213/ad.servicebindservice D/YAYUN_BIND: onCreate:
01-21 08:47:22.152 21213-21213/ad.servicebindservice D/YAYUN_BIND: onBind:
```

图 6.50　BINDSERVICE 之 Log 信息

可以看出，onCreate 和 onBind 方法被依次调用，再次单击这个 Button 也不会调用 onBind 方法，也就是说 onBind 方法只会调用一次。单击 UNBINDSERVICE 按钮，查看 Log 如图 6.51 所示。

```
01-21 08:47:51.475 21213-21213/ad.servicebindservice D/YAYUN_BIND: onUnbind:
01-21 08:47:51.475 21213-21213/ad.servicebindservice D/YAYUN_BIND: onDestroy:
```

图 6.51　UNBINDSERVICE 之 Log 信息

可以看出，解绑时会依次调用 onUnbind 和 onDestroy 方法。细心的读者可以看出，虽然在 ServiceConnection 的抽象方法中添加了 Log，但是 Log 并没有打印，原因是 MyServiceToBind 的 onBind 方法返回了 null。

修改 MyServiceToBind，代码如下：

```java
public class MyServiceToBind extends Service {
    private static final String TAG = "YAYUN_BIND";
    private MyBinder myBinder = new MyBinder();

    class MyBinder extends Binder {
        private void test() {
            Log.d(TAG, "test: ");
        }
    }

    @Override
    public void onCreate() {
        super.onCreate();
        Log.d(TAG, "onCreate: ");
    }

    @Nullable
    @Override
    public IBinder onBind(Intent intent) {
        Log.d(TAG, "onBind: ");
        return myBinder;
    }
    // 省略部分相同代码
}
```

这里创建的一个内部类 MyBinder 继承自 Binder，同时通过 new 的方式创建了一个 MyBinder 对象，修改 onBind 方法的返回值为这个新建的 MyBinder 对象。这时再次运行实例，并单击 BINDSERVICE 按钮，查看 Log，如图 6.52 所示。

```
01-21 08:51:24.003 24907-24907/ad.servicebindservice D/YAYUN_BIND: onCreate:
01-21 08:51:24.003 24907-24907/ad.servicebindservice D/YAYUN_BIND: onBind:
01-21 08:51:24.017 24907-24907/ad.servicebindservice D/YAYUN_MAIN_ACTIVITY: onServiceConnected:
```

图 6.52　BINDSERVICE 之 Log 信息

可以看出，MainActivity 中的 onServiceConnected 方法被回调了，单击 UNBINDSERVICE 查看 Log，如图 6.53 所示。

```
01-21 08:51:49.151 24907-24907/ad.servicebindservice D/YAYUN_BIND: onUnbind:
01-21 08:51:49.152 24907-24907/ad.servicebindservice D/YAYUN_BIND: onDestroy:
```

图 6.53　UNBINDSERVICE 之 Log 信息

可以看出，仅回调了 onUnbind 和 onDestroy 方法，并没有回调 onServiceDisconnected 方法，其原因官网给出了如下说明：

> Called when a connection to the Service has been lost. This typically happens when the process hosting the service has crashed or been killed. This does not remove the ServiceConnection itself -- this binding to the service will remain active, and you will receive a call to onServiceConnected(ComponentName, IBinder) when the Service is next running.

也就是说，这个方法在正常情况下不会被回调，只有在 Service 因异常而断开（Service 所在的进程 crash 或者被 kill）时才会被回调。

上面讲到了通过 Bind 方式启动的 Service 在其启动的 Activity 销毁时，这个 Service 也会销毁，下面我们来验证一下。首先单击 BINDSERVICE 按钮绑定服务，然后单击返回键销毁 Activity，这时应用 crash 了，Log 信息如图 6.54 所示。

```
Activity ad.servicebindservice.MainActivity has leaked ServiceConnection ad.servicebindservice.MainActivity$1@11ee8f9 that was originally bound here
android.app.ServiceConnectionLeaked: Activity ad.servicebindservice.MainActivity has leaked ServiceConnection ad.servicebindservice
    at android.app.LoadedApk$ServiceDispatcher.<init>(LoadedApk.java:1336)
    at android.app.LoadedApk.getServiceDispatcher(LoadedApk.java:1231)
    at android.app.ContextImpl.bindServiceCommon(ContextImpl.java:1450)
    at android.app.ContextImpl.bindService(ContextImpl.java:1422)
    at android.content.ContextWrapper.bindService(ContextWrapper.java:636)
    at ad.servicebindservice.MainActivity.bindService(MainActivity.java:50) <1 internal calls>
    at android.support.v7.app.AppCompatViewInflater$DeclaredOnClickListener.onClick(AppCompatViewInflater.java:288)
    at android.view.View.performClick(View.java:5637)
    at android.view.View$PerformClick.run(View.java:22429)
    at android.os.Handler.handleCallback(Handler.java:751)
    at android.os.Handler.dispatchMessage(Handler.java:95)
    at android.os.Looper.loop(Looper.java:154)
    at android.app.ActivityThread.main(ActivityThread.java:6119) <1 internal calls>
    at com.android.internal.os.ZygoteInit$MethodAndArgsCaller.run(ZygoteInit.java:886)
    at com.android.internal.os.ZygoteInit.main(ZygoteInit.java:776)
```

图 6.54　应用 crash 信息

其原因是在 Activity 销毁时没有解绑 Service。

在 MainActivity 覆写 onDestroy 方法，代码如下：

```java
    @Override
protected void onDestroy() {
    super.onDestroy();
    if (mIsUnBind) return;
    unbindService(serviceConnection);
    mIsUnBind = true;
}
```

在 onDestroy 方法中首先判断有没有没解绑的服务,若存在则调用 unbindService 方法解绑服务,并将标志位设置为 true。这时单击 BINDSERVICE 按钮之后再单击返回退出 Activity 将不会出现 crash。Log 信息如图 6.55 所示。

```
01-21 09:07:11.336 16087-16087/ad.servicebindservice D/YAYUN_BIND: onCreate:
01-21 09:07:11.336 16087-16087/ad.servicebindservice D/YAYUN_BIND: onBind:
01-21 09:07:11.350 16087-16087/ad.servicebindservice D/YAYUN MAIN ACTIVITY: onServiceConnected:
01-21 09:07:14.109 16087-16087/ad.servicebindservice D/YAYUN_BIND: onUnbind:
01-21 09:07:14.110 16087-16087/ad.servicebindservice D/YAYUN_BIND: onDestroy:
```

图 6.55　退出 Activity 之 Log 信息

第 7 章　Android 存储操作实战

保存数据和文件也是智能机最基本的一项功能，本章将介绍 Android 平台数据存储的几种方式：
- 使用 SharedPreferences 存储数据；
- 文件存储数据；
- SQLite 数据库存储数据；
- 使用 ContentProvider 存储数据。

这几种存储方式都是开发和学习中经常遇到的，应聘面试时也经常会被问到，下面通过实例来学习这几种存储方式。

7.1　轻型存储器——SharedPreferences

适用场景和特点：一种轻型的数据存储方式，适用于数据量少、数据类型单一的情形。例如应用程序的配置信息，用户名密码的本地保存等。

保存原理：保存基于 xml 文件存储的键值对数据，通常用来存储一些简单的配置信息。通过 DDMS 的 File Explorer 面板，展开文件浏览树，很明显，SharedPreferences 数据总是存储在 /data/data/< 当前项目包名 package name>/shared_prefs 目录下。

实现 SharedPreferences 存储的步骤如下：
- 获得 SharedPreferences 对象：通过 Context 提供的 getSharedPreferences(String name, int mode) 方法来获取 SharedPreferences 对象实例，该方法中 name 表示要操作的 xml 文件名。第二个参数有三种，分别如下：
  - Context.MODE_PRIVATE：指定该 SharedPreferences 数据只能被本应用序读、写。
  - Context.MODE_WORLD_READABLE：指定该 SharedPreferences 数据能被其他应用程序读，但不能写。
  - Context.MODE_WORLD_WRITEABLE：指定该 SharedPreferences 数据能被其他应用程序读、写。
- 获得 SharedPreferences.Editor 对象：SharedPreferences 的对象调用 edit 方法可以获得 SharedPreferences.Editor 对象。
- 通过 Editor 接口的 putXXX 方法保存键值（Key-Value）对：例如保存用户名和密码信息，"editor.putString ("username", "user1"); editor.putString ("password", "123456")"。
- 通过 Editor 接口的 commit 方法提交键值对：调用 commit 方法即 editor.commit。

7.1.1　SharedPreferences 基本用法

下面通过简单的实例来学习上面的方法。

主布局文件（activity_main.xml）代码如下：

```
<?xml version="1.0" encoding="utf-8"?>
<LinearLayout xmlns:android="http://schemas.android.com/apk/res/android"
    android:layout_width="match_parent"
```

```xml
    android:layout_height="match_parent"
    android:orientation="vertical">

    <RelativeLayout
        android:layout_width="match_parent"
        android:layout_height="wrap_content">

        <TextView
            android:id="@+id/tv_username"
            android:layout_width="80dp"
            android:layout_height="50dp"
            android:gravity="center"
            android:text="用户名："  />

        <EditText
            android:id="@+id/et_username"
            android:layout_width="match_parent"
            android:layout_height="50dp"
            android:layout_toRightOf="@+id/tv_username" />
    </RelativeLayout>

    <RelativeLayout
        android:layout_width="match_parent"
        android:layout_height="wrap_content">

        <TextView
            android:id="@+id/tv_password"
            android:layout_width="80dp"
            android:layout_height="50dp"
            android:gravity="center"
            android:padding="10dp"
            android:text="密码：" />

        <EditText
            android:id="@+id/et_password"
            android:layout_width="match_parent"
            android:layout_height="50dp"
            android:layout_toRightOf="@+id/tv_password"
            android:padding="10dp" />
    </RelativeLayout>

    <Button
        android:layout_width="match_parent"
        android:layout_height="wrap_content"
        android:layout_margin="10dp"
        android:onClick="commit"
        android:text="提交" />
</LinearLayout>
```

上述代码中，最外层采用线性布局，设置 orientation 属性值为 vertical（垂直方式），里面引入了两个 RelativeLayout（相对布局），每一个 RelativeLayout 中都添加了一个 TextView 和一个 EditText，TextView 在左边，用于显示提示信息，EditText 用于输入信息。最下面的 Button 设置了 onClick 属性值，单击这个按钮时将上面 EditText 中的信息保存到 SharedPreferences 中。

MainActivity.java 代码如下：

```java
public class MainActivity extends AppCompatActivity {
    private EditText mEditTextUserName;
    private EditText mEditTextPassword;
    private String mUserName, mPassword;
    SharedPreferences.Editor mEditor;

    @Override
    protected void onCreate(Bundle savedInstanceState) {
        super.onCreate(savedInstanceState);
        setContentView(R.layout.activity_main);
        mEditTextPassword = (EditText) findViewById(R.id.et_password);
        mEditTextUserName = (EditText) findViewById(R.id.et_username);
        // 获得 SharedPreferences 对象
        SharedPreferences sharedPreferences = getSharedPreferences("userinfo", MODE_PRIVATE);
        // 获得 SharedPreferences.Editor
        mEditor = sharedPreferences.edit();
    }

    public void commit(View view) {
        mUserName = mEditTextUserName.getText().toString();
        mPassword = mEditTextPassword.getText().toString();
        // 保存数据
        mEditor.putString("userName", mUserName);
        mEditor.putString("password", mPassword);
        // 提交
        mEditor.commit();
    }
}
```

上述代码中使用 getSharedPreferences("userinfo"，MODE_PRIVATE) 方法获得 SharedPreferences 对象，第一个是参数是保存的文件名，第二个是读写权限。使用 sharedPreferences.edit 方法获得 SharedPreferences.Editor 对象 editor。使用 SharedPreferences.Editor 类的 putString("userName",userName) 方法保存数据，第一个参数是保存数据的 key 值，第二个是 Value 值。最后调用 commit 方法，提交保存。

运行实例，输入用户名和密码，如图 7.1 所示。单击"提交"按钮，如图 7.2 所示。

可以看出，EditText 被清空并 Toast 显示提示信息。下面介绍如何在 Android Studio 上查看 SharedPreferences 保存的文件。单击 Tools 菜单，如图 7.3 所示。

再选择 Android → Android Device Monitor，打开 DDMS 并选择 data → data，找到项目所在目录，如图 7.4 所示。

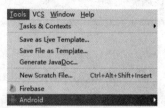

图 7.1　SharedPreferences 保存数据一　　图 7.2　SharedPreferences 保存数据二　　图 7.3　Tools 菜单

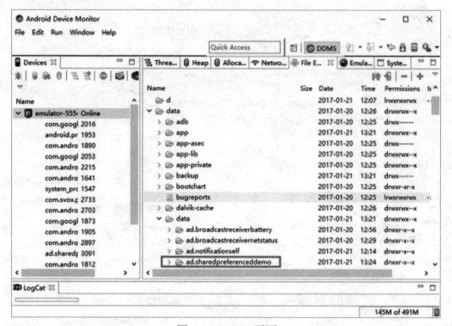

图 7.4　DDMS 页面

然后打开包名路径，选择 shared_prefs 文件夹，可以看到 userinfo.xml 文件，如图 7.5 所示。

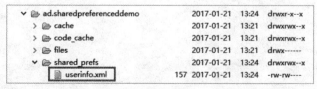

图 7.5　userinfo.xml 文件

选择右上方的小手机标识导出文件到桌面，打开文件，如图 7.6 所示。

可以看到数据以 map 标签包裹，key 和 value 也都是指定的。这样一条 SharedPreferences 记录就完成了。

查看动态图，请扫描图 7.7 中的二维码。

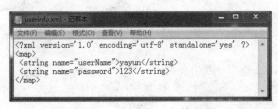
图 7.6 userinfo.xml 文件内容

图 7.7 SharedPreferences 保存数据二维码

7.1.2 SharedPreferences 实现自动登录功能

SharedPreferences 可以用来记录用户名和密码，因此，可以通过本地保存的 SharedPreferences 来解决用户每次登录都要重输用户名和密码的问题。

本实例主要有三个界面，即登录界面（LoginActivity）、中转界面（JumpActivity）、欢迎界面（Welcome）。

LoginActivity 是起始页，所以修改 AndroidManifest.xml 代码如下：

```xml
// 省略部分相同代码
<application
    // 省略部分相同代码
    <activity android:name=".LoginActivity">
        <intent-filter>
            <action android:name="android.intent.action.MAIN" />

            <category android:name="android.intent.category.
            LAUNCHER" />
        </intent-filter>
    </activity>
    <activity android:name=".JumpActivity" />
    <activity android:name=".WelcomeActivity" />
</application>
</manifest>
```

上述代码将默认生成的 MainActivity 替换成 LoginActivity，则 LoginActivity 就变成了应用默认的起始页，另外两个 Activity（JumpActivity 和 WelcomeActivity）要记得在文件中配置。

登录界面的布局文件（login_layout.xml）代码如下：

```xml
<?xml version="1.0" encoding="utf-8"?>
<RelativeLayout xmlns:android="http://schemas.android.com/apk/res/android"
    android:layout_width="fill_parent"
    android:layout_height="wrap_content">

    <TextView
        android:id="@+id/tv_zh"
        android:layout_width="wrap_content"
        android:layout_height="35dip"
        android:layout_marginLeft="12dip"
```

```xml
        android:layout_marginTop="10dip"
        android:gravity="bottom"
        android:text="用户名:"
        android:textColor="#000000"
        android:textSize="18sp" />

    <EditText
        android:id="@+id/et_userName"
        android:layout_width="fill_parent"
        android:layout_height="40dip"
        android:layout_below="@id/tv_zh"
        android:layout_marginLeft="12dip"
        android:layout_marginRight="10dip" />

    <TextView
        android:id="@+id/tv_password"
        android:layout_width="wrap_content"
        android:layout_height="35dip"
        android:layout_below="@id/et_userName"
        android:layout_marginLeft="12dip"
        android:layout_marginTop="10dip"
        android:gravity="bottom"
        android:text="密码:"
        android:textColor="#000000"
        android:textSize="18sp" />

    <EditText
        android:id="@+id/et_password"
        android:layout_width="fill_parent"
        android:layout_height="40dip"
        android:layout_below="@id/tv_password"
        android:layout_marginLeft="12dip"
        android:layout_marginRight="10dip"
        android:maxLines="200"
        android:password="true"
        android:scrollHorizontally="true" />

    <CheckBox
        android:id="@+id/cb_auto"
        android:layout_width="wrap_content"
        android:layout_height="wrap_content"
        android:layout_below="@id/et_password"
        android:layout_marginLeft="12dip"
        android:text="自动登录"
        android:textColor="#000000" />

    <Button
        android:id="@+id/btn_login"
        android:layout_width="match_parent"
        android:layout_height="50dip"
        android:layout_alignRight="@+id/et_password"
```

```xml
            android:layout_below="@+id/cb_auto"
            android:gravity="center"
            android:text=" 登录 "
            android:textColor="#000000"
            android:textSize="18sp" />
</RelativeLayout>
```

上述代码中最外层采用相对布局，相对布局可以灵活地控制控件的位置。为了记录用户的行为（是否需要记住密码），这里添加了一个 CheckBox 控件，用户选中 CheckBox 时记录密码。

中转界面（JumpActivity.java）的布局文件（jump_layout.xml）代码如下：

```xml
<?xml version="1.0" encoding="utf-8"?>
<RelativeLayout xmlns:android="http://schemas.android.com/apk/res/android"
    android:layout_width="match_parent"
    android:layout_height="match_parent">

    <ProgressBar
        android:id="@+id/pgBar"
        android:layout_width="wrap_content"
        android:layout_height="wrap_content"
        android:layout_centerInParent="true" />

    <TextView
        android:id="@+id/tv1"
        android:layout_width="wrap_content"
        android:layout_height="wrap_content"
        android:layout_below="@id/pgBar"
        android:layout_centerHorizontal="true"
        android:text=" 正在登录 ..."
        android:textColor="#000000"
        android:textSize="18sp" />
</RelativeLayout>
```

为了模拟加载效果，这里添加了一个 ProgressBar 并在其下方添加了一个 TextView 显示提示信息。

欢迎页面的布局文件（welcome_layout.xml）代码如下：

```xml
<?xml version="1.0" encoding="utf-8"?>
<LinearLayout xmlns:android="http://schemas.android.com/apk/res/android"
    android:layout_width="match_parent"
    android:layout_height="match_parent"
    android:layout_gravity="center"
    android:orientation="vertical">

    <TextView
        android:layout_width="match_parent"
        android:layout_height="wrap_content"
        android:gravity="center"
```

```xml
        android:padding="20dp"
        android:text=" 登录成功,进入用户界面 "
        android:textColor="#000000"
        android:textSize="20sp" />
</LinearLayout>
```

上述代码仅添加一个 TextView 用于表明当前页面为用户欢迎界面。

LoginActivity.java 代码如下:

```java
public class LoginActivity extends Activity {
    private EditText mUserNameET, mPasswordET;
    private CheckBox mCheckBox;
    private Button mButtonLogin;
    private String mUserNameValue, mPasswordValue;
    private SharedPreferences mSharedPreferences;
    private boolean mIsLogined;

    public void onCreate(Bundle savedInstanceState) {
        super.onCreate(savedInstanceState);
        this.requestWindowFeature(Window.FEATURE_NO_TITLE);
        setContentView(R.layout.login_layout);
        // 获取 SharedPreferences 对象
        mSharedPreferences = getSharedPreferences("userInfo",
                Context.MODE_PRIVATE);
        mUserNameET = (EditText) findViewById(R.id.et_userName);
        mPasswordET = (EditText) findViewById(R.id.et_password);
        mCheckBox = (CheckBox) findViewById(R.id.cb_auto);
        mButtonLogin = (Button) findViewById(R.id.btn_login);
        // 获取 AUTO_ISCHECK 的值
        mIsLogined = mSharedPreferences.getBoolean("AUTO_ISCHECK", false);
        // 假如 SharedPreferences 保存过用户名和密码
        if (mIsLogined) {
            // 跳转到中转页面
            Intent intent = new Intent(LoginActivity.this, JumpActivity.
            class);
            LoginActivity.this.startActivity(intent);
            LoginActivity.this.finish();
        } else {
            mUserNameET.setText(mSharedPreferences.getString("USER_NAME",
            ""));
            mPasswordET.setText(mSharedPreferences.getString("PASSWORD",
            ""));
        }
        mButtonLogin.setOnClickListener(new View.OnClickListener() {
            public void onClick(View v) {
                mUserNameValue = mUserNameET.getText().toString();
                mPasswordValue = mPasswordET.getText().toString();
                if (mUserNameValue.equals("yayun") && mPasswordValue.
                equals("123")) {
                    SharedPreferences.Editor editor = mSharedPreferences.
                    edit();
                    // 记录用户名
```

```
                    editor.putString("USER_NAME", mUserNameValue);
                    // 记录密码
                    editor.putString("PASSWORD", mPasswordValue);
                    // 记录 CheckBox 是否被选中
                    editor.putBoolean("AUTO_ISCHECK", mCheckBox.isChecked())
                            .commit();
                    // 提交
                    editor.commit();
                    Intent intent = new Intent(LoginActivity.this,
                    JumpActivity.class);
                    LoginActivity.this.startActivity(intent);
                    LoginActivity.this.finish();
                } else {
                    Toast.makeText(LoginActivity.this, "用户名或密码错误",
                            Toast.LENGTH_LONG).show();
                }
            }
        });
    }
}
```

每次登录之前，首先在 SharedPreferences 中获取 key 为 AUTO_ISCHECK 的布尔型变量的值，若其值为 true，说明成功存储了用户名和密码，这时直接跳转到中转界面，否则显示登录界面。

第一次登录的时候如果输入了正确的用户名和密码，则使用 putString 方法记录用户名和密码并记录"自动登录"按钮是否被选中，若被选中则下次再登录时不需要输入用户名密码，直接跳转到中转界面。

对于登录中转界面（JumpActivity.java）代码如下：

```
public class JumpActivity extends Activity {
    @Override
    protected void onCreate(Bundle savedInstanceState) {
        super.onCreate(savedInstanceState);
        setContentView(R.layout.jump_layout);
        myThread.start();
    }

    private Thread myThread = new Thread(new Runnable() {
        @Override
        public void run() {
            try {
                // 模拟登录耗时
                Thread.sleep(1000);
            } catch (InterruptedException e) {
                e.printStackTrace();
            }

            Intent intent = new Intent(JumpActivity.this, WelcomeActivity.
            class);
            startActivity(intent);
            finish();
        }
```

```
        });
    }
```

这里在一个新的线程中启动欢迎界面,为了模拟登录的耗时过程,这里调用了 Thread 的 sleep 方法让线程休眠 1s 后才启动欢迎界面。最后记得再调用 Thread 的 start 方法开始线程。

运行实例,输入正确的用户名和密码(用户名:yayun,密码:123),如图 7.8 所示。单击"登录"按钮,如图 7.9 所示。

跳转到登录中转界面,1s 后进入到欢迎界面,如图 7.10 所示。

图 7.8 SharedPreferences 实现自动登录一

图 7.9 SharedPreferences 实现自动登录二

图 7.10 SharedPreferences 实现自动登录三

查看动态图,请扫描图 7.11 中的二维码。

打开 SharedPreference 文件,如图 7.12 所示。

图 7.11 SharedPreferences 实现自动登录二维码

图 7.12 SharedPreferences 实现自动登录保存文件内容

可以看出已经记录了 AUTO_ISCHECK 为 true。

7.2 Android 数据库 SQLite

SQLite 是一款基于嵌入式系统的轻型数据库,支持 Windows/Linux/UNIX 等主流的操作系统,同时能够跟很多程序语言相结合,例如 Tcl、PHP、Java、C++、.Net 等,还有 ODBC 接口,同样与 MySQL、PostgreSQL 这两款开源世界著名的数据库管理系统相比,它的处理速度更快。

SQLite 特点:

- 轻量级：SQLite 和 C/S 模式的数据库软件不同，它是进程内的数据库引擎，因此不存在数据库的客户端和服务器。使用 SQLite 一般只需要带上它的一个动态库，就可以享受它的全部功能。
- 跨平台／可移植性：除了主流操作系统 Windows、Linux、SQLite 外，还支持其他一些不常用的操作系统。
- 弱类型的字段：同一列中的数据可以是不同类型。
- 不需要安装：Android 系统内置的数据库，不需要额外安装。
- 开源：一款开源免费的轻量级数据库。

一般数据采用的固定的静态数据类型，而 SQLite 采用的是动态数据类型，会根据存入值自动判断。SQLite 支持以下几种常用的数据类型：
- VARCHAR(n)：长度不固定且其最大长度为 n 的字串，n 不能超过 4000。
- INTEGER：值被标识为整数，依据值的大小可以依次被存储为 1,2,3,4,5,6,7,8。
- DATA：包含了年份、月份、日期。
- TEXT：值为文本字符串，使用数据库编码存储（TUTF-8，UTF-16BE 或 UTF-16-LE）。

SQLite 有两个常用的类：SQLiteOpenHelper 和 SQLiteDatabase。下面分别对这两个类的使用方法进行讲解。

7.2.1　SQLiteOpenHelper 类

这是一个用来创建和升级数据库的辅助类，其继承结构如下：

public abstract class

SQLiteOpenHelper

extends Object

java.lang.Object

　　　　□　　　android.database.sqlite.SQLiteOpenHelper

由继承结构可以看出，SQLiteOpenHelper 类是一个抽象类，因此要想使用它必须通过继承的方式，同时还要覆写它的两个常用的抽象方法 onCreate（SQLiteDatabase db）和 onUpgrade（SQLiteDatabase db, int oldVersion, int newVersion）。此外，还必须重载其构造函数，构造函数中有四个参数：上下文对象、数据库名、自定义的 Cursor（这里一般传入 null）、当前数据库的版本号。

这个类中还有两个重要的方法：
- getReadableDatabase：返回一个 SQLiteDatabase 对象，只读的方式，如果没有则创建。
- getWritableDatabase：返回一个 SQLiteDatabase 对象，读写的方式，如果没有则创建。

下面通过实例讲解一下如何使用辅助类创建一个数据库。

首先，需要创建一个这样的辅助类：

```java
public class MySQLiteHelper extends SQLiteOpenHelper {
    // 创建数据库的SQL 语句
    public static final String CREATE_DATABASE = "CREATE TABLE " +
            "user(id INTEGER PRIMARY KEY AUTOINCREMENT," +
            "username TEXT,sex TEXT,age INTEGER)";
    private Context context;
    // 构造方法
```

```java
    public MySQLiteHelper(Context context, String name,
                    SQLiteDatabase.CursorFactory factory, int
                    version) {
        super(context, name, factory, version);
        this.context = context;
    }
    // 覆写的方法 onCreate
    @Override
    public void onCreate(SQLiteDatabase db) {
        // 调用 SQLiteDatabase 类的 exeSQL 方法创建数据库
        db.execSQL(CREATE_DATABASE);
        Toast.makeText(context, "数据库创建成功", Toast.LENGTH_SHORT).
        show();
    }
    // 覆写的方法 onUpgrade
    @Override
    public void onUpgrade(SQLiteDatabase db, int oldVersion, int
    newVersion) {
    }
}
```

上述代码中新建的辅助类 MySQLiteHelper 继承自 SQLiteOpenHelper，重载了构造方法，覆写了抽象方法 onCreate 和 onUpgrade，自定义了 SQL 语句，在 onCreate 方法中调用了 SQLiteDatabase 的 execSQL 方法创建数据库。

MainActivity.java 代码如下：

```java
public class MainActivity extends AppCompatActivity {
    private MySQLiteHelper mMySQLiteHelper;
    private Button mButton;

    @Override
    protected void onCreate(Bundle savedInstanceState) {
        super.onCreate(savedInstanceState);
        final LinearLayout layout = new LinearLayout(this);
        layout.setOrientation(LinearLayout.VERTICAL);
        mButton = new Button(this);
        mButton.setText("创建数据库");
        layout.addView(mButton);
        // 这个方法在初始化控件后调用
        setContentView(layout);
        // 实例化辅助类
        mMySQLiteHelper = new MySQLiteHelper(this, "users.db", null, 1);
        mButton.setOnClickListener(new View.OnClickListener() {
            @Override
            public void onClick(View v) {
                // 创建或得到数据库（已经创建数据库则返回已存在的，否则创建）
                mMySQLiteHelper.getWritableDatabase();
            }
        });
    }
}
```

这里采用了动态的方式添加 Button，动态添加 Button 首先创建一个线性布局对象，然后创建一个 Button 对象，而后调用线性布局对象的 addView 方法将这个创建的 Button 对象加入进去，最后调用 setContentView 方法显示出来。

这里还初始化了数据库辅助类 MySQLiteHelper，初始化这个对象传入四个参数：上下文对象、数据库名称、工厂类（这里传入 null 即可）版本号。在 Button 的单击监听中调用 getWritableDatabase 方法创建数据库。

运行实例，如图 7.13 所示。

单击"创建数据库"按钮，Toast 显示"数据库创建成功"。

打开 DDMS，查找 data/data/ 包名 /databases 文件夹，其下有两个文件：users.db（创建的数据库）和 users.db-journal（日志文件），如图 7.14 所示。

图 7.13　Android 数据库创建　　　　图 7.14　DDMS 查看数据库

图 7.14 说明数据库创建成功。

数据库创建成功了，但是没有数据，下面通过 SQLiteDatabase 中的方法实现数据的增、删、改、查等操作。

7.2.2　SQLiteDatabase 类

SQLiteDatabase 类代表一个数据库对象，提供了操作数据库的一些方法。在 Android 的 SDK 目录下有 sqlite3 工具，可以利用它创建数据库、创建表和执行一些 SQL 语句。这个类的常用方法如表 7.1 所示。

表 7.1　SQLiteDatabase 类的常用方法

方　　法	说　　明
openOrCreateDatabase(String path, SQLiteDatabase.CursorFactory factory)	打开或创建一个数据库
insert(String table, String nullColumnHack,ContentValues values)	插入一条数据
delete(String table,String whereClause, String[] whereArgs)	删除一条数据

续表

方法	说明
update(String table,ContentValues values, String whereClause,String[] whereArgs)	修改一条数据
query(String table,String[] columns, String selection,String[] selectionArgs, String groupBy,String having,String orderBy)	查询一条记录
execSQL(String sql)	执行一条 SQL 语句
close	关闭数据库连接

下面通过一个小实例对上面方法的使用进行讲解，首先向前面创建的 user 表中插入几条数据，同时为了验证插入的正确性，查询插入的数据并打印出来。

主布局文件（activity_main.xml）代码如下：

```xml
<?xml version="1.0" encoding="utf-8"?>
<LinearLayout xmlns:android="http://schemas.android.com/apk/res/android"
    android:layout_width="match_parent"
    android:layout_height="match_parent"
    android:orientation="vertical">

    <Button
        android:id="@+id/btn_create"
        android:layout_width="match_parent"
        android:layout_height="wrap_content"
        android:gravity="center"
        android:text=" 创建数据库 " />

    <Button
        android:id="@+id/btn_add"
        android:layout_width="match_parent"
        android:layout_height="wrap_content"
        android:gravity="center"
        android:text=" 添加一条数据 " />

    <Button
        android:id="@+id/btn_query"
        android:layout_width="match_parent"
        android:layout_height="wrap_content"
        android:gravity="center"
        android:text=" 查询数据 " />
</LinearLayout>
```

上述代码在线性布局中添加了三个 Button 并为每一个 Button 添加了 id 属性。

数据库辅助类和上一小节中的一致，这里就不再介绍。

MainActivity.java 代码如下：

```java
public class MainActivity extends Activity implements View.OnClickListener {
    private MySQLiteHelper mMySQLiteHelper;
    private Button mButtonCreate, mButtonAdd, mButtonQuery;

    @Override
```

```java
protected void onCreate(Bundle savedInstanceState) {
    super.onCreate(savedInstanceState);
    setContentView(R.layout.activity_main);
    initViews();
    // 实例化辅助类
    mMySQLiteHelper = new MySQLiteHelper(this, "users.db", null, 1);
}

// 初始化View
private void initViews() {
    mButtonCreate = (Button) findViewById(R.id.btn_create);
    mButtonAdd = (Button) findViewById(R.id.btn_add);
    mButtonQuery = (Button) findViewById(R.id.btn_query);
    mButtonAdd.setOnClickListener(this);
    mButtonCreate.setOnClickListener(this);
    mButtonQuery.setOnClickListener(this);
}

@Override
public void onClick(View v) {
    switch (v.getId()) {
        // 创建数据库
        case R.id.btn_create:
            // 调用getWritableDatabase方法创建数据库
            mMySQLiteHelper.getWritableDatabase();
            break;
        // 插入数据
        case R.id.btn_add:
            SQLiteDatabase db = mMySQLiteHelper.getWritableDatabase();
            // 数据集合
            ContentValues cv = new ContentValues();
            cv.put("username", "亚运");
            cv.put("sex", "男");
            cv.put("age", 26);
            db.insert("user", null, cv);
            Toast.makeText(MainActivity.this, "插入数据成功",
                    Toast.LENGTH_SHORT).show();
            break;
        // 数据查询
        case R.id.btn_query:
            SQLiteDatabase dbquery = mMySQLiteHelper.
            getWritableDatabase();
            //query返回一个Cursor对象
            Cursor cursor = dbquery.query("user", null, null, null,
            null, null, null);
            if (cursor.moveToFirst()) {
                do {
                    String username = cursor.getString(
                            cursor.getColumnIndex("username"));
                    String sex = cursor.getString(cursor.
                    getColumnIndex("sex"));
```

```
                            int age = cursor.getInt(cursor.
                            getColumnIndex("age"));
                        Log.d("datas", "姓名:" + username +
                            "、性别:" + sex + "、年龄:" + age);
                    } while (cursor.moveToNext());
                }
                cursor.close();
                break;
        }
    }
}
```

数据插入时,需要借助 ContentValues 类对数据进行封装,这个类类似于 Map 类,以 key-value 键值对保存数据,其中 key 表示数据库中字段名称,value 表示对应的值,然后调用 insert 方法插入到数据表中。其中 insert 方法有三个参数:第一个是数据表名;第二个参数表示如果插入的数据每一列都为空的话,需要指定此行中某一列的名称,一般传入 null;第三个是要插入的数据集(用 ContentValues 类进行封装)。

query 方法需要传入七个参数,方法如下:

```
db.query(String table, String[] columns, String selection, String[]
selectionArgs, String groupBy, String having, String orderBy);
```

table:数据表名;columns:要查询列的名称数组;selection:条件字句,相当于 where;selectionArgs:条件字句,参数数组;groupBy:分组列;having:分组条件;orderBy:排序列。这里查询所有数据,只需第一个参数传入数据表名,其余参数全部传入 null 即可。

数据查询方法 query 会返回一个 Cursor 类游标对象,这个对象的常用方法如表 7.2 所示。

表 7.2 Cursor 类的常用方法

方　法	说　明
close	关闭 Cursor,释放资源
getColumnCount	返回总列数
getColumnIndex(String columnName)	返回指定列名,若不存在返回 -1
getColumnName(int columnIndex)	返回索引列名
getCount	返回 Cursor 中的行数
moveToFirst	移动光标到第一行
moveToLast	移动光标到最后一行
moveToNext	移动光标到下一行
moveToPosition(int position)	移动光标到指定位置
moveToPrevious	移动光标到上一行

运行实例,如图 7.15 所示。

连续单击"添加一条数据"按钮三次,插入三条数据,然后单击"查询数据"按钮,查出三条已插入的数据,如图 7.16 所示。

图 7.15 Android 数据库
增加和查询

图 7.16 Android 数据库增加和查询 Log

可以看出，三条数据被成功插入。

上面对数据的插入和数据查询进行了讲解，下面研究数据更新和数据删除的操作。

在上面主布局文件（activity_main.xml）中添加两个按钮，代码如下：

```xml
<Button
    android:id="@+id/btn_update"
    android:layout_width="match_parent"
    android:layout_height="wrap_content"
    android:gravity="center"
    android:text=" 更新数据 " />

<Button
    android:id="@+id/btn_delete"
    android:layout_width="match_parent"
    android:layout_height="wrap_content"
    android:gravity="center"
    android:text=" 删除数据 " />
```

修改 MainActivity.java 代码如下：

```java
public class MainActivity extends Activity implements View.OnClickListener {
    private MySQLiteHelper mMySQLiteHelper;
    private Button mButtonCreate, mButtonAdd, mButtonQuery,
            mButtonUpdate, mButtonDelete;

    @Override
    protected void onCreate(Bundle savedInstanceState) {
    // 省略部分相同代码
    }

    // 初始化 View
    private void initViews() {
        // 省略部分相同代码
        mButtonQuery.setOnClickListener(this);
```

```java
            mButtonUpdate = (Button) findViewById(R.id.btn_update);
            mButtonDelete = (Button) findViewById(R.id.btn_delete);
            mButtonDelete.setOnClickListener(this);
            mButtonUpdate.setOnClickListener(this);
    }

    @Override
    public void onClick(View v) {
        switch (v.getId()) {
            // 创建数据库
            case R.id.btn_create:
                // 省略部分相同代码
                break;
            // 插入数据
            case R.id.btn_add:
                // 省略部分相同代码
                break;
            // 数据查询
            case R.id.btn_query:
                // 省略部分相同代码
                break;
            // 数据更新
            case R.id.btn_update:
                SQLiteDatabase dbupdate = mMySQLiteHelper.
                        getWritableDatabase();
                ContentValues cvupdate = new ContentValues();
                cvupdate.put("username", " 奥运 ");
                dbupdate.update("user", cvupdate,
                        "username=?", new String[]{" 亚运 "});
                break;
            // 数据删除
            case R.id.btn_delete:
                SQLiteDatabase dbdelete = mMySQLiteHelper.
                getWritableDatabase();
                dbdelete.delete("user", "id>?", new String[]{"1"});
                break;
        }
    }
}
```

上面的代码中涉及到了两个关键方法，具体如下：

- update(String table, ContentValues values, String whereClause, String[] whereArgs)：数据更新。第一个参数表示数据表名；第二个参数表示条件字句，相当于 where；第三个参数是条件字句的参数数组。
- delete(String table, String whereClause, String[] whereArgs)：数据删除。第一个参数表示数据表名；第二个参数是条件字句，相当于 where；第三个参数是条件字句的参数数组。

运行实例，如图 7.17 所示。

首先单击"数据更新"按钮，然后单击"查询数据"按钮，数据更新成功，如图 7.18 所示。

```
01-23 14:09:42.097  30533-30533/ad.sqliteaddquery D/datas: 姓名：奥运、性别：男、年龄：26
01-23 14:09:42.097  30533-30533/ad.sqliteaddquery D/datas: 姓名：奥运、性别：男、年龄：26
01-23 14:09:42.097  30533-30533/ad.sqliteaddquery D/datas: 姓名：奥运、性别：男、年龄：26
```

图 7.17　Android 数据库　　　　图 7.18　Android 数据库更新 Log
　　　　更新和删除

可以看出，所有为用户名为"亚运"的数据都变成了"奥运"。然后单击"删除数据"按钮，再次单击"查询数据"按钮，如图 7.19 所示。

```
01-23 14:11:59.680  30533-30533/ad.sqliteaddquery D/datas: 姓名：奥运、性别：男、年龄：26
```

图 7.19　Android 数据库删除

可以看出，只剩下一条数据，说明所有 id>1 的记录都被删除了，删除操作成功。

7.3　数据中心——ContentProvider

严格意义上来讲，ContentProvider 并不能算数据存储的方式，它一般被用来在不同的应用之间共享数据。例如我们在使用 QQ 或微信时都会遇到是否匹配手机联系人的提示信息，匹配手机联系人这一操作就需要用到本节要讲解的 ContentProvider 组件。

其使用方式一般有两种：一种是读取和操作其他应用程序（例如电话簿、信息、多媒体等），这种在开发中较为常用；另一种是创建自己的 ContentProvider，把应用程序的数据提供外部访问接口，这种在开发中比较少用。

下面主要研究第一种方式的用法，即通过 ContentProvider 操作其他应用程序数据。Android 提供了一个用于数据操作的操作类 ContentResolver，当外部应用需要对 ContentProvider 中的数据进行添加、删除、修改和查询操作时，可以使用 ContentResolver 类中的方法来完成。要获取 ContentResolver 对象，可以使用 Context 提供的 getContentResolver 方法。ContentResolver 提供了如下几个常用的操作方法：

- public Uri insert(Uri uri, ContentValues values)：添加数据到指定 Uri 的 ContentProvider 中，values 参数是对要插入数据的封装。
- public int delete(Uri uri, String selection, String[] selectionArgs)：从指定 Uri 的 ContentProvider 中删除数据，selection 表示约束条件，selectionArgs 是约束条件参数。
- public int update(Uri uri, ContentValues values, String selection, String[] selectionArgs)：更新指定 Uri 的 ContentProvider 中的数据。

- public Cursor query(Uri uri, String[] projection, String selection, String[] selectionArgs, String sortOrder)：查询指定 Uri 的 ContentProvider。需要传入五个参数：第一个参数是 Uri 对象；第二个参数是指定查询的列名；第三个参数是约束条件；第四个参数是约束条件参数值；第五个参数表示排序方式。

这里涉及到了 Uri 类，Uri 是统一资源定位符的意思，表示一个资源的唯一地址。可以把一个 Uri 地址分成三个部分：第一个部分是协议头" content://"，其格式固定，类似 Url 地址中的" http://"。第二个部分是权限（authority）部分，可以唯一标识这个 ContentProvider，可以看作是 Uri 地址中的主机名部分。第三个部分是路径名，表示要操作的数据表，可以看作 Uri 地址中的具体页面。常用写法如下：

content://com.yayun.demo.provider/table1

content://com.yayun.demo.provider/table2

这里需要传入的参数都是 Uri 对象，由 Uri 字符串借助 Uri 类的 parse 方法即可将 Uri 字符串转变成 Uri 对象。

下面以查询系统联系人为例，对 ContentProvider 的用法进行学习。

主布局文件（activity_main.xml）代码如下：

```xml
<?xml version="1.0" encoding="utf-8"?>
<LinearLayout xmlns:android="http://schemas.android.com/apk/res/android"
    android:layout_width="match_parent"
    android:layout_height="match_parent"
    android:orientation="vertical">

    <Button
        android:layout_width="match_parent"
        android:layout_height="wrap_content"
        android:onClick="btnQuery"
        android:text="查询数据" />

    <ListView
        android:id="@+id/lv"
        android:layout_width="match_parent"
        android:layout_height="wrap_content"/>
</LinearLayout>
```

上述代码中采用线性布局，添加了一个 Button 控件并设置了其 onClick 属性为 btnQuery，在 Button 控件的下方添加了一个 ListView 控件用来显示联系人。

MainActivity.java 代码如下：

```java
public class MainActivity extends Activity {
    private ListView mListView;
    ContentResolver mContentResolver;
    // 数据源集合
    private List<String> mDatas = new ArrayList<String>();
    Cursor cursor = null;

    @Override
    protected void onCreate(Bundle savedInstanceState) {
        super.onCreate(savedInstanceState);
```

```java
        setContentView(R.layout.activity_main);
        mListView = (ListView) findViewById(R.id.lv);
        mContentResolver = getContentResolver();
    }

    public void btnQuery(View view) {
        // 运行时权限申请
        if (ActivityCompat.checkSelfPermission(this, Manifest.
        permission.READ_CONTACTS)
                != PackageManager.PERMISSION_GRANTED) {
            if (Build.VERSION.SDK_INT >= 23) {
                ActivityCompat.requestPermissions(
                        this, new String[]{Manifest.permission.READ_
                        CONTACTS}, 123);
            }
        } else {
            // 查询所有联系人
            requestDatas();
        }
    }


    @Override
    public void onRequestPermissionsResult(int requestCode,
                                           @NonNull String[] permissions,
                                           @NonNull int[] grantResults) {
        super.onRequestPermissionsResult(requestCode, permissions,
        grantResults);
        if (requestCode == 123 && ActivityCompat.checkSelfPermission(this,
                Manifest.permission.READ_CONTACTS)
                == PackageManager.PERMISSION_GRANTED) {
            cursor = mContentResolver.query(
                    ContactsContract.CommonDataKinds.Phone.CONTENT_URI,
                    null, null, null, null);
            requestDatas();
        }
    }

    private void requestDatas() {
        if (cursor != null) {
            while (cursor.moveToNext()) {
                int id = cursor.getInt(cursor.getColumnIndex(
                        ContactsContract.CommonDataKinds.Phone._ID));
                String displayName = cursor.getString(cursor.
                getColumnIndex(
                        ContactsContract.CommonDataKinds.Phone.DISPLAY_
                        NAME));
                String number = cursor.getString(cursor.getColumnIndex(
                        ContactsContract.CommonDataKinds.Phone.NUMBER));
                mDatas.add(id + "- 姓名: " + displayName + "- 电话: " + number);
            }
        }
```

```
            if (cursor != null) {
                cursor.close();
                // 设置适配器
                mListView.setAdapter(new ArrayAdapter<>(this,
                        android.R.layout.simple_list_item_1, mDatas));
            }
        }
    }
```

这里要注意，读取联系人在 Android 6.0 以上要动态获取权限，否则单击 Button 时应用会 crash。对于动态申请权限，首先判断是否已获取了 Manifest.permission.READ_CONTACTS 权限，如果已获取则直接调用 requestDatas 方法读取联系人信息。如果没有则调用 requestPermission 方法请求权限。需要传入三个参数：上下文对象、权限、返回码。onRequestPermissionsResult 方法是回调方法，根据返回码判断是否是读取联系人请求权限的返回，若是，则读取联系人信息。

查询所有联系人用到了 ContentResolver 类的 query 方法，传入五个参数，第一个是 Uri，这里传入系统内置的常量，因为要查询所有联系人，所以后面的条件传入 null 即可。Query 方法返回一个 Cursor 对象，对这个对象进行遍历，读出其中的联系人和电话信息，最后设置适配器，将数据在 ListView 中显示出来。

注意对于 SDK 小于 23 的情况，要在 AndroidManifest.xml 中注册权限，代码如下：

```
<uses-permission android:name="android.permission.READ_CONTACTS"/>
```

运行实例，如图 7.20 所示。单击"允许"按钮后，如图 7.21 所示。

所有联系人信息将在 ListView 中显示出来。查看动态图，请扫描图 7.22 中的二维码。

图 7.20　ContentProvider 查询联系人一　　图 7.21　ContentProvider 查询联系人二　　图 7.22　ContentProvider 查询联系人二维码

这时再次单击"查询数据"按钮应用将会出现 crash，Log 信息如图 7.23 所示。

原因是一次查询之后我们调用了 Cursor 的 close 方法将其关闭了。修改代码如下：

```
public class MainActivity extends Activity {
    // 省略部分相同代码
    private void requestDatas() {
```

```
            mDatas.clear();
            cursor = mContentResolver.query(
                    ContactsContract.CommonDataKinds.Phone.CONTENT_URI,
                    null, null, null, null);
            if (cursor != null) {
            // 省略部分相同代码
```

```
Caused by: android.database.StaleDataException: Attempted to access a cursor after it has been closed.
    at android.database.BulkCursorToCursorAdaptor.throwIfCursorIsClosed(BulkCursorToCursorAdaptor.java:63)
    at android.database.BulkCursorToCursorAdaptor.getCount(BulkCursorToCursorAdaptor.java:69)
    at android.database.AbstractCursor.moveToPosition(AbstractCursor.java:219)
    at android.database.AbstractCursor.moveToNext(AbstractCursor.java:268)
    at android.database.CursorWrapper.moveToNext(CursorWrapper.java:202)
    at ad.contentproviderdemo.MainActivity.requestDatas(MainActivity.java:67)
    at ad.contentproviderdemo.MainActivity.btnQuery(MainActivity.java:45)
    at java.lang.reflect.Method.invoke(Native Method)
```

图 7.23 ContentProvider 查询联系人 crash

每次调用 requestDatas 方法时调用 List 的 clear 方法清空 List 集合，并调用 ContentResolver 的 query 方法获得 Cursor 对象。这时再次单击"查询数据"按钮就不会再出现 crash 的问题。

第 8 章　Android 动画操作实战

Android 动画可以让应用界面更友好，用户体验更好，现在的应用基本或多或少都引入了 Android 动画。Android 给开发者们提供了两个动画类：Animation 和 Animator。

Animation 是一个实现 Android UI 界面动画效果的 API。Animation 提供了一系列的动画效果，可以进行旋转、缩放、淡入淡出等，这些效果可以应用在绝大多数的控件中。

Animation 可认为是传统动画，3.0 以后 Android 引入了属性动画 Animator 类，和 Animation 动画不同，Animator 类直接改变控件的属性值。下面介绍 Animator 相对于 Animation 的劣势。

1. 版本兼容

Android 1.0 时就提供了 Animation 类，3.0 以后才引入 Animator，也就是说无法满足目前开发环境 2.x 的兼容支持的，而且现在的 support 包中也没有对于低版本的 Animator 进行支持，所以 Animation 兼容性更强。不过考虑到如今 Android 已经升级到了 7.0，这个问题就不再存在了。

2. 实现效率

由于 Animator 是直接通过设置对象的 Setter、Getter 方法来实现动画效果，因此为了满足对任意对象调用正确方法，Animator 使用了 Java 反射机制，而 Animation 则是直接通过代码对矩阵进行处理，所以就效率这一方面而言，Animator 比不上 Animation。

下面介绍 Animator 相较于 Animation 的优势。

1. 适用性

上面讲到了由于 Animator 使用了反射机制导致其效率偏低，但是这也带来了其适用的对象范围的增加，Animation 仅对 View 这一种对象有用，但是 Animator 可以设置任意对象的属性。

2. 使用效果

Animation 仅仅是对 View 的显示进行改变，其实本身的属性并没有改变，在开发中可能会遇到一些问题，而 Animator 属性动画则不存在这个问题。

8.1　Android 传统动画——Tween（补间动画）

Animation 类实现的类可以称为传统动画。对于传统动画，Android 提供了两类动画：Tween（补间动画）和 Frame（帧动画）。Tween 有四种动画形式，即 AlphaAnimation（渐变动画）、RotateAnimation（旋转动画）、ScaleAnimation（尺寸动画）和 TranslateAnimation（位移动画），当然这些动画形式还可以随意进行组合，构成组合动画 AnimationSet。常用构造方法如表 8.1 所示。

表 8.1 传统动画的常用构造方法

构 造 方 法	参 数 说 明
AlphaAnimation(Context context, AttributeSet attrs)	上下文对象，xml 格式动画文件
AlphaAnimation(float fromAlpha, float toAlpha)	起始透明度值，结束透明度值
RotateAnimation(float fromDegrees, float toDegrees, int pivotXType, float pivotXValue, int pivotYType, float pivotYValue)	起始角度，结束角度，X 轴旋转基准，旋转基准值，Y 轴旋转基准，旋转基准值
ScaleAnimation(float fromX, float toX, float fromY, float toY, int pivotXType, float pivotXValue, int pivotYType, float pivotYValue)	起始 X 轴大小，结束 X 轴大小，起始 Y 轴大小，结束 Y 轴大小，X 轴基准，基准值，Y 轴基准，基准值
TranslateAnimation(float fromXDelta, float toXDelta, float fromYDelta, float toYDelta)	起始 X 坐标，结束 X 坐标，起始 Y 坐标，结束 Y 坐标

Animation 中包含了很多属性和方法，这些属性和方法是操作动画的重要手段，要予以理解，常用属性和方法如表 8.2 所示。

表 8.2 Animaiton 的常用属性和方法

配 置 属 性	相 关 方 法	说　　明
android:duration	setDuration(long)	动画持续时间，单位为毫秒
android:fillAfter	setFillAfter(boolean)	参数为布尔型，传入 true，表示动画转化在动画结束后应用
android:interpolator	setInterpolator(Interpolator)	设置动画速率变化
android:repeatCount	setRepeatCount(int)	设置动画重复次数
android:repeatMode	setRepeatMode(int)	设置动画重复方式
android:startOffset	setStartOffset(long)	动画延迟多少秒后执行，一般可以用在动画集合中

对于帧动画比较好理解，其原理和动画片相似，一张张图片按照某种规则进行排序，然后按照一定速度切换起来，由于人眼的视觉暂留特性就会在大脑中形成连贯的动画画面，电影院中的电影就是这种形式的动画。

8.1.1 AlphaAnimation——渐变动画

下面对 AlphaAnimation 的用法进行讲解，其继承结构如下：
public class
AlphaAnimation
extends Animation
java.lang.Object
　□ android.view.animation.Animation
　□ android.view.animation.AlphaAnimation
由继承结构可以看出，AlphaAnimation 继承自 Animation 类，常用构造方法如下：

```
AlphaAnimation(float fromAlpha, float toAlpha)
```

上述方法需要传入两个参数：第一个是起始透明度值；第二个是结束透明度值，取值在 0～1 之间。每一种补间动画都可以通过两种方式实现：一种是代码方式实现；另一种是 XML 方式实现。下面分别做介绍。

1. 代码实现

新建项目，主布局文件（activity_main.xml）代码如下：

```xml
<?xml version="1.0" encoding="utf-8"?>
<LinearLayout xmlns:android="http://schemas.android.com/apk/res/android"
    android:layout_width="match_parent"
    android:layout_height="match_parent"
    android:orientation="vertical">

    <ImageView
        android:id="@+id/image"
        android:layout_width="wrap_content"
        android:layout_height="wrap_content"
        android:layout_gravity="center_horizontal" />

    <Button
        android:id="@+id/start"
        android:layout_width="match_parent"
        android:layout_height="wrap_content"
        android:text=" 开始动画 " />

    <Button
        android:id="@+id/cancel"
        android:layout_width="match_parent"
        android:layout_height="wrap_content"
        android:onClick="cancelAnimation"
        android:text=" 取消动画 " />
</LinearLayout>
```

上述代码采用线性布局，设置 orientation 属性值为 vertical，最上面添加一个 ImageView 控件用来显示演示的图片，在下面添加两个 Button 用来控制动画的开始和取消。

MainActivity.java 代码如下：

```java
public class MainActivity extends Activity {
    private ImageView mImageView;
    private Button mButtonStart;
    private Button mButtonCancel;

    @Override
    public void onCreate(Bundle savedInstanceState) {
        super.onCreate(savedInstanceState);
        setContentView(R.layout.activity_main);
        mImageView = (ImageView) findViewById(R.id.image);
        mImageView.setBackgroundResource(R.drawable.android);
        mButtonStart = (Button) findViewById(R.id.start);
        mButtonCancel = (Button) findViewById(R.id.cancel);
        final AlphaAnimation animation = new AlphaAnimation(1, 0);
        // 设置动画持续时间
        animation.setDuration(2000);
        // 设置重复次数
        animation.setRepeatCount(3);
```

```
        // 执行前的等待时间
        animation.setStartOffset(1000);
        mButtonStart.setOnClickListener(new View.OnClickListener() {
            public void onClick(View arg0) {
                mImageView.startAnimation(animation);
            }
        });
        mButtonCancel.setOnClickListener(new View.OnClickListener() {
            public void onClick(View v) {
                animation.cancel();
            }
        });
    }
}
```

在 onCreate 方法中通过 new 的方式创建了一个 AlphaAnimation 对象，实例化 AlphaAnimation 对象时传入两个参数，第一个参数为初始的 alpha 值，第二个参数为结束后的 alpha 值。

这里涉及了几个方法，下面做一下说明：

- setDuration 方法：动画对象调用此方法，传入毫秒数作为参数，表示动画持续的时间。
- setRepeatCount 方法：这个方法是可以控制动画重复的次数，调用方法时传入一个 int 值，这个值表示动画重复的次数。
- setStartOffset 方法：这个方法可以控制动画开始前等待的时间，传入的参数表示等待的时长，单位为毫秒。
- startAnimation 方法：控件调用该方法，传入一个动画对象，可以为控件设置动画效果。
- cancel 方法：这个方法可以取消动画的执行。

运行实例，如图 8.1 所示。查看动态图，请扫描图 8.2 中的维码。

图 8.1　Android 动画之渐变动画　　　　图 8.2　Android 动画之渐变动画二维码

2. XML 方式实现

XML 方式实现需要先创建一个放置 xml 文件的文件夹，在 res 文件夹下右击，在弹出的快捷菜单中选择 New → Android Resource Directory，在 Resource type 下拉列表框中选择 anim，如图 8.3 所示。

选择 anim 后，会在 res 目录下新建一个 anim 文件夹。在 anim 文件夹下单击新建一个 Animation Resource File，输入动画文件名，如图 8.4 所示。

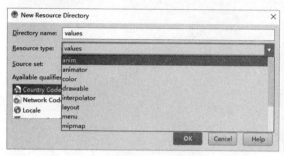

图 8.3　创建 anim 文件夹一

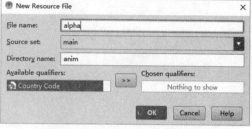

图 8.4　创建 anim 文件夹二

单击 OK 按钮即可新建一个动画文件，新建动画文件后默认在文件中已有一个 set 标签，这个标签是动画文件的根节点，代码如下：

```xml
<?xml version="1.0" encoding="utf-8"?>
<set xmlns:android="http://schemas.android.com/apk/res/android">

</set>
```

在 set 标签里新建一个 alpha 标签并添加一些属性，代码如下：

```xml
<?xml version="1.0" encoding="utf-8"?>
<set xmlns:android="http://schemas.android.com/apk/res/android">
    <alpha
        android:duration="200"
        android:fromAlpha="1.0"
        android:repeatCount="3"
        android:toAlpha="0.0" />
</set>
```

上述代码添加了 duration 属性，设置其值为 200，动画持续 200ms；添加了 fromAlpha 和 toAlpha 属性，设置其值为 1.0 到 0.0；添加了 repeatCount 属性，设置其值为 3，动画重复 3 次。

主布局文件（activity_main.xml）代码如下，

```xml
<?xml version="1.0" encoding="utf-8"?>
<LinearLayout xmlns:android="http://schemas.android.com/apk/res/android"
    android:layout_width="match_parent"
    android:layout_height="match_parent"
    android:orientation="vertical">

    <ImageView
        android:id="@+id/image"
```

```xml
            android:layout_width="wrap_content"
            android:layout_height="wrap_content"
            android:layout_gravity="center_horizontal" />

    <Button
        android:id="@+id/start"
        android:layout_width="match_parent"
        android:layout_height="wrap_content"
        android:text=" 开始动画 " />
</LinearLayout>
```

MainActivity.java 代码如下：

```java
public class MainActivity extends Activity {
    private ImageView mImageView;
    private Button mButtonStart;

    @Override
    public void onCreate(Bundle savedInstanceState) {
        super.onCreate(savedInstanceState);
        setContentView(R.layout.activity_main);
        mImageView = (ImageView) findViewById(R.id.image);
        mImageView.setBackgroundResource(R.drawable.android);
        mButtonStart = (Button) findViewById(R.id.start);
        mButtonStart.setOnClickListener(new View.OnClickListener() {
            public void onClick(View arg0) {
                Animation animation = AnimationUtils.loadAnimation(
                        MainActivity.this, R.anim.alpha);
                mImageView.startAnimation(animation);
            }
        });
    }
}
```

上述代码通过 AnimationUtils 静态方法 loadAnimation 可以将一个动画文件转换成一个 Animation 对象，loadAnimation 需要传入两个参数，第一个为上下文对象，第二个为动画文件的 id，最后调用 startAnimation 方法开始动画。

运行实例，如图 8.5 所示。

图 8.5 文件方式实现 Android 渐变动画

单击"开始动画"按钮动画开始执行。

8.1.2 RotateAnimation——旋转动画

RotateAnimation 类的继承结构如下：
public abstract class
Animation
extends Object
implements Cloneable
java.lang.Object
　└ Android.view.animation.Animation
Known Direct Subclasses
　　AlphaAnimation, AnimationSet, RotateAnimation, ScaleAnimation, TranslateAnimation
　　由继承结构可以看出，同 AlphaAnimation 和 TranslateAnimation 一样，RotateAnimation 和 ScaleAnimation 动画也都是 Animation 类的子类，这里一起进行介绍。
　　RotateAnimation 常用构造方法如下：

```
RotateAnimation(float fromDegrees, float toDegrees,
                int pivotXType, float pivotXValue,
                int pivotYType, float pivotYValue)
```

常用属性如下：
- fromDegrees：起始角度值。
- toDegrees：结束角度值。
- pivotXType：转动点 X 轴的转动标准，共三种，RELATIVE_TO_SELF 以自己为标准，RELATIVE_TO_PARENT 以父组件为标准，ABSOLUTE 表示绝对位置。
- pivotXValue：针对上面标准的值，取值为 0 ～ 1。
- pivotYType：转动点 Y 轴的转动标准，也是三种，RELATIVE_TO_SELF 以自己为标准，RELATIVE_TO_PARENT 以父组件为标准，ABSOLUTE 表示绝对位置。
- pivotYValue：针对上面标准的值，取值为 0 ～ 1。

下面通过实例来看一下 RotateAnimation 的用法。

1. 代码方式实现

新建项目，布局文件（activity_main.xml）代码如下：

```xml
<?xml version="1.0" encoding="utf-8"?>
<LinearLayout xmlns:android="http://schemas.android.com/apk/res/android"
    android:layout_width="match_parent"
    android:layout_height="match_parent"
    android:orientation="vertical">

    <Button
        android:id="@+id/btn_rotate"
        android:layout_width="match_parent"
        android:layout_height="wrap_content"
        android:layout_margin="10dp"
        android:gravity="center"
```

```
            android:onClick="rotate"
            android:text=" 旋转动画 "
            android:textSize="20sp" />

    <ImageView

        android:id="@+id/imgview"
        android:layout_width="100dp"
        android:layout_height="100dp"
        android:layout_gravity="center"
        android:src="@drawable/compass90" />

</LinearLayout>
```

MainActivity.java 代码如下：

```
public class MainActivity extends AppCompatActivity {
    private ImageView mImageView;

    @Override
    protected void onCreate(Bundle savedInstanceState) {
        super.onCreate(savedInstanceState);
        setContentView(R.layout.activity_main);
        mImageView = (ImageView) findViewById(R.id.imgview);
    }

    public void rotate(View view) {
        RotateAnimation rotateAnimation = new RotateAnimation(0, 270,
                Animation.RELATIVE_TO_SELF, 0.5f,
                Animation.RELATIVE_TO_SELF, 0.5f);
        rotateAnimation.setDuration(2000);
        rotateAnimation.setFillAfter(true);
        mImageView.setAnimation(rotateAnimation);
        mImageView.startAnimation(rotateAnimation);
    }
}
```

上述代码通过 new 的方式创建了一个 RotateAnimation 对象，需要传入五个参数（参照上文），然后调用 setDuration 方法设置动画持续的时间；调用 setFillAfter 方法传入 true 表示动画结束后保持状态；调用 setAnimation 方法将 RotateAnimation 对象设置给了 ImgeView 控件，最后调用 startAnimation 方法开始动画。运行实例，如图 8.6 所示。单击"旋转动画"按钮，如图 8.7 所示。

可以看出，旋转 270°后动画结束，并保持 270°的状态。

查看动态图，扫描图 8.8 中的二维码。

2. XML 方式实现

动画文件（rotate.xml）代码如下：

```
<?xml version="1.0" encoding="utf-8"?>
<set xmlns:android="http://schemas.android.com/apk/res/android">
    <rotate
```

```
        android:duration="200"
        android:fromDegrees="0"
        android:pivotX="50%"
        android:pivotY="50%"
        android:repeatCount="infinite"
        android:toDegrees="180" />
</set>
```

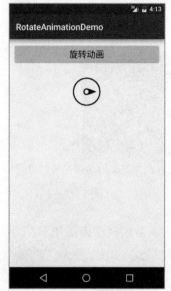

图 8.6　Android 旋转动画实例一　　图 8.7　Android 旋转动画实例二　　图 8.8　Android 旋转动画实例二维码

其中的 repeatCount 属性值设置为 infinite，也就是动画重复次数为无穷次。
MainActivity.java 代码如下：

```
public class MainActivity extends AppCompatActivity {
    private ImageView mImageView;

    @Override
    protected void onCreate(Bundle savedInstanceState) {
        super.onCreate(savedInstanceState);
        setContentView(R.layout.activity_main);
        mImageView = (ImageView) findViewById(R.id.imgview);
    }
    public void rotate(View view) {
        Animation rotateAnimation = AnimationUtils.loadAnimation(
                MainActivity.this, R.anim.rotate);
        mImageView.startAnimation(rotateAnimation);
    }
}
```

上述代码同样采用 loadAnimation 方法引入了布局文件，并调用 startAnimation 方法开始动画。

运行实例，如图 8.9 所示。查看动态图，请扫描图 8.10 中的二维码。

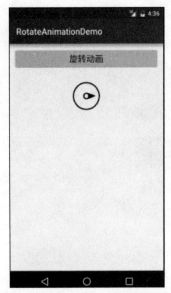

图 8.9　XML 方式实现旋转动画实例

图 8.10　XML 方式实现旋转动画实例二维码

8.1.3　ScaleAnimation——尺寸动画

尺寸动画，顾名思义，改变添加动画的对象的尺寸大小。其常用构造方法如下：

```
public ScaleAnimation (float fromX, float toX,
                       float fromY, float toY,
                       int pivotXType, float pivotXValue,
                       int pivotYType, float pivotYValue)
```

常用属性如下：
- fromX：表示起始时 X 轴方向上的大小，取值 0～1。
- toX：表示结束是 X 轴方向上的大小，取值 0～1。
- fromY：表示起始时 Y 轴方向上的大小，取值 0～1。
- toY：表示结束时 Y 轴方向上的大小，取值 0～1。
- pivotXType：缩放点 X 轴的缩放标准，共三种，RELATIVE_TO_SELF 以自己为标准，RELATIVE_TO_PARENT 以父组件为标准，ABSOLUTE 表示绝对位置。
- pivotXValue：针对上面标准的值，取值 0～1。
- pivotYType：缩放点 Y 轴的缩放标准，也是三种，RELATIVE_TO_SELF 以自己为标准，RELATIVE_TO_PARENT 以父组件为标准，ABSOLUTE 表示绝对位置；
- pivotYValue：针对上面标准的值，取值 0～1。

下面通过一个实例进行演示：

1. 代码方式实现

主布局文件（activity_main.xml）代码如下：

```
<?xml version="1.0" encoding="utf-8"?>
<LinearLayout xmlns:android="http://schemas.android.com/apk/res/android"
    android:layout_width="match_parent"
    android:layout_height="match_parent"
```

```xml
        android:orientation="vertical">

    <Button
        android:id="@+id/btn_rotate"
        android:layout_width="match_parent"
        android:layout_height="wrap_content"
        android:layout_margin="10dp"
        android:gravity="center"
        android:onClick="scale"
        android:text=" 尺寸动画 "
        android:textSize="20sp" />

    <ImageView
        android:id="@+id/imgview"
        android:layout_width="100dp"
        android:layout_height="100dp"
        android:layout_gravity="center"
        android:src="@drawable/magnify" />

</LinearLayout>
```

MainActivity.java 代码如下：

```java
public class MainActivity extends AppCompatActivity {
    private ImageView mImageView;

    @Override
    protected void onCreate(Bundle savedInstanceState) {
        super.onCreate(savedInstanceState);
        setContentView(R.layout.activity_main);
        mImageView = (ImageView) findViewById(R.id.imgview);
    }

    public void scale(View view) {
        ScaleAnimation scaleAnimation = new ScaleAnimation(
                1f, 0f, 1f, 0f, Animation.RELATIVE_TO_SELF, 0.5f,
                Animation.RELATIVE_TO_SELF, 0.5f);
        scaleAnimation.setDuration(2000);
        scaleAnimation.setRepeatCount(2);
        mImageView.setAnimation(scaleAnimation);
        mImageView.startAnimation(scaleAnimation);
    }
}
```

构造方法需要传入八个参数，每个参数的含义参考实例前的说明。

运行实例，如图 8.11 所示。

2. XML 方式实现

动画文件（scale.xml）代码如下：

```xml
<?xml version="1.0" encoding="utf-8"?>
<set xmlns:android="http://schemas.android.com/apk/res/android">
```

```xml
<scale
    android:duration="2000"
    android:fillAfter="true"
    android:fromXScale="1"
    android:fromYScale="1"
    android:pivotX="50%"
    android:pivotY="50%"
    android:repeatCount="1"
    android:repeatMode="reverse"
    android:toXScale="2.0"
    android:toYScale="2.0" />
</set>
```

上述代码添加了 repeatMode 属性表示重复的行为，属性值有 restart 和 reverse 两种，这里设置了其值为 reverse，具体含义可以通过代码实例运行看出。

在主布局文件（activity_main.xml）中新添加一个 Button 用于触发 XML 方式的动画，代码如下：

```xml
<Button
    android:layout_width="match_parent"
    android:layout_height="wrap_content"
    android:layout_margin="10dp"
    android:gravity="center"
    android:onClick="scaleByLayout"
    android:text="尺寸动画XML方式"
    android:textSize="20sp" />
```

MainActivity.java 中添加代码如下：

```java
public void scaleByLayout(View view) {
    Animation animation = AnimationUtils.loadAnimation(MainActivity.
    this,R.anim.scale);
    mImageView.startAnimation(animation);
}
```

运行实例，如图 8.12 所示。单击"尺寸动画 XML 方式"按钮，运行效果请扫描图 8.13 中的二维码。

图 8.11　代码方式实现 Android 尺寸动画实例

图 8.12　Android 尺寸动画实例

图 8.13　Android 尺寸动画实例二维码一

修改动画文件的 repeatMode 属性值为 restart，代码如下：

```
android:repeatMode="restart"
```

再次运行实例，扫描图 8.14 中的二维码。

8.1.4　TranslateAnimation——位移动画

本节将介绍另一种形式的动画——TranslateAnimation（位移动画），TranslateAnimation 类继承结构如下：

public class TranslateAnimation extends Animation

java.lang.Object

图 8.14　Android 尺寸动画实例二维码二

□ anndroid.view.animation.Animation

□ android.view.animation.TranslateAnimation

位移动画也是 Animation 类的子类，常用构造方法如下：

```
TranslateAnimation(float fromXDelta, float toXDelta, float fromYDelta,
float toYDelta)
```

有四个参数，具体如下：
- fromXDelta：起始 X 坐标。
- toXDelta：结束 X 坐标。
- fromYDelta：起始 Y 坐标。
- toYDelta：结束 Y 坐标。

同样可以通过两种方式实现位移动画，下面通过实例进行演示。

1. 代码方式实现

主布局文件（activity_main.xml）代码如下：

```
<?xml version="1.0" encoding="utf-8"?>
<LinearLayout xmlns:android="http://schemas.android.com/apk/res/android"
    android:layout_width="match_parent"
    android:layout_height="match_parent"
    android:orientation="vertical">

    <Button
        android:id="@+id/btn"
        android:layout_width="match_parent"
        android:layout_height="wrap_content"
        android:layout_margin="10dp"
        android:gravity="center"
        android:onClick="translate"
        android:text="位移动画"
        android:textSize="20sp" />

    <ImageView
        android:id="@+id/imgview"
```

```
            android:layout_width="100dp"
            android:layout_height="100dp"
            android:src="@mipmap/ic_launcher" />

</LinearLayout>
```

MainActivity.java 代码如下：

```
public class MainActivity extends AppCompatActivity {
    private ImageView imageView;

    @Override
    protected void onCreate(Bundle savedInstanceState) {
        super.onCreate(savedInstanceState);
        setContentView(R.layout.activity_main);
        imageView = (ImageView) findViewById(R.id.imgview);

    }

    public void translate(View view) {
        TranslateAnimation translateAnimation = new TranslateAnimation(
                0, 200, 0, 200);
        translateAnimation.setDuration(2000);
        translateAnimation.setFillAfter(true);
        imageView.setAnimation(translateAnimation);
        imageView.startAnimation(translateAnimation);
    }
}
```

上述代码通过 new 的方式创建一个 TranslateAnimation 对象，传入四个参数：起始 x 轴坐标、起始 y 轴坐标、结束 x 轴坐标、结束 y 轴坐标。

运行实例，如图 8.15 所示。单击"位移动画"按钮，如图 8.16 所示。

图 8.15　Android 位移动画实例一

图 8.16　Android 位移动画实例二

结果图片从（0,0）移动到了（200,200）。

2. XML 方式实现

动画文件（translate.xml）代码如下：

```xml
<?xml version="1.0" encoding="utf-8"?>
<set xmlns:android="http://schemas.android.com/apk/res/android"
    android:fillAfter="true">
    <translate
        android:duration="2000"
        android:fromXDelta="0"
        android:fromYDelta="0"
        android:repeatCount="3"
        android:toXDelta="0"
        android:toYDelta="100" />
</set>
```

在主布局文件（activity_main.xml）中添加 Button 控件，代码如下：

```xml
<Button
    android:layout_width="match_parent"
    android:layout_height="wrap_content"
    android:layout_margin="10dp"
    android:gravity="center"
    android:onClick="translateByXML"
    android:text=" 位移动画 XML 方式 "
    android:textSize="20sp" />
```

添加了 onClick 属性，在 MainActivity.java 中添加代码如下：

```java
public void translateByXML(View view) {
    Animation translateAnimation = AnimationUtils.loadAnimation(
            MainActivity.this,R.anim.translate);
    imageView.startAnimation(translateAnimation);
}
```

上述代码在 Button 按钮的单击事件监听中，调用 AnimationUtils 的 loadAnimation 方法创建一个 Animation 对象，调用 ImageView 的 startAnimation 方法将这个 Animation 作用在 Imageiew 上。运行实例，如图 8.17 所示。

图 8.17　Android 位移动画实例 XML 方式

8.2 Android 传统动画进阶

8.2.1 动画插值器 Interpolator

由 API 文档可以看出，Interpolator 有比较多的实现子类，如下：

public interface
Interpolator
implements TimeInterpolator
android.view.animation.Interpolator
Known Indirect Subclasses
AccelerateDecelerateInterpolator, AccelerateInterpolator,
AnticipateInterpolator, AnticipateOvershootInterpolator,
BounceInterpolator, CycleInterpolator, DecelerateInterpolator,
LinearInterpolator, OvershootInterpolator

常用的动画插值器如表 8.3 所示。

表 8.3　常用的动画插值器

插 值 器	说 明
AccelerateDecelerateInterpolator	先加速后减速，中间速率最高
AccelerateInterpolator	一直加速
BounceInterpolator	动画回弹
CycleInterpolator	速率改变正弦曲线
DecelerateInterpolator	减速
LinearInterpolator	匀速

下面通过实例看一下这些插值器的用法和效果。

新建项目，主布局文件（activity_main.xml）代码如下：

```xml
<?xml version="1.0" encoding="utf-8"?>
<LinearLayout xmlns:android="http://schemas.android.com/apk/res/android"
    android:layout_width="match_parent"
    android:layout_height="match_parent"
    android:orientation="vertical">

    <Button
        android:layout_width="match_parent"
        android:layout_height="wrap_content"
        android:onClick="AccelerateDecelerateInterpolator"
        android:text="AccelerateDecelerateInterpolator" />

    <Button
        android:layout_width="match_parent"
        android:layout_height="wrap_content"
        android:onClick="AccelerateInterpolator"
        android:text="AccelerateInterpolator" />

    <Button
        android:layout_width="match_parent"
```

```xml
        android:layout_height="wrap_content"
        android:onClick="BounceInterpolator"
        android:text="BounceInterpolator" />

    <Button
        android:layout_width="match_parent"
        android:layout_height="wrap_content"
        android:onClick="CycleInterpolator"
        android:text="CycleInterpolator" />

    <Button
        android:layout_width="match_parent"
        android:layout_height="wrap_content"
        android:onClick="DecelerateInterpolator"
        android:text="DecelerateInterpolator" />

    <Button
        android:layout_width="match_parent"
        android:layout_height="wrap_content"
        android:onClick="LinearInterpolator"
        android:text="LinearInterpolator" />

    <ImageView
        android:id="@+id/image"
        android:layout_width="wrap_content"
        android:layout_height="wrap_content"
        android:src="@drawable/compass90" />
</LinearLayout>
```

上述代码添加了六个 Button 控件，每个 Button 对应设置了 onClick 属性，添加了一个 ImageView 用来显示插值器效果。

MainActivity.java 代码如下：

```java
public class MainActivity extends AppCompatActivity {
    private ImageView mImageView;
    private RotateAnimation mRotateAnimation;

    @Override
    protected void onCreate(Bundle savedInstanceState) {
        super.onCreate(savedInstanceState);
        setContentView(R.layout.activity_main);
        mImageView = (ImageView) findViewById(R.id.image);
        mRotateAnimation = new RotateAnimation(0, 360,
                Animation.RELATIVE_TO_SELF, 0.5f,
                Animation.RELATIVE_TO_SELF, 0.5f);
        mRotateAnimation.setDuration(2000);
    }

    public void AccelerateDecelerateInterpolator(View view) {
        mRotateAnimation.setInterpolator(new AccelerateDecelerateInterp
        olator());
```

```
        mImageView.startAnimation(mRotateAnimation);
    }

    public void AccelerateInterpolator(View view) {
        mRotateAnimation.setInterpolator(new AccelerateInterpolator());
        mImageView.startAnimation(mRotateAnimation);
    }

    public void BounceInterpolator(View view) {
        mRotateAnimation.setInterpolator(new BounceInterpolator());
        mImageView.startAnimation(mRotateAnimation);
    }

    public void CycleInterpolator(View view) {
        mRotateAnimation.setInterpolator(new CycleInterpolator(3.0f));
        mImageView.startAnimation(mRotateAnimation);
    }

    public void DecelerateInterpolator(View view) {
        mRotateAnimation.setInterpolator(new DecelerateInterpolator());
        mImageView.startAnimation(mRotateAnimation);
    }

    public void LinearInterpolator(View view) {
        mRotateAnimation.setInterpolator(new LinearInterpolator());
        mImageView.startAnimation(mRotateAnimation);
    }
}
```

上述代码添加了一个 RotateAnimation 对象，在不同的 onClick 监听事件中都调用了 RotateAnimation 的 setInterpolator 方法传入不同的插值器对象，最后调用 startAnimation 方法开始动画。

运行实例，如图 8.18 所示。查看动态图，请扫描图 8.19 中的二维码。

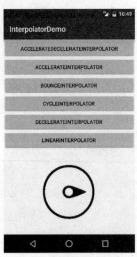

图 8.18　Android 动画插值器实例

图 8.19　Android 动画插值器实例二维码

8.2.2 动画监听器 AnimationListener

查看源码：

```java
public static interface AnimationListener {
    void onAnimationStart(Animation animation);      // 动画开始
    void onAnimationEnd(Animation animation);        // 动画结束
    void onAnimationRepeat(Animation animation);     // 动画重复
}
```

可以看出，AnimationListener 是一个内部接口，实现这个接口必须实现三个抽象方法，分别在动画开始、动画结束和动画重复时回调。下面通过实例来看一下动画监听器的用法。

主布局文件（activity_main.xml）代码如下：

```xml
<?xml version="1.0" encoding="utf-8"?>
<LinearLayout xmlns:android="http://schemas.android.com/apk/res/android"
    android:layout_width="match_parent"
    android:layout_height="match_parent"
    android:orientation="vertical">

    <Button
        android:layout_width="match_parent"
        android:layout_height="wrap_content"
        android:onClick="animationListener"
        android:text="AnimationListenerDemo" />

    <TextView
        android:id="@+id/textView"
        android:gravity="center_horizontal"
        android:layout_width="match_parent"
        android:layout_height="38dp"
        android:padding="4dp" />

    <ImageView
        android:id="@+id/image"
        android:layout_width="200dp"
        android:layout_height="200dp"
        android:layout_gravity="center_horizontal"
        android:layout_marginTop="50dp"
        android:src="@drawable/cardiology" />
</LinearLayout>
```

上述代码添加了一个 Button 控件用于控制动画的开始，添加了一个 TextView 用来显示动画的当前状态，添加了一个 ImageView 用来演示动画效果。

MainActivity.java 代码如下：

```java
public class MainActivity extends AppCompatActivity {
    private ImageView mImageView;
    private TextView mTextView;

    @Override
    protected void onCreate(Bundle savedInstanceState) {
        super.onCreate(savedInstanceState);
```

```
        setContentView(R.layout.activity_main);
        mImageView = (ImageView) findViewById(R.id.image);
        mTextView=(TextView)findViewById(R.id.textView);
}

public void animationListener(View view) {
    ScaleAnimation scaleAnimation = new ScaleAnimation(1f,
            0.9f, 1f, 0.9f, Animation.RELATIVE_TO_SELF,
            0.5f, Animation.RELATIVE_TO_SELF, 0.5f);
    scaleAnimation.setDuration(1000);
    scaleAnimation.setRepeatCount(3);
    scaleAnimation.setAnimationListener(new Animation.
    AnimationListener() {
        @Override
        public void onAnimationStart(Animation animation) {
            mTextView.setText("动画开始...");
            mTextView.setTextSize(20);
        }

        @Override
        public void onAnimationEnd(Animation animation) {
            mTextView.setText("动画结束...");
            mTextView.setTextSize(20);
        }

        @Override
        public void onAnimationRepeat(Animation animation) {
            mTextView.setText("动画重复...");
            mTextView.setTextSize(20);
        }
    });
    mImageView.startAnimation(scaleAnimation);
}
}
```

上述代码创建了一个 ScaleAnimation 对象用来演示动画监听事件，调用 ScaleAnimation 方法的 setAnimationListener，这个方法需要传入一个 AnimationListener 对象，创建这个对象需要覆写三个方法，在每个方法中都调用 TextView 的 setText 方法显示当前的状态。

运行实例，如图 8.20 所示。查看动态图，请扫描图 8.21 中的二维码。

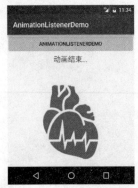

图 8.20　Android 动画监听器实例

图 8.21　Android 动画监听器实例二维码

8.2.3 动画集 AnimationSet

顾名思义，动画集也就是动画集合，不同种类的动画结合在一起显示运行。它同样也有两种方式实现动画集效果：XML 方式实现和代码方式实现。

1. XML 方式实现

这个实例用来模仿淘宝的"添加到购物车"效果，也就是不断旋转、不断缩小、不断移动的动画。

动画文件（animation_set.xml）代码如下：

```xml
<?xml version="1.0" encoding="utf-8"?>
<set xmlns:android="http://schemas.android.com/apk/res/android">
    <rotate
        android:duration="500"
        android:fromDegrees="0"
        android:pivotX="50%"
        android:pivotY="50%"
        android:repeatCount="3"
        android:toDegrees="360" />
    <scale
        android:duration="2000"
        android:fromXScale="1"
        android:fromYScale="1"
        android:pivotX="50%"
        android:pivotY="50%"
        android:repeatMode="reverse"
        android:toXScale="0.0"
        android:toYScale="0.0" />
    <translate
        android:duration="2000"
        android:fromXDelta="0"
        android:fromYDelta="0"
        android:toXDelta="720"
        android:toYDelta="1080" />
</set>
```

上述代码添加了三种样式的动画：rotate（旋转动画）、scale（尺寸动画）、translate（位移动画）。

主布局文件（activity_main.xml）代码如下：

```xml
<?xml version="1.0" encoding="utf-8"?>
<RelativeLayout xmlns:android="http://schemas.android.com/apk/res/android"
    android:layout_width="match_parent"
    android:layout_height="match_parent">

    <ImageView
        android:id="@+id/imageViewCart"
        android:layout_width="60dp"
        android:layout_height="60dp"
        android:layout_alignParentBottom="true"
        android:layout_alignParentRight="true"
        android:layout_margin="10dp"
        android:src="@drawable/shopping_cart" />
```

```xml
    <ImageView
        android:id="@+id/imageViewShirt"
        android:layout_width="80dp"
        android:layout_height="80dp"
        android:layout_gravity="center_horizontal"
        android:layout_margin="80dp"
        android:src="@drawable/tshirt22" />
</RelativeLayout>
```

上述代码添加了两个 ImageView 控件,一个表示购物商品,一个表示购物车。

MainActivity.java 代码如下:

```java
public class MainActivity extends AppCompatActivity {
    private ImageView mImageViewCart, mImageTshirt;

    @Override
    protected void onCreate(Bundle savedInstanceState) {
        super.onCreate(savedInstanceState);
        setContentView(R.layout.activity_main);
        mImageTshirt = (ImageView) findViewById(R.id.imageViewShirt);
        mImageViewCart = (ImageView) findViewById(R.id.imageViewCart);

        mImageTshirt.setOnClickListener(new View.OnClickListener() {
            @Override
            public void onClick(View v) {
                AnimationSet animationSet = (AnimationSet)
                        AnimationUtils.loadAnimation(MainActivity.this,
                                R.anim.animation_set);
                animationSet.setFillAfter(true);
                animationSet.setInterpolator(new
                AccelerateInterpolator());
                mImageTshirt.startAnimation(animationSet);
            }
        });
    }
}
```

AnimationSet 对象是调用了 AnimationUtils 的静态方法 loadAnimation 获得,并添加了加速的动画插值器。运行实例并单击 T-shirt 图片,如图 8.22 所示。可以看出,商品一边旋转一边缩小地移动到购物车中。查看动态图,请扫描图 8.23 中的二维码。

2. 代码方式实现

修改 MainActivity.java 代码如下:

```java
public class MainActivity extends AppCompatActivity {
    // 省略部分相同代码
    mImageTshirt.setOnClickListener(new View.OnClickListener() {
        @Override
        public void onClick(View v) {
```

```java
            AnimationSet animationSet = new AnimationSet(true);
            RotateAnimation rotateAnimation = new RotateAnimation
            (0, 360,
                    Animation.RELATIVE_TO_SELF, 0.5f,
                    Animation.RELATIVE_TO_SELF, 0.5f);
            AlphaAnimation alphaAnimation = new AlphaAnimation(1f, 0f);
            TranslateAnimation translateAnimation = new
            TranslateAnimation(0, 720, 0, 1080);
            animationSet.addAnimation(rotateAnimation);
            animationSet.addAnimation(translateAnimation);
            animationSet.addAnimation(alphaAnimation);
            animationSet.setFillAfter(true);
            animationSet.setDuration(2000);
            animationSet.setInterpolator(new
            AccelerateInterpolator());
            mImageTshirt.startAnimation(animationSet);
        }
    });
}
```

上述代码首先通过 new 的方式创建一个 AnimationSet 对象，实例化时需要传入一个布尔值作为参数，表示是否所有的动画子项共用插值器。这里为动画集准备了三种动画：RotateAnimation（旋转动画）、AlphaAnimation（透明度动画）和 TranslateAnimation（位移动画）。调用 AnimationSet 的 addAnimation 方法即可成功地将一个动画子项添加到动画集当中。

运行实例，如图 8.24 所示。动画目标对象将一边旋转、一边移动、一边渐隐着到"购物车"中。查看动态图，请扫描图 8.25 中的二维码。

图 8.22 Android 组合
动画实例

图 8.23 Android 组合动画
实例二维码

图 8.24 Android 组合动画
代码方式实现

图 8.25　Android 组合动画代码方式实现二维码

8.2.4　LayoutAnimationController 组件动画

LayoutAnimationController 类继承结构如下：

public class

LayoutAnimationController

extends Object

java.lang.Object

　　□　　android.view.animation.LayoutAnimationController

由继承结构可以看出，LayoutAnimationController 类直接继承 Object 类，用于在组件上添加动画效果，这些组件包括常用的 ListView 和 GridView 等。同样，可以通过配置文件和代码两种方式实现。

LayoutAnimationController 主要构造方法如下：

```
LayoutAnimationController(Animation animation)
```

其参数为动画对象。其属性和方法如表 8.4 所示。

表 8.4　组件动画的属性和方法

属　　性	方　　法	说　　明
android:animation	setAnimation(Animation)	设置动画，用于组件中的每一个子项
android:animationOrder	setOrder(int)	动画执行顺序，主要有三种
android:delay		动画执行延迟

子项动画加载的顺序主要有以下三种：

- ORDER_NORMAL：正常顺序，由上到下。
- ORDER_RANDOM：随机顺序。
- ORDER_REVERSE：倒序。

下面通过实例来看一下这个类的用法。

主布局文件（activity_main.xml）代码如下：

```xml
<?xml version="1.0" encoding="utf-8"?>
<RelativeLayout xmlns:android="http://schemas.android.com/apk/res/android"
    android:layout_width="match_parent"
    android:layout_height="match_parent">

    <ListView
        android:id="@+id/lv"
        android:layout_width="match_parent"
        android:layout_height="match_parent" />
</RelativeLayout>
```

上述代码在布局中添加一个 ListView 用来演示组件动画。

添加一个动画文件（animation_layout.xml）代码如下：

```xml
<?xml version="1.0" encoding="utf-8"?>
<set xmlns:android="http://schemas.android.com/apk/res/android">
    <scale
        android:duration="2000"
        android:fromXScale="0"
        android:fromYScale="0"
        android:pivotX="50%"
        android:pivotY="50%"
        android:repeatMode="reverse"
        android:toXScale="1.0"
        android:toYScale="1.0" />
    <alpha
        android:duration="1000"
        android:fromAlpha="0"
        android:toAlpha="1" />
</set>
```

上述代码在动画文件中添加了一个尺寸动画、添加了一个渐变动画。

MainActivity.java 代码如下：

```java
public class MainActivity extends AppCompatActivity {
    private ListView mListView;

    @Override
    protected void onCreate(Bundle savedInstanceState) {
        super.onCreate(savedInstanceState);
        setContentView(R.layout.activity_main);
        mListView = (ListView) findViewById(R.id.lv);
        ArrayList list = new ArrayList();
        for (int i = 0; i < 10; i++) {
            list.add("Android百战经典第" + i + "战");
        }
        ArrayAdapter<String> adapter = new ArrayAdapter<String>(this,
                android.R.layout.simple_list_item_1, list);
        mListView.setAdapter(adapter);
        // 获得 Animation 对象
        Animation animation = AnimationUtils.loadAnimation(this,
                R.anim.animation_layout);
        // 获得 LayoutAnimationController 对象
        LayoutAnimationController layoutAnimationController = new
                LayoutAnimationController(animation);
        layoutAnimationController.setOrder(LayoutAnimationController.ORDER_RANDOM);
        // 设置布局动画
        mListView.setLayoutAnimation(layoutAnimationController);
        // 开始动画
        mListView.startLayoutAnimation();
    }
}
```

上述代码调用 AnimationUtils 的 loadAnimation 方法传入两个参数获得 Animation 对象，通过 new 的方式传入这个 Animation 对象而得到一个 LayoutAnimationController 对象；然后调用 setOrder 方法传入一个常量，这里传入的是随机值（子项动画随机开始执行）；最后调用 setLayoutAnimation 方法设置组件动画，调用 startLayoutAnimation 方法开始组动画。

运行实例，如图 8.26 所示。查看动态图，请扫描图 8.27 中的二维码。

图 8.26　Android 组件动画实例

图 8.27　Android 组件动画实例二维码

8.3　Android 传统动画——Frame Animation（帧动画）

AnimationDrawable 类的继承结构如下：
public class
AnimationDrawable
extends DrawableContainer
implements Animatable Runnable
java.lang.Object
　　□　android.graphics.drawable.Drawable
　　　　□　android.graphics.drawable.DrawableContainer
　　　　　　□　android.graphics.drawable.AnimationDrawable

帧动画就像动画片一样，一张张图片有序排列，并通过一定的速度播放出来，由于人眼的视觉暂留效应，就会呈现出连贯的画面感。电影院里看的电影实际上就是高速播放的连续性图片，现在电影放映的标准是每秒 24 帧，即每秒播放 24 张图片。若要使用 Frame Animation 动画则必须用到 AnimationDrawable 类。根据 API 文档可以看出，此类并不在 Animation 类下。下面就研究其主要属性，如表 8.5 所示。

表 8.5　帧动画属性

属　　性	简　　介
android:drawable	每一帧图片
android:duration	每一帧之间间隔
android:oneshot	是否显示一次，false 为重复显示

API 中演示了如何配置一个帧动画文件，代码如下：

```xml
<!-- Animation frames are wheel0.png -- wheel5.png files inside the
res/drawable/ folder -->
<animation-list android:id="@+id/selected" android:oneshot="false">
    <item android:drawable="@drawable/wheel0" android:duration="50" />
    <item android:drawable="@drawable/wheel1" android:duration="50" />
    <item android:drawable="@drawable/wheel2" android:duration="50" />
    <item android:drawable="@drawable/wheel3" android:duration="50" />
</animation-list>
```

上述代码中文件用一个 animation-list 的标签包裹，一个 oneshot 属性（设置成 false 表示重复播放），数个 item 标签。一个 item 标签表示一张帧图，其中包含了 drawable 属性，表示帧图资源的位置。duration 属性表示一帧的持续时间，单位是毫秒。若每一帧是 50ms，则一秒播放 20 帧图像。

帧动画文件（frame_animation.xml）在 drawable 文件夹下，代码如下：

```xml
<?xml version="1.0" encoding="utf-8"?>
<animation-list xmlns:android="http://schemas.android.com/apk/res/android"
    android:oneshot="false">
    <item
        android:drawable="@drawable/meituan1"
        android:duration="150" />
    <item
        android:drawable="@drawable/meituan2"
        android:duration="150" />
</animation-list>
```

可以看出，两个图片每隔 150ms 切换一次。

这个实例用到了自定义 View，将提示框继承了 ProgressDialog，因此要自定义一个布局文件（dialog.xml），代码如下：

```xml
<?xml version="1.0" encoding="utf-8"?>
<RelativeLayout xmlns:android="http://schemas.android.com/apk/res/android"
    android:layout_width="wrap_content"
    android:layout_height="wrap_content"
    android:layout_gravity="center">

    <ImageView
        android:id="@+id/loadingIv"
        android:layout_width="wrap_content"
        android:layout_height="wrap_content"
        android:background="@drawable/frame_animation" />

    <TextView
        android:id="@+id/loadingTv"
        android:layout_width="wrap_content"
        android:layout_height="wrap_content"
        android:layout_alignBottom="@+id/loadingIv"
        android:layout_centerHorizontal="true"
        android:text=" 正在加载中 .."
```

```
            android:textSize="20sp" />
</RelativeLayout>
```

自定义的 dialog 布局文件中添加了一个 ImageView 用来显示 FrameAnimation 动画，下方的 TextView 用来显示提示文字。

主布局文件（activity_main.xml）代码如下：

```xml
<?xml version="1.0" encoding="utf-8"?>
<RelativeLayout xmlns:android="http://schemas.android.com/apk/res/android"
    android:layout_width="match_parent"
    android:layout_height="match_parent">

    <Button
        android:id="@+id/btn_test"
        android:layout_width="match_parent"
        android:layout_height="wrap_content"
        android:onClick="test"
        android:text=" 模拟请求 " />
</RelativeLayout>
```

上述代码添加了一个 Button 控件用来控制自定义 Dialog 的显示与否。

自定义的 ProgressDialog 的代码如下：

```java
public class ProgressDialogDemo extends ProgressDialog {
    private AnimationDrawable mAnimation;
    private ImageView mImageView;
    private String mLoadingTip;
    private TextView mLoadingTv;
    private int mResid;

    public ProgressDialogDemo(Context context, String content, int id) {
        super(context);
        this.mLoadingTip = content;
        this.mResid = id;
        setCanceledOnTouchOutside(true);
    }

    @Override
    protected void onCreate(Bundle savedInstanceState) {
        super.onCreate(savedInstanceState);
        initView();
        initData();
    }

    private void initData() {
        mImageView.setBackgroundResource(mResid);
        // 通过 ImageView 对象拿到背景显示的 AnimationDrawable
        mAnimation = (AnimationDrawable) mImageView.getBackground();
        mImageView.post(new Runnable() {
            @Override
            public void run() {
                // 调用 AnimationDrawable 的 start 方法开始动画
```

```
            mAnimation.start();
        }
    });
    mLoadingTv.setText(mLoadingTip);
}

private void initView() {
    // 显示界面
    setContentView(R.layout.dialog);
    mLoadingTv = (TextView) findViewById(R.id.loadingTv);
    mImageView = (ImageView) findViewById(R.id.loadingIv);
}
```

上述代码创建了一个含有三个参数的构造方法：第一个是上下文对象；第二个是显示内容；第三个是显示的图片。在 onCreate 方法中初始化要显示的控件（initView）和控件的数据（initData）。

MainActivity.java 代码如下：

```
public class MainActivity extends AppCompatActivity {
    ProgressDialogDemo dialog;

    @Override
    protected void onCreate(Bundle savedInstanceState) {
        super.onCreate(savedInstanceState);
        setContentView(R.layout.activity_main);
    }

    public void test(View v) {
        dialog = new ProgressDialogDemo(this, "正在加载中",
                R.drawable.frame_animation);
        dialog.show();
        Handler handler = new Handler();
        // 此方法常用于延迟操作
        handler.postDelayed(new Runnable() {
            @Override
            public void run() {
                dialog.dismiss();
            }
        }, 3000);//3 秒钟后调用 dismiss 方法隐藏；
    }
}
```

上述代码在单击事件的监听方法 test 中，通过 new 方法初始化了自定义的 ProgressDialogDemo，传入了三个参数，调用 Dialog 的 show 方法显示自定义的 ProgressDialog，创建了一个 Handler 对象并调用其 postDelayed 方法。这个方法中需要传入两个参数：第一个是 Runnable 对象；第二个是延迟时间（3000ms）。

运行实例，单击"模拟请求"按钮，如图 8.28 所示。查看动态图，请扫描图 8.29 中的二维码。

图 8.28　Android 帧动画实例　　　　图 8.29　Android 帧动画实例二维码

8.4　Android 属性动画——ObjectAnimator

8.4.1　属性动画与传统动画的区别

属性动画和传统动画有什么不同呢？下面可以通过这个经典实例来看一个最基本的不同点。

主布局文件（activity_main.xml）代码如下：

```xml
<?xml version="1.0" encoding="utf-8"?>
<LinearLayout xmlns:android="http://schemas.android.com/apk/res/android"
    android:layout_width="match_parent"
    android:layout_height="match_parent"
    android:orientation="vertical">

    <ImageView
        android:id="@+id/imageview"
        android:layout_width="wrap_content"
        android:layout_height="wrap_content"
        android:src="@mipmap/ic_launcher" />

    <Button
        android:layout_width="match_parent"
        android:layout_height="wrap_content"
        android:onClick="translate"
        android:text="translate" />
</LinearLayout>
```

MainActivity.java 代码如下：

```
public class MainActivity extends AppCompatActivity {
    ImageView mImageView;

    @Override
```

```java
protected void onCreate(Bundle savedInstanceState) {
    super.onCreate(savedInstanceState);
    setContentView(R.layout.activity_main);
    mImageView = (ImageView) findViewById(R.id.imageview);
    mImageView.setOnClickListener(new View.OnClickListener() {
        @Override
        public void onClick(View v) {
            Toast.makeText(MainActivity.this,
                "点我", Toast.LENGTH_SHORT).show();
        }
    });
}

public void translate(View view) {
    TranslateAnimation translateAnimation = new
        TranslateAnimation(0, 200, 0, 0);
    translateAnimation.setDuration(2000);
    translateAnimation.setFillAfter(true);
    mImageView.startAnimation(translateAnimation);
}
}
```

上述代码为 ImageView 添加了单击事件监听，在覆写的 onClick 方法中添加了一个 Toast 信息；对于 Button 的单击事件，translate 方法中添加了一个 TranslateAnimation 位移动画，设置了 setFillAfter 方法传入 true 表示移动后停留下来。

运行实例，移动后单击图片不会再次弹出 Toast，单击图片原来的地方会再次弹出 Toast，也就是说传统的动画仅仅只是改变 View 的显示，并没有改变 View 的实际位置属性。查看动态图，请扫描图 8.30 中的二维码。

图 8.30　Android 传统动画之位移动画二维码

下面用属性动画来实现一下上面相同的功能，修改 MainActivity.java 代码如下：

```java
public class MainActivity extends AppCompatActivity {
    // 省略部分相同代码
    public void translate(View view) {
        // 动画类型及参数值
        ObjectAnimator.ofFloat(mImageView, "translationX", 0, 200)
                // 动画时长
                .setDuration(1000)
                // 开始动画
                .start();
    }
}
```

查看 ofFloat 源码如下：

```java
public static ObjectAnimator ofFloat(Object target, String propertyName,
float... values) {
    ObjectAnimator anim = new ObjectAnimator(target, propertyName);
```

```
        anim.setFloatValues(values);
        return anim;
}
```

ofFloat 方法为 ObjectAnimator 的静态方法，调用这个方法需要传入几个参数：第一个参数为动画目标对象；第二个参数为要修改的属性值；第三个为可变长度的长度，这里传入了 0 和 200，表示在 X 轴方向上从 0 移动到 200。

再次运行实例，可以看出，单击移动后的图片也会弹出 Toast，也就是说属性动画不仅移动了 View 的显示，还移动了 View 本身。

查看动态图，请扫描图 8.31 中的二维码。

除了上面的平移动画，属性动画还提供了其他多种动画，通过下面实例学习。

图 8.31 Android 属性动画之位移动画二维码

8.4.2 旋转动画

为了便于演示，添加一个按钮，其单击事件响应的方法 rotate 代码如下：

```
public void rotate(View view) {
    // 以 X 轴为轴旋转一圈
    ObjectAnimator.ofFloat(mImageView, "rotationX", 0, 360)
            .setDuration(1000).start();
    // 以 Y 轴为轴旋转一圈
    ObjectAnimator.ofFloat(mImageView, "rotationY", 0, 360)
            .setDuration(1000).start();
}
```

上述代码同样也是调用 ObjectAnimator 的 ofFloat 方法，第二个参数传入了 rotationX 和 rotationY，表示绕 X 轴、Y 轴旋转，0 和 360 表示旋转一周。

8.4.3 尺寸动画

再次添加一个按钮用于演示尺寸动画，其单击事件响应的方法 scale 代码如下：

```
public void scale(View view) {
    ObjectAnimator.ofFloat(mImageView, "scaleX", 1, 2.0f).setDuration
    (1000).start();
    ObjectAnimator.ofFloat(mImageView, "scaleY", 1, 2.0f).setDuration
    (1000).start();
}
```

上述代码调用了 ObjectAnimator 的 ofFloat 方法，第二个参数传入了 scaleX 和 scaleY，表示沿 X 轴和 Y 轴方向缩放，1 和 2.0f 表示由原来大小扩大到原来的两倍。

8.4.4 渐变动画

再次添加一个按钮用于演示渐变动画，其单击事件响应的方法 alpha 代码如下：

```
public void alpha(View view) {
    ObjectAnimator.ofFloat(mImageView, "alpha", 1, 0.5f).setDuration
    (1000).start();
}
```

上述代码调用了 ObjectAnimator 的 ofFloat 方法，第二个参数传入了 alpha，表示了透明度，1 和 0.5f 表示由不透明到半透明。

下面看一下完整的代码。

主布局文件（activity_main.xml）代码如下：

```xml
<?xml version="1.0" encoding="utf-8"?>
<LinearLayout xmlns:android="http://schemas.android.com/apk/res/android"
    android:layout_width="match_parent"
    android:layout_height="match_parent"
    android:orientation="vertical">

    <ImageView
        android:id="@+id/imageview"
        android:layout_width="wrap_content"
        android:layout_height="wrap_content"
        android:src="@mipmap/ic_launcher" />

    <Button
        android:layout_width="match_parent"
        android:layout_height="wrap_content"
        android:onClick="translate"
        android:text="translate" />

    <Button
        android:layout_width="match_parent"
        android:layout_height="wrap_content"
        android:onClick="rotate"
        android:text="rotate" />

    <Button
        android:layout_width="match_parent"
        android:layout_height="wrap_content"
        android:onClick="scale"
        android:text="scale" />

    <Button
        android:layout_width="match_parent"
        android:layout_height="wrap_content"
        android:onClick="alpha"
        android:text="alpha" />
</LinearLayout>
```

上述代码添加了四个 Button 控件，为每个 Button 添加了不同的 onClick 属性值来响应和处理不同的单击事件。

MainActivity.java 代码如下：

```java
public class MainActivity extends Activity {
    ImageView mImageView;
```

```java
@Override
protected void onCreate(Bundle savedInstanceState) {
    super.onCreate(savedInstanceState);
    setContentView(R.layout.activity_main);
    mImageView = (ImageView) findViewById(R.id.imageview);
    mImageView.setOnClickListener(new View.OnClickListener() {
        @Override
        public void onClick(View v) {
            Toast.makeText(MainActivity.this, "点我",
                    Toast.LENGTH_SHORT).show();
        }
    });
}

public void translate(View view) {
    // 动画类型及参数值
    ObjectAnimator.ofFloat(mImageView, "translationX", 0, 200)
            // 动画时长
            .setDuration(1000)
            // 开始动画
            .start();
}

public void rotate(View view) {
    // 以 X 轴为轴旋转一圈
    ObjectAnimator.ofFloat(mImageView, "rotationX", 0, 360)
            .setDuration(1000).start();
    // 以 Y 轴为轴旋转一圈
    ObjectAnimator.ofFloat(mImageView, "rotationY", 0, 360)
            .setDuration(1000).start();
}

public void scale(View view) {
    ObjectAnimator.ofFloat(mImageView, "scaleX", 1, 2.0f).setDuration
    (1000).start();
    ObjectAnimator.ofFloat(mImageView, "scaleY", 1, 2.0f).setDuration
    (1000).start();
}

public void alpha(View view) {
    ObjectAnimator.ofFloat(mImageView, "alpha", 1, 0.5f).setDuration
    (1000).start();
}
```

上述代码每个 Button 的监听中调用不同的接口添加属性动画，单击 TRANSLATE 按钮，如图 8.32 所示，查看其他动画效果，请扫描图 8.33 中的二维码。

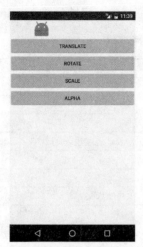

图 8.32　Android 属性动画之位移动画实例

图 8.33　Android 属性动画之位移动画实例二维码

8.4.5　XML 方式实现属性动画

上面是使用代码的方法操作属性动画，那么能不能像传统动画那样通过 XML 文件定义属性动画呢？当然是可以的，下面通过一个实例看一下如何实现。

同样，新建一个文件夹存储属性动画文件，在 res 文件夹下新建一个文件夹，右击，在弹出快捷菜单中选择 New → Android Resource Directory，然后在 Resource type 下拉列表框中选择 animator，如图 8.34 所示。

在这个文件夹右击，在弹出快捷菜单中选择 New → Animator Resource File，在 File name 下拉列表框中 background，如图 8.35 所示。

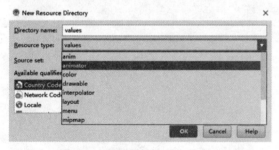

图 8.34　创建属性动画文件夹一

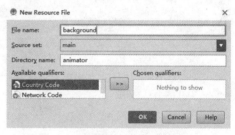

图 8.35　创建属性动画文件夹二

这个动画文件的代码如下：

```xml
<?xml version="1.0" encoding="utf-8"?>
<objectAnimator xmlns:android="http://schemas.android.com/apk/res/android"
    android:duration="5000"
    android:propertyName="backgroundColor"
    android:repeatCount="infinite"
    android:repeatMode="reverse"
    android:valueFrom="#000000"
    android:valueTo="#ffffff"
    android:valueType="intType" >
</objectAnimator>
```

上述代码用一个 objectAnimator 标签包裹。属性动画的常用属性如表 8.6 所示。

表 8.6 属性动画的常用属性

属 性	说 明
duration	动画持续时间
propertyName	属性名
repeatCount	动画重复次数，-1 表示无限循环
repeatMode	重复模式，还有一种模式 restart，表示连续重复
valueFrom	动画属性起始值
valueTo	动画属性结束值
valueType	属性值的类型，有两种 intType 和 floatType

主布局文件（activity_main.xml）代码如下：

```xml
<?xml version="1.0" encoding="utf-8"?>
<LinearLayout xmlns:android="http://schemas.android.com/apk/res/android"
    android:layout_width="match_parent"
    android:layout_height="match_parent"
    android:orientation="vertical">

    <ImageView
        android:id="@+id/image"
        android:layout_width="80dp"
        android:layout_height="80dp"
        android:layout_gravity="center_horizontal"
        android:layout_margin="10dp"
        android:src="@mipmap/ic_launcher" />

    <Button
        android:layout_width="match_parent"
        android:layout_height="wrap_content"
        android:onClick="fromXML"
        android:text="fromXML" />
</LinearLayout>
```

ImageView 是动画作用的对象，单击 Button 触发动画的执行。

MainActivity.java 代码如下：

```java
public class MainActivity extends AppCompatActivity {
    private ImageView mImageView;

    @Override
    protected void onCreate(Bundle savedInstanceState) {
        super.onCreate(savedInstanceState);
        setContentView(R.layout.activity_main);
        mImageView = (ImageView) findViewById(R.id.image);
    }

    public void fromXML(View view) {
        // 加载属性动画文件需要用到 AnimatorInflater 类
        ObjectAnimator objectAnimator = (ObjectAnimator) AnimatorInflater
```

```
            .loadAnimator(MainActivity.this, R.animator.background);
        // 用于动画计算的需要,如果开始和结束的值不是基本类型的时候,这个方法是需要的
        objectAnimator.setEvaluator(new ArgbEvaluator());
        // 设置动画的设置目标
        objectAnimator.setTarget(mImageView);
        objectAnimator.start();
    }
}
```

这里使用了 AnimatorInflater 的静态方法 loadAnimator 将动画文件转换成 ObjectAnimator 对象,这个方法和 Animations 的静态方法 loadAnimation 方法类似,同样需要传入两个参数,第一个为上下文对象,第二个为动画文件的 id。

这里用到了 setEvaluator 方法,这个方法用来设置属性动画值的类型,这里传入了 ArgbEvaluator 对象,在 API 中,这个类的解释如下:

```
This evaluator can be used to perform type interpolation between integer
values that represent ARGB colors.
```

也就是说这个类是用来设置颜色属性,实例化时传入起始和结束的颜色值。除了这个类之外,系统还提供了 FloatValue 和 IntValue,分别用来设置 float 型和 int 型的值。

setTarget 方法用来设置属性动画作用的对象,最后调用 start 方法开始属性动画。

运行实例,如图 8.36 所示。单击图中的 FROMXML 按钮后,颜色在 #000000 纯黑和 #ffffff 纯白之间不断变化。查看动态图,请扫描图 8.37 中的二维码。

图 8.36 属性动画 XML 方式实现

图 8.37 属性动画 XML 方式实现二维码

8.5 Android 属性动画——ValueAnimator

上一节介绍了结合 ObjectAnimator 类来实现属性动画,除了使用这个类之外,还可以使用 ObjectAnimator 类的父类 ValueAnimator 来实现,它的使用将更加灵活。下面介绍使用这个类来实现属性动画的几个步骤。

- 调用 ValueAnimator 的静态方法 ofInt 创建一个 ValueAnimator 对象。
- 调用 addUpdateListener 方法为动画添加属性值更新监听,这个方法需要传入一

个 AnimatorUpdateListener，这是一个接口，实现这个接口必须覆写其抽象方法 onAnimationUpdate，这个方法将在动画属性更新时回调。
- 调用 setInterpolator 方法为动画添加插值器（可选）。
- 调用 setDuration 方法为动画添加动画持续时间。
- 调用 start 方法开始动画。

下面通过实例看一下如何使用 ValueAnimator 实现类似抽屉组件的效果。什么是抽屉组件呢？即需要时拉出，不需要时收起，此功能在开发中很常用。

主布局文件（activity_main.xml）代码如下：

```xml
<?xml version="1.0" encoding="utf-8"?>
<RelativeLayout xmlns:android="http://schemas.android.com/apk/res/android"
    android:layout_width="match_parent"
    android:layout_height="match_parent">

    <TextView
        android:id="@+id/text"
        android:layout_width="match_parent"
        android:layout_height="80dp"
        android:gravity="center|bottom"
        android:text=" 单击我可以收缩 " />

    <WebView
        android:id="@+id/webview"
        android:layout_width="match_parent"
        android:layout_height="match_parent"
        android:layout_alignParentBottom="true"
        android:layout_below="@+id/text" />
</RelativeLayout>
```

上述代码添加了一个 RelativeLayout 作为父布局组件，在上方添加了一个 TextView 组件，在下方添加了一个 WebView 组件并设置其宽、高属性值为 match_parent，还需要设置 layout_alignParentBottom 属性值为 true，将 WebView 放置在屏幕底部。

MainActivity.java 代码如下：

```java
public class MainActivity extends Activity {
    private TextView mTextView;
    private boolean show = false;
    private WebView mWebView;
    private int height = 0;

    @Override
    protected void onCreate(Bundle savedInstanceState) {
        super.onCreate(savedInstanceState);
        requestWindowFeature(Window.FEATURE_NO_TITLE);
        setContentView(R.layout.activity_main);
        mTextView = (TextView) findViewById(R.id.text);
        new Thread(new Runnable() {
            @Override
            public void run() {
                try {
```

```java
                    Thread.sleep(100);
                } catch (InterruptedException e) {
                    e.printStackTrace();
                }
                height = mTextView.getMeasuredHeight();
            }
        }).start();

        mWebView = (WebView) findViewById(R.id.webview);
        mWebView.loadUrl("http://www.baidu.com");
        mTextView.setOnClickListener(new View.OnClickListener() {
            @Override
            public void onClick(View v) {
                performAnim();
            }
        });
    }

    private void performAnim() {
        show = !show;
        ValueAnimator valueAnimator;
        if (show) {
            valueAnimator = ValueAnimator.ofInt(height, 40);
            mTextView.setText("单击我可以展开");
        } else {
            valueAnimator = ValueAnimator.ofInt(40, height);
            mTextView.setText("单击我可以收缩");
        }
        valueAnimator.addUpdateListener(new ValueAnimator.
        AnimatorUpdateListener() {
            @Override
            public void onAnimationUpdate(ValueAnimator valueAnimator) {
                int heightTemp = (Integer) valueAnimator.
                getAnimatedValue();
                mTextView.getLayoutParams().height = heightTemp;
                mTextView.requestLayout();
            }
        });
        valueAnimator.setInterpolator(new BounceInterpolator());
        valueAnimator.setDuration(1000);
        valueAnimator.start();
    }
}
```

要对高度做动画，首先要获得 TextView 控件的高度，若直接在 onCreate 方法中调用 getHeight 方法不能获得真实的控件高度（可以测试一下，这个值为0），原因是过早地调用了获得宽度的方法。解决办法很多，这里新开启了一个线程，延迟100ms执行 getMeasuredHeight 方法获得控件的高度。

对于动画方面，根据是否隐藏（show 值）初始化两个 ValueAniamtor 对象，若 TextView 显示时，则设定其参数值为 height → 40，反之设定为 40 → height，同时改变 TextView 的显示。调用 addUpdateListener 方法为 ValueAnimator 添加动画值更新的监听器，动画值更新就会回调 onAnimationUpdate 方法，这时调用 getAnimatedValue 方法获得更新后的高度值,并将这个高度值赋给 TextView,这里通过改变其 LayoutParams 的属性 height 来实现,最后不要忘记调用 requestLayout 方法刷新 Layout。

要想让浏览器 WebView 成功连接网络,必须要配置网络权限:

```
<uses-permission android:name="android.permission.INTERNET"/>
```

运行实例,如图 8.38 所示。单击最上方的 TextView,如图 8.39 所示,TextView 收缩了上去,同样 WebView 也就跟了上去。这里是通过改变 TextView 的高度来实现这一效果,同样还可以改变 WebView 的高度或改变其他属性值来实现,原理比较相近,都是在 onAnimationUpdate 中动态改变控件中 LayoutParams 的属性值,读者可以自行测试,举一反三。

查看动态图,请扫描图 8.40 中的二维码。

图 8.38　属性动画 ValueAnimator 实现实例一

图 8.39　属性动画 ValueAnimator 实现实例二

图 8.40　属性动画 ValueAnimator 实现实例二维码

8.6　Android 属性动画集

上一节通过 ObjectAnimator 类了解了 Android 属性动画,不过这些都是单一的动画,下面通过例子学习组合动画,也就是动画集是如何实现的。主要可以通过以下几种方式来实现,下面一一介绍。

8.6.1　简单的组合方式

主布局文件(activity_main.xml)代码如下:

```
<?xml version="1.0" encoding="utf-8"?>
<RelativeLayout xmlns:android="http://schemas.android.com/apk/res/android"
    android:layout_width="match_parent"
    android:layout_height="match_parent">

    <ImageView
        android:id="@+id/imageView"
        android:layout_width="100dp"
        android:layout_height="100dp"
```

```
            android:layout_centerInParent="true"
            android:onClick="animatorset"
            android:src="@mipmap/ic_launcher" />

</RelativeLayout>
```

上述代码采用相对布局作为父布局，在父布局中添加一个 ImageView 用来作为属性动画集的演示对象，设置其 layout_centerInPartent 属性值为 true，将 ImageView 放在手机屏幕的正中，为其添加了 onClick 属性监听单击事件。

MainActivity.java 代码如下：

```
public class MainActivity extends AppCompatActivity {
    private ImageView mImageView;

    @Override
    protected void onCreate(Bundle savedInstanceState) {
        super.onCreate(savedInstanceState);
        setContentView(R.layout.activity_main);
        mImageView = (ImageView) findViewById(R.id.imageView);
    }

    public void animatorset(View view) {
        ObjectAnimator.ofFloat(mImageView, "rotation", 0F, 360F)
                .setDuration(1000).start();
        ObjectAnimator.ofFloat(mImageView, "scaleX", 1F, 2F)
                .setDuration(1000).start();
        ObjectAnimator.ofFloat(mImageView, "translationX", 0F, 200F)
                .setDuration(1000).start();
    }
}
```

上述代码在单击监听的方法中添加了三个 ObjectAnimator 对象，所有动画作用的对象都是这个 ImageView，第一个属性动画为旋转动画（0°~360°旋转），第二个属性动画为尺寸动画（一倍到两倍尺寸），第三个属性动画为位移动画（由 0F 移动到 200F）。

运行实例，如图 8.41 所示。单击这个图片后，执行属性动画如图 8.42 所示。

查看动态图，请扫描图 8.43 中的二维码。

图 8.41　简单组合方式实现属性动画集实例一　　图 8.42　简单组合方式实现属性动画集实例二　　图 8.43　简单组合方式实现属性动画集实例二维码

通过动态图可以看出，简单的属性动画集合在一起时，所有动画一起执行。

8.6.2 PropertyValuesHolder 方式

Android 提供了 PropertyValuesHolder 类来包装属性动画对象，下面通过一个实例来看一下这个类的用法。

修改 MainActivity.java 代码如下：

```java
public class MainActivity extends AppCompatActivity {
    private ImageView mImageView;

    @Override
    protected void onCreate(Bundle savedInstanceState) {
        super.onCreate(savedInstanceState);
        setContentView(R.layout.activity_main);
        mImageView = (ImageView) findViewById(R.id.imageView);
    }

    public void animatorset(View view) {
        PropertyValuesHolder propertyValuesHolder1 = PropertyValuesHolder
                .ofFloat("rotation", 0F, 360F);
        PropertyValuesHolder propertyValuesHolder2 = PropertyValuesHolder
                .ofFloat("scaleX", 1F, 2F);
        PropertyValuesHolder propertyValuesHolder3 = PropertyValuesHolder
                .ofFloat("translationX", 0F, 200F);
        ObjectAnimator.ofPropertyValuesHolder(mImageView,
        propertyValuesHolder1,
                propertyValuesHolder2, propertyValuesHolder3).
                setDuration(1000).start();
    }
}
```

上述代码在单击事件监听方法中，首先创建了三个 PropertyValuesHolder 对象，这里调用了 PropertyValuesHolder 的静态方法 ofFloat，这个方法源码如下：

```
ofFloat(String propertyName, float... values)
```

该方法需要传入两个参数，第一个为属性名，第二个为可变长度的参数。

最后调用 ObjectAnimator 类的静态方法 ofPropertyValuesHolder，这个方法源码如下：

```java
public static ObjectAnimator ofPropertyValuesHolder(Object target,
        PropertyValuesHolder... values) {
    ObjectAnimator anim = new ObjectAnimator();
    anim.setTarget(target);
    anim.setValues(values);
    return anim;
}
```

可以看出，这个方法需要传入两个参数，第一个为属性集作用的对象，第二个参数同样是可变长度参数，这里传入的是所有的 PropertyValuesHolder 对象。

最后调用 ObjectAnimator 对象的 setDuration 设置动画时长，调用 ObjectAnimator 的 start 方法开始动画。

运行实例可以获得和上一部分同样的效果，所有动画也是同时执行的。

8.6.3 AnimatorSet 方式

上面的方式中所有的动画都是同时执行的，Android 还提供了 AnimatorSet 类来更灵活地操作动画。

修改 MainActivity.java 代码如下：

```java
public class MainActivity extends AppCompatActivity {
    private ImageView mImageView;

    @Override
    protected void onCreate(Bundle savedInstanceState) {
        super.onCreate(savedInstanceState);
        setContentView(R.layout.activity_main);
        mImageView = (ImageView) findViewById(R.id.imageView);
    }

    public void animatorset(View view) {
        AnimatorSet animatorSet = new AnimatorSet();
        ObjectAnimator objectAnimator1 = ObjectAnimator
                .ofFloat(mImageView, "rotation", 0F, 360F);
        ObjectAnimator objectAnimator2 = ObjectAnimator
                .ofFloat(mImageView, "scaleX", 1F, 2F);
        ObjectAnimator objectAnimator3 = ObjectAnimator
                .ofFloat(mImageView, "translationX", 0F, 200F);
        animatorSet.playTogether(objectAnimator1, objectAnimator2,
                objectAnimator3);
        animatorSet.setDuration(1000);
        animatorSet.start();
    }
}
```

上述代码首先实例化一个 AnimatorSet 对象，然后用三个 ObjectAnimator 对象包装三个属性动画，最后调用 AnimatorSet 的 playTogether 方法，传入三个包装的属性动画，运行实例得到一起执行的动画效果。当然除了 playTogether 方法之外还有 playSequentially 方法，将上面的 playTogether 方法替换成 playSequentially 方法，代码如下：

```java
animatorSet.playSequentially(objectAnimator1, objectAnimator2,
        objectAnimator3);
```

这个方法会按照参数的顺序依次执行所有属性动画，查看动态图，请扫描图 8.44 中的二维码。

可以看出，首先运行旋转动画，然后进行尺寸动画，最后运行了位移动画。其实还可以更多方式控制动画的运行，修改代码如下：

```java
public class MainActivity extends AppCompatActivity {
    private ImageView mImageView;

    @Override
    protected void onCreate(Bundle savedInstanceState) {
```

```java
        super.onCreate(savedInstanceState);
        setContentView(R.layout.activity_main);
        mImageView = (ImageView) findViewById(R.id.imageView);
    }

    public void animatorset(View view) {
        AnimatorSet animatorSet = new AnimatorSet();
        ObjectAnimator objectAnimator1 = ObjectAnimator
                .ofFloat(mImageView, "rotation", 0F, 360F);
        ObjectAnimator objectAnimator2 = ObjectAnimator
                .ofFloat(mImageView, "scaleX", 1F, 2F);
        ObjectAnimator objectAnimator3 = ObjectAnimator
                .ofFloat(mImageView, "translationX", 0F, 200F);
        animatorSet.play(objectAnimator3).with(objectAnimator2);
        animatorSet.play(objectAnimator1).after(objectAnimator2);
        animatorSet.setDuration(1000);
        animatorSet.start();
    }
}
```

这里用到了 with 和 after 方法，单从方法的命名上就可以看出，with 表示一起运行，after 表示在后面运行，也就是说在本项目中尺寸动画和位移动画一起运行，旋转动画在它们后面运行。查看动态图，请扫描图 8.45 中的二维码。

图 8.44　Animator 方式实现组合动画二维码一　　图 8.45　Animator 方式实现组合动画二维码二

8.7　Android 属性动画实现浮动菜单

前几节对属性动画的知识进行了介绍，本节将对前几节的知识进行总结，同样也是通过一个实例的方式来实现。百度阅读的浮动菜单如图 8.46 所示。

这个控件可以通过简单的属性动画来模拟和实现。

主布局文件（activity_main.xml）代码如下：

```xml
<?xml version="1.0" encoding="utf-8"?>
<FrameLayout xmlns:android="http://schemas.android.com/apk/res/android"
    android:layout_width="match_parent"
    android:layout_height="match_parent">

    <ImageView
        android:id="@+id/img1"
        android:layout_width="60dp"
        android:layout_height="60dp"
        android:layout_gravity="bottom|right"
```

```xml
        android:layout_margin="20dp"
        android:src="@drawable/wifi" />

    <ImageView
        android:id="@+id/img2"
        android:layout_width="60dp"
        android:layout_height="60dp"
        android:layout_gravity="bottom|right"
        android:layout_margin="20dp"
        android:src="@drawable/music" />

    <ImageView
        android:id="@+id/img3"
        android:layout_width="60dp"
        android:layout_height="60dp"
        android:layout_gravity="bottom|right"
        android:layout_margin="20dp"
        android:src="@drawable/more" />

    <ImageView
        android:id="@+id/img_more"
        android:layout_width="60dp"
        android:layout_height="60dp"
        android:layout_gravity="bottom|right"
        android:layout_margin="20dp"
        android:src="@drawable/rabbit" />
</FrameLayout>
```

图 8.46　百度阅读的浮动菜单

上述代码采用了 FrameLayout 帧布局，这个布局最适合层叠布局，添加了四个 ImageView，并设置了所有 ImageView 的 layout_gravity 都为 bottom|right（右边底部），同时也设置了相同的 layout_margin 值，这时所有的 ImageView 都会重叠在一起。

MainActivity.java 代码如下：

```java
public class MainActivity extends AppCompatActivity implements View.
OnClickListener {
    private ImageView mImageView1, mImageView2, mImageView3,
    mImageViewMore;
    private boolean mIsSelected = false;
    private List<View> mList = new ArrayList<>();

    @Override
    protected void onCreate(Bundle savedInstanceState) {
        super.onCreate(savedInstanceState);
        setContentView(R.layout.activity_main);
        initViews();
    }

    private void initViews() {
        mImageView1 = (ImageView) findViewById(R.id.img1);
        mImageView2 = (ImageView) findViewById(R.id.img2);
        mImageView3 = (ImageView) findViewById(R.id.img3);
        mImageViewMore = (ImageView) findViewById(R.id.img_more);
        mImageViewMore.setOnClickListener(this);
        mImageView1.setOnClickListener(this);
        mImageView2.setOnClickListener(this);
        mImageView3.setOnClickListener(this);
        mList.add(mImageView1);
        mList.add(mImageView2);
        mList.add(mImageView3);
    }

    private void endAnimator() {
        mIsSelected = false;
        ObjectAnimator animator = ObjectAnimator
                .ofFloat(mImageViewMore, "rotation", 0F, 360F).
                setDuration(300);
        // 添加动画插值器
        animator.setInterpolator(new BounceInterpolator());
        // 开始动画
        animator.start();
        for (int i = 0; i < 3; i++) {
            ObjectAnimator.ofFloat(mList.get(i), "translationY", -200 *
            (i + 1), 0F)
                    .setDuration(1000).start();
        }
    }

    private void startAnimator() {
        mIsSelected = true;
        for (int i = 0; i < 3; i++) {
            ObjectAnimator animator = ObjectAnimator
                    .ofFloat(mList.get(i), "translationY", 0F, -200 *
                    (i + 1))
                    .setDuration(1000);
```

```
                animator.setInterpolator(new BounceInterpolator());
                animator.start();
            }
            ObjectAnimator.ofFloat(mImageViewMore, "rotation", 0F, 360F)
                    .setDuration(300).start();
        }

        @Override
        public void onClick(View v) {
            switch (v.getId()) {
                case R.id.img_more:
                    if (!mIsSelected) {
                        startAnimator();
                    } else {
                        endAnimator();
                    }
                    break;
                case R.id.img1:
                    Toast.makeText(MainActivity.this, "img1",
                            Toast.LENGTH_SHORT).show();
                    break;
                case R.id.img2:
                    Toast.makeText(MainActivity.this, "img2",
                            Toast.LENGTH_SHORT).show();
                    break;
                case R.id.img3:
                    Toast.makeText(MainActivity.this, "img3",
                            Toast.LENGTH_SHORT).show();
                    break;
            }
        }
    }
```

上述代码在 onCreate 方法中调用 initViews 方法，这个方法中初始化了四个 ImageView 并为每个 ImageView 注册了单击事件监听，同时调用了 List 的 add 方法将三个 ImageView 添加到了 List 集合中。

上述代码添加了两个方法——startAnimator 和 endAnimator，单击 image_more 时必调用一个（通过标志位 mIsSelected 来决定调用哪一个）。对于 startAnimator 方法，为每个 ImageView 都添加了不同的位移属性动画，同时为了更好地显示效果，这里还设置了动画插值器，最后调用 start 方法开始动画。

对于 endAnimator 方法，首先将标志位 mIsSelected 设置成 false，而后调用相反的位移动画将所有 ImageView 归位。

运行实例，如图 8.47 所示。单击右下角 Button，如图 8.48 所示。

查看动态图，请扫描图 8.49 中的二维码。

图 8.47　Android 属性动画实现
　　　　浮动菜单一

图 8.48　Android 属性动画实现浮动菜单二　　图 8.49　Android 属性动画实现浮动菜单二维码

第 9 章　Android 网络操作实战

智能手机用户对网络十分依赖，每到一个陌生的地方首先想到的就是如何获取 WiFi 密码，这也充分体现了网络对 Android 手机的重要性，脱离了网络的 Android 手机就像鱼儿离开了水。现如今市面上几乎所有的应用都需要联网才能发挥它的最大功能，因此开发者要对 Android 网络开发进行学习。

9.1 Android 网络核心控件 WebView

Android 内部提供了一个能加载显示网页的控件——WebView，它使用 WebKit 引擎渲染显示网页，这个控件内部提供了一些方法，我们可以借助这些接口和方法实现自己的浏览器。一般来讲，实现一个最基本的 WebView 浏览器需要三个步骤：

- 可以通过 <WebView> 标签在布局文件中添加一个 WebView 控件；
- 在 Activity 中实例化这个 WebView，并调用 WebView 的 loadUrl 方法加载显示网页；
- 在 AndroidManifest.xml 中配置网络访问的权限：

```
<uses-permission android:name="android.permission.INTERNET"/>
```

9.1.1 简单的 WebView

下面通过实例实践一下上面的步骤，来实现一个最简单的浏览器。
主布局文件（activity_main.xml）代码如下：

```xml
<?xml version="1.0" encoding="utf-8"?>
<RelativeLayout xmlns:android="http://schemas.android.com/apk/res/android"
    android:layout_width="match_parent"
    android:layout_height="match_parent">

    <EditText
        android:id="@+id/edit_web"
        android:layout_width="match_parent"
        android:layout_height="wrap_content" />

    <Button
        android:id="@+id/btn_load"
        android:layout_width="wrap_content"
        android:layout_height="wrap_content"
        android:layout_alignParentRight="true"
        android:background="@android:color/transparent"
        android:text="load" />

    <WebView
        android:id="@+id/webView"
        android:layout_width="match_parent"
        android:layout_height="match_parent"
```

```
            android:layout_below="@+id/edit_web" />
</RelativeLayout>
```

上述代码添加了一个 EditText 供用户输入网址信息，添加了一个 Button 监听单击事件加载 EditText 中输入的网址，WebView 负责渲染这个网址页面。

MainActivity.java 代码如下：

```
public class MainActivity extends AppCompatActivity {
    private EditText mEditText;
    private Button mButton;
    private WebView mWebView;

    @Override
    protected void onCreate(Bundle savedInstanceState) {
        super.onCreate(savedInstanceState);
        setContentView(R.layout.activity_main);
        initViews();
        mButton.setOnClickListener(new View.OnClickListener() {
            @Override
            public void onClick(View v) {
                mWebView.loadUrl(mEditText.getText().toString());
            }
        });
    }

    private void initViews() {
        mEditText = (EditText) findViewById(R.id.edit_web);
        mButton = (Button) findViewById(R.id.btn_load);
        mWebView = (WebView) findViewById(R.id.webView);
    }
}
```

initViews 方法负责实例化控件，同时在 onCreate 方法中为 Button 添加了单击事件的监听，调用 WebView 的 loadUrl 方法加载 EditText 中输入的网址。

最后不要忘记在 AndroidManifest.xml 中添加网络权限，代码如下：

```
<uses-permission android:name="android.permission.INTERNET"/>
```

运行实例，在 EditText 中输入网址并单击 LOAD 按钮，如图 9.1 所示。可以看出，百度页面被成功地加载到 WebView 中。查看动态图，请扫描图 9.2 中的二维码。

图 9.1　WebView 加载网页实例

图 9.2　WebView 加载网页实例二维码

9.1.2 丰富 WebView 功能

上面仅仅实现了一个网页加载的功能，一个完整的浏览器还应该有翻页、刷新、进度提示等功能。

修改主布局文件（activity_main.xml）代码如下：

```xml
<?xml version="1.0" encoding="utf-8"?>
<RelativeLayout xmlns:android="http://schemas.android.com/apk/res/android"
    android:layout_width="match_parent"
    android:layout_height="match_parent">
    // 省略部分相同代码
    <ProgressBar
        android:id="@+id/progressBar"
        style="@android:style/Widget.Holo.Light.ProgressBar.Horizontal"
        android:layout_width="match_parent"
        android:layout_height="wrap_content"
        android:layout_below="@id/btn_load" />

    <WebView
        android:id="@+id/webView"
        android:layout_width="match_parent"
        android:layout_height="match_parent"
        android:layout_above="@+id/ll_bottom"
        android:layout_below="@+id/progressBar" />

    <LinearLayout
        android:id="@+id/ll_bottom"
        android:layout_width="match_parent"
        android:layout_height="wrap_content"
        android:layout_alignParentBottom="true"
        android:orientation="horizontal">

        <ImageButton
            android:layout_width="wrap_content"
            android:layout_height="wrap_content"
            android:onClick="reload"
            android:src="@android:drawable/ic_menu_rotate" />

        <Button
            android:id="@+id/left"
            android:layout_width="wrap_content"
            android:layout_height="wrap_content"
            android:onClick="left"
            android:text="上一页" />

        <Button
            android:id="@+id/right"
            android:layout_width="wrap_content"
            android:layout_height="wrap_content"
            android:onClick="right"
            android:text="下一页" />

    </LinearLayout>
</RelativeLayout>
```

上述代码在文本框的下方添加了一个 ProgressBar 用来指示网页加载的进度，在最下方添加了一个LinearLayout并在这个线性布局中添加了三个按钮，分别是："重载""上一页"和"下一页"。设置LinearLayout的orientation属性为horizontal，子控件水平放置。

修改MainActivity.java代码如下：

```java
public class MainActivity extends Activity {
    private EditText mEditText;
    private Button mButton, mButtonLeft, mButtonRight;
    private WebView mWebView;
    private ProgressBar mProgressBar;

    @Override
    protected void onCreate(Bundle savedInstanceState) {
        super.onCreate(savedInstanceState);
        requestWindowFeature(Window.FEATURE_NO_TITLE);
        setContentView(R.layout.activity_main);
        initViews();
        mEditText.setText("http://www.baidu.com");
        mButton.setOnClickListener(new View.OnClickListener() {
            @Override
            public void onClick(View v) {
                mWebView.loadUrl(mEditText.getText().toString());
            }
        });
        mWebView.setWebViewClient(new WebViewClient() {
          @Override
          public boolean shouldOverrideUrlLoading(WebView view, String url) {
                view.loadUrl(url);
                return true;
            }
        });
        mWebView.setWebChromeClient(new WebChromeClient() {
          @Override
          public void onProgressChanged(WebView view, int newProgress) {
                super.onProgressChanged(view, newProgress);
                mProgressBar.setProgress(newProgress);
            }
        });
    }

    private void initViews() {
        mEditText = (EditText) findViewById(R.id.edit_web);
        mButton = (Button) findViewById(R.id.btn_load);
        mButtonLeft = (Button) findViewById(R.id.left);
        mButtonRight = (Button) findViewById(R.id.right);
        mWebView = (WebView) findViewById(R.id.webView);
        mProgressBar = (ProgressBar) findViewById(R.id.progressBar);
    }

    public void reload(View view) {
        mWebView.reload();
    }
```

```java
    public void right(View view) {
        mWebView.goForward();
    }

    public void left(View view) {
        mWebView.goBack();
    }

    @Override
    public boolean onKeyDown(int keyCode, KeyEvent event) {
        if (keyCode == KeyEvent.KEYCODE_BACK && mWebView.canGoBack()) {
            mWebView.goBack();
            return true;
        }
        return super.onKeyDown(keyCode, event);
    }
}
```

这里添加了几个功能点，下面一一进行讲解：

- 网页加载进度条：在地址栏下方添加了一个进度条提示网页加载进度，这里调用 WebView 的 setWebChromeClient 方法并传入 WebChromeClient 对象作为参数，同时覆写了 onProgressChanged 方法，通过这个方法中的参数 newProgress 即可获得当前网页加载的进度，调用 ProgressBar 的 setProgress 方法将这个进度设置到进度条中。
- 上一页功能：单击"上一页"按钮将会回到前一次加载的页面，这里通过调用 WebView 的 goBack 方法实现。
- 下一页功能：单击"下一页"按钮将会跳转到刚才已经打开的页面，这里通过调用 WebView 的 goForward 方法来实现。
- 重载功能：单击"重载"按钮，将刷新当前页面，调用 WebView 的 reload 方法实现。
- "返回键"返回上一页功能：这里覆写了 onKeyDown 方法，判断当 keycode=KeyEvent.KEYCODEBACK(返回键) 并且 WebView 可返回（调用 WebView 的 canGoBack 方法）时调用 WebView 的 goBack 方法返回上一页，否则退出应用。

运行实例，如图 9.3 所示。查看动态图，请扫描图 9.4 中的二维码。

图 9.3　WebView 加载网页进阶实例

图 9.4　WebView 加载网页进阶实例二维码

9.2 WebView 滚动事件

WebView 最常用的浏览方式是上下滚动,可不可以监听滚动事件并添加一些逻辑,让应用实用性、交互性更好呢?WebView 提供了 OnScrollChangeListener 接口供我们实现监听 WebView 的滚动事件,实现这个接口需要覆写其抽象方法 onScrollChange 方法,这个方法在滚动时会回调。除了实现这个监听之外,WebView 还提供了 getScrollX 和 getScrollY 方法来监听当前滚动到的位置。结合一下这些方法,实现监听 WebView 划动到底部和顶部的功能。

首先介绍 WebView 的常用方法,如表 9.1 所示。

表 9.1 WebView 的常用方法

方 法	说 明
getContentHeight	获得整个网页的高度
getScale	获得当前网页的缩放比
getScrollY	获得当前滚动的位置的 Y 坐标值

9.2.1 WebView 滚动监听的实现

下面通过一个实例来监听 WebView 的滚动事件。

在主布局文件(activity_main.xml)中添加一个 WebView 控件,代码如下:

```xml
<?xml version="1.0" encoding="utf-8"?>
<RelativeLayout xmlns:android="http://schemas.android.com/apk/res/android"
    android:layout_width="match_parent"
    android:layout_height="match_parent">

    <WebView
        android:id="@+id/webview"
        android:layout_width="match_parent"
        android:layout_height="match_parent" />
</RelativeLayout>
```

上述代码设置 WebView 的宽高属性都为 match_parent。

MainActivity.java 代码如下:

```java
public class MainActivity extends Activity {
    private WebView mWebView;

    @Override
    protected void onCreate(Bundle savedInstanceState) {
        super.onCreate(savedInstanceState);
        requestWindowFeature(Window.FEATURE_NO_TITLE);
        setContentView(R.layout.activity_main);
        initViews();
        mWebView.loadUrl("http://blog.csdn.net/yayun0516");
        mWebView.setWebViewClient(new WebViewClient() {
            @Override
            public boolean shouldOverrideUrlLoading(WebView view,
                String url) {
```

```java
            view.loadUrl(url);
            return true;
        }
    });

    mWebView.setOnScrollChangeListener(new View.
OnScrollChangeListener() {
        @Override
        public void onScrollChange(View v, int scrollX, int scrollY,
                            int oldScrollX, int oldScrollY) {
            float webViewHeight =
                    mWebView.getContentHeight() * mWebView.getScale();
            float nowHeight =
                    mWebView.getHeight() + mWebView.getScrollY();
            if (nowHeight == webViewHeight) {
                Toast.makeText(MainActivity.this, "已经处于底端",
                        Toast.LENGTH_SHORT).show();
            } else if (mWebView.getScrollY() == 0) {
                Toast.makeText(MainActivity.this, "已经处于顶端",
                        Toast.LENGTH_SHORT).show();
            }else {

            }
        }
    });
}

private void initViews() {
    mWebView = (WebView) findViewById(R.id.webview);
}

@Override
public boolean onKeyDown(int keyCode, KeyEvent event) {
    if (keyCode == KeyEvent.KEYCODE_BACK && mWebView.canGoBack()) {
        mWebView.goBack();
        return true;
    }
    return super.onKeyDown(keyCode, event);
}
}
```

上述代码调用 setOnScrollChangeListener 为 WebView 方法添加滚动事件监听，采用匿名内部类的方式，实现了 OnScrollChangeListener 接口并覆写了 onScrollChange 方法，这个方法在 WebView 滚动时会不断被回调。至于如何判断何时滚动到顶部和底部，这里需要获得两个值：

- 整个网页的长度：调用 WebView 的 getContentHeight 可以得到整个网页的长度。由于网页可能被缩放，因此要获得准确的长度要乘以网页的缩放比例（通过 getScale 方法获得缩放比例）。
- 当前滚动到的位置：由 getScrollY 方法可以获得当前滚动到 Y 方向上的位置。getHeight 方法可以得到 WebView 控件的高度。

这里认为若控件高度和 Y 方向上位置之和等于 WebView 中网页长度时滚动到了底部，若 Y 方向上的位置为 0 时滚动到了顶部。

最后不要忘记在 AndroidManifest.xml 中配置网络权限，代码如下：

```
<uses-permission android:name="android.permission.INTERNET"/>
```

运行实例，如图 9.5 所示。滚动网页到最底部会弹出"已经处于低端"的提示信息，滚动网页到最顶部弹出提示信息，如图 9.6 所示。

查看动态图，请扫描图 9.7 中的二维码。

图 9.5　Android WebView 底端监听　　　图 9.6　Android WebView 顶端监听　　　图 9.7　Android WebView 顶端底端监听二维码

9.2.2　WebView 一键回到顶部功能实现

监听滚动到底部或顶部在实际开发中有什么实际用处呢？不要着急，下面实现一个有用的功能，通过监听滚动到底部显示"回到顶部"按钮，单击这个按钮回到顶部。

在主布局文件（activity_main.xml）中再添加一个 Button，代码如下：

```
<?xml version="1.0" encoding="utf-8"?>
<RelativeLayout xmlns:android="http://schemas.android.com/apk/res/android"
    android:layout_width="match_parent"
    android:layout_height="match_parent">

    <WebView
        android:id="@+id/webview"
        android:layout_width="match_parent"
        android:layout_height="match_parent" />

    <Button
        android:id="@+id/btn_up"
        android:layout_width="70dp"
        android:layout_height="70dp"
```

```xml
            android:layout_alignParentBottom="true"
            android:layout_alignParentRight="true"
            android:background="@drawable/up"
            android:onClick="up"
            android:visibility="gone" />
</RelativeLayout>
```

上述代码添加一个 Button，置于右下角并设置其 visibility 属性值为 gone，即一开始不显示这个按钮。

修改 MainActivity.java 代码如下：

```java
public class MainActivity extends Activity {
    private WebView mWebView;
    private Button mButton;

    @Override
    protected void onCreate(Bundle savedInstanceState) {
        super.onCreate(savedInstanceState);
        // 省略部分相同代码
        mWebView.setOnScrollChangeListener(new View.
        OnScrollChangeListener() {
            @Override
            public void onScrollChange(View v, int scrollX, int scrollY,
                                       int oldScrollX, int oldScrollY) {
                float webViewHeight =
                        mWebView.getContentHeight() * mWebView.
                        getScale();
                float nowHeight =
                        mWebView.getHeight() + mWebView.getScrollY();
                if (nowHeight == webViewHeight) {
                    mButton.setVisibility(View.VISIBLE);
                    Toast.makeText(MainActivity.this, "已经处于底端",
                            Toast.LENGTH_SHORT).show();
                } else if (mWebView.getScrollY() == 0) {
                    Toast.makeText(MainActivity.this, "已经处于顶端",
                            Toast.LENGTH_SHORT).show();
                } else {
                    mButton.setVisibility(View.GONE);
                }
            }
        });
    }

    private void initViews() {
        mWebView = (WebView) findViewById(R.id.webview);
        mButton = (Button) findViewById(R.id.btn_up);
    }

    // 省略部分相同代码
    public void up(View view) {
        mButton.setVisibility(View.GONE);
```

```
        mWebView.scrollTo(0, 0);
    }
}
```

在代码中控制 Button 的显示与否，滚动到底部的时候调用 setVisibility 方法并传入 View.VISIBLE 方法显示 Button 按钮，否则隐藏这个 Button 按钮。在 Button 的单击监听方法 up 中调用 WebView 的 scrollTo 方法滚动顶部（两个参数，即 x 和 y 的坐标，这里都传入 0）。

运行实例，如图 9.8 所示。当滚动到底部时，Button 按钮显示出来了，单击这个按钮会迅速回到顶部，如图 9.9 所示。

查看动态图，请扫描图 9.10 中的二维码。

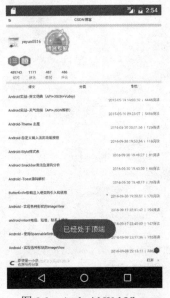

图 9.8　Android WebView 一键回到顶部一　　图 9.9　Android WebView 一键回到顶部二　　图 9.10　Android WebView 一键回到顶部二维码

9.2.3　WebView 退出记忆功能实现

scrollTo 方法除了在这里使用之外，还有一个十分常用的场景。使用过微信公众号的朋友应该都会注意到，查看一篇文章到某个位置关闭这个页面后，再一次进入这篇文章的时候会记忆上次翻到的记录，这个功能十分实用而其实现也十分简单，下面通过实例来学习一下。

修改 MainActivity.java 代码如下：

```java
public class MainActivity extends Activity {
    private WebView mWebView;
    private Button mButton;

    @Override
    protected void onCreate(Bundle savedInstanceState) {
        super.onCreate(savedInstanceState);
        requestWindowFeature(Window.FEATURE_NO_TITLE);
```

```java
        setContentView(R.layout.activity_main);
        initViews();
        mWebView.loadUrl("http://blog.csdn.net/yayun0516");
        int loc = getLoc();
        mWebView.scrollTo(0, loc);
        mWebView.setWebViewClient(new WebViewClient() {
            @Override
            public boolean shouldOverrideUrlLoading(WebView view,
            String url) {
                view.loadUrl(url);
                return true;
            }
        });

    // 省略部分相同代码
    private void saveLoc() {
        SharedPreferences sharedPreferences = getSharedPreferences("loc",
                Activity.MODE_PRIVATE);
        SharedPreferences.Editor editor = sharedPreferences.edit();
        editor.putInt("scrollY", mWebView.getScrollY());
        editor.commit();
    }

    private int getLoc() {
        SharedPreferences sharedPreferences = getSharedPreferences("loc",
                Activity.MODE_PRIVATE);
        int loc = sharedPreferences.getInt("scrollY", 0);
        return loc;
    }

    @Override
    protected void onDestroy() {
        super.onDestroy();
        saveLoc();
    }
}
```

这里用到了 SharedPreferences 来保存临时数据，添加了两个方法：saveLoc 方法用来保存网页的滚动位置，在 onDestroy 中调用这个方法；getLoc 方法用来获取保存的位置数据，在 onCreate 方法中调用。

运行实例，查看动态图，请扫描图 9.11 中的二维码。

通过动态图可以发现，滚动到某一位置退出应用，再次进入应用时会自动滚动到刚才退出的位置。

9.2.4　WebView 联合滚动实现

上面是 WebView 单一控件的滚动，若两个控件一起滚动可不可以实现呢？这里使用 ScrollView 来实现联合滚动。

新建一个项目，其主布局文件（activity_main.xml）代码如下：

图 9.11　Android WebView 退出位置记忆功能实例

```xml
<?xml version="1.0" encoding="utf-8"?>
<ScrollView xmlns:android="http://schemas.android.com/apk/res/android"
    android:layout_width="match_parent"
    android:layout_height="match_parent">

    <RelativeLayout
        android:layout_width="match_parent"
        android:layout_height="match_parent">

        <ImageView
            android:id="@+id/image"
            android:layout_width="match_parent"
            android:layout_height="230dp"
            android:src="@drawable/image" />

        <WebView
            android:id="@+id/webView"
            android:layout_width="match_parent"
            android:layout_height="match_parent"
            android:layout_below="@+id/image"
            android:minHeight="400dp">

        </WebView>
    </RelativeLayout>

</ScrollView>
```

上述代码最外层用 ScrollView 滚动布局来包裹，由于 ScrollView 中只能添加一个组件，因此 ImageView 和 WebView 用 RelativeLayout 来包裹。

MainActivity.java 代码如下：

```java
public class MainActivity extends Activity {
    private WebView mWebView;

    @Override
    protected void onCreate(Bundle savedInstanceState) {
        super.onCreate(savedInstanceState);
        setContentView(R.layout.activity_main);
        mWebView = (WebView) findViewById(R.id.webView);
        mWebView.loadUrl("http://blog.csdn.net/yayun0516");
        mWebView.setWebViewClient(new WebViewClient() {
            @Override
            public boolean shouldOverrideUrlLoading(WebView view,
            String url) {
                view.loadUrl(url);
                return true;
            }
        });
    }
}
```

注意添加网络权限，代码如下：

```
<uses-permission android:name="android.permission.INTERNET"/>
```

运行实例，如图9.12所示。查看动态图，请扫描图9.13中的二维码。

图 9.12　Android WebView 联合滚动实例　　图 9.13　Android WebView 联合滚动实例二维码

9.3　网络连接类——HttpURLConnection

Android 提供了两个类来实现 Http 请求：HttpURLConnection 和 HttpClient。HttpClient 在 Android 6.0 版本后被直接删除了，这里就不再介绍。HttpURLConneciton 的使用方式比较固定，主要步骤如下：

将要访问的 URL 路径包装成 URL 对象，代码如下：

```
URL url=new URL(path);
```

通过 URL 获取 HttpURLConnection 对象，代码如下：

```
HttpURLConnection conn = (HttpURLConnection) url.openConnection();
```

设置请求方式，代码如下：

```
conn.setRequestMethod("GET");
```

设置连接超时时长，代码如下：

```
conn.setConnectTimeout(5000);
```

获取请求返回的输入流，代码如下：

```
InputStream inputStream =conn.getInputStream();
```

最后，从输入流获得返回的字符串信息。

9.3.1　HttpURLConnection 打印网页

下面通过一个实例看一下如何实现 HttpURLConnection 请求。

主布局文件（activity_main.xml）代码如下：

```xml
<?xml version="1.0" encoding="utf-8"?>
<RelativeLayout xmlns:android="http://schemas.android.com/apk/res/android"
    android:layout_width="match_parent"
    android:layout_height="match_parent">

    <Button
        android:id="@+id/btn"
        android:layout_width="match_parent"
        android:layout_height="wrap_content"
        android:onClick="get"
        android:text=" 请求网络 " />

    <TextView
        android:id="@+id/text"
        android:layout_width="match_parent"
        android:layout_height="wrap_content"
        android:layout_below="@+id/btn"
        android:text="Hello World!" />
</RelativeLayout>
```

上述代码在相对布局中添加了两个控件，Button 按钮响应单击事件发起网络请求，TextView 用来显示网络请求的内容。

MainActivity.java 代码如下：

```java
public class MainActivity extends AppCompatActivity {
    private static final String URLSTRING = "http://www.baidu.com";
    private HttpURLConnection mHttpURLConnection;
    private TextView mTextView;

    @Override
    protected void onCreate(Bundle savedInstanceState) {
        super.onCreate(savedInstanceState);
        setContentView(R.layout.activity_main);
        mTextView = (TextView) findViewById(R.id.text);
    }

    private String getConnecitonContent() {
        InputStream inputStream = null;
        try {
            URL url = new URL(URLSTRING);
            mHttpURLConnection = (HttpURLConnection) url.openConnection();
            mHttpURLConnection.setConnectTimeout(5 * 1000);
            mHttpURLConnection.setReadTimeout(5 * 1000);
            mHttpURLConnection.setRequestMethod("GET");
            inputStream = mHttpURLConnection.getInputStream();
            String response = convertStreamToString(inputStream);
```

```
            return response;
        } catch (MalformedURLException e) {
            e.printStackTrace();
            return "";
        } catch (IOException e) {
            e.printStackTrace();
            return "";
        } finally {
            if (inputStream != null) {
                try {
                    inputStream.close();
                } catch (IOException e) {
                    e.printStackTrace();
                }
            }
            if (mHttpURLConnection != null) {
                mHttpURLConnection.disconnect();
            }
        }
    }

    private String convertStreamToString(InputStream is) throws
    IOException {
        BufferedReader reader = new BufferedReader(new InputStreamReader
        (is));
        StringBuffer sb = new StringBuffer();
        String line;
        while ((line = reader.readLine()) != null) {
            sb.append(line + "\n");
        }
        String respose = sb.toString();
        if (reader != null) {
            reader.close();
        }
        return respose;
    }

    public void get(View view) {
        mTextView.setText(getConnecitonContent());
    }
}
```

在 getConnectionContent 方法中首先获得一个 HttpURLConnection 对象，获得这个对象需要调用 URL 的 openConnection 方法，然后调用几个方法为这个对象添加几个属性。这些基本属性有通过 setConnectTimeOut 方法设置的"请求超时"属性；通过 setReadTimeout 方法设置的"读取超时"，这两个方法参数的单位都为毫秒；通过 setRequestMethod 方法设置的请求方式属性，这里设置了请求方式为 GET。调用 HttpURLConnection 的 getInputStream 方法可以得到请求返回的输入流，这个输入流包含请求返回的内容，这里用自定义的 convertStreamToString 方法得到返回的字符串。最后在 finally 方法中调用 close 方法关闭流，调用 disconnect 方法关闭网络连接。

convertStreamToString 方法可以将一个输入流转换成一个字符串信息，这个方法的写法比较固定，可以抽取出来作为工具类使用。首先创建一个 BufferedReader 对象，创建这个对象需要传入一个 InputStreamReader 对象（同样，创建 InputStreamReader 对象需要传入 InputStream 对象），创建一个 StringBuffer 对象用来保存返回的字符串信息，然后使用一个 while 循环一行一行读取，将一行数据追加保存在 StringBuffer 对象，最后调用 close 方法关闭这个字符读取流。

网络请求不要忘记在 AndroidManifest.xml 中添加网络权限，代码如下：

```
<uses-permission android:name="android.permission.INTERNET"/>
```

运行实例并单击"请求网络"按钮，应用 crash，Log 信息如下：

```
Caused by: android.os.NetworkOnMainThreadException
at android.os.StrictMode$AndroidBlockGuardPolicy.onNetwork(StrictMode.
java:1273)
```

可以看出，Log 信息提示网络请求在主线程异常，也就说网络请求、文件读取等耗时操作在主线程时可能会出现异常，因此必须创建一个子线程用于网络请求。同样 Android 还有一个"行规"，即"子线程不能更新主线程 UI"，TextView 属于"主线程 UI"，子线程中获取的请求信息不能直接更新到"主线程 UI"，Android 提供了 Handler 来实现子线程间接更新主线程 UI 的功能。

修改 MainActivity.java 代码如下：

```java
public class MainActivity extends AppCompatActivity {
    private static final int RESPONSE = 1;
    private static final String URLSTRING = "http://www.baidu.com";
    private HttpURLConnection mHttpURLConnection;
    private TextView mTextView;
    private Handler mHandler = new Handler() {
        @Override
        public void handleMessage(Message msg) {
            super.handleMessage(msg);
            switch (msg.what) {
                case RESPONSE:
                    mTextView.setText((CharSequence) msg.obj);
                    break;
            }
        }
    };
    // 省略部分相同代码
    public void get(View view) {
        new Thread(new Runnable() {
            @Override
            public void run() {
                Message message = new Message();
                message.what = RESPONSE;
                message.obj = getConnecitonContent();
                mHandler.sendMessage(message);
            }
        }).start();
    }
}
```

上述代码在单击监听事件中，创建一个 Thread 对象，创建这个 Thread 需要传入一个 Runnable 对象，这个 Runnable 是一个接口，实现这个借口必须覆写其抽象方法 run，在这个 run 方法中创建了一个 Message 对象并设置其 what 属性为 RESPONSE（这个属性用于分辨不同的 Message 对象），设置其 obj 属性为网络请求返回的字符串（Message 内容），最后调用 Handler 的 sendMessage 方法将这个包装了网络返回内容的 Message 对象发送出去。

上面将 Message 对象发送出去，下面就要创建一个 Handler 对象来接收和处理这个消息，这里通过 new 的方式创建了这个 Handler 对象，为了处理接收到的 Message 对象，这里覆写了 handleMessage 方法，这个方法的参数 Message 即为发送的 Message 对象，根据 Message 的 what 属性可以区别不同的 Message 对象，调用 TextView 的 setText 方法将 Message 中的内容（Message 的 obj 属性）显示出来。

运行实例，如图 9.14 所示。

可以看出，整个百度首页的源码被打印出来，这些都是 html 源码，用户看不懂这些源码，借助浏览器就可以将这些 html 代码变成网页了。

图 9.14 Android 网络之下载网页

9.3.2 HttpURLConnection 下载图片

上面通过 HttpURLConneciton 类获得了一个字符串信息，除此之外，还可以获取图片信息。

新建项目，主布局文件（activity_main.xml）代码如下：

```
<?xml version="1.0" encoding="utf-8"?>
<RelativeLayout xmlns:android="http://schemas.android.com/apk/res/android"
    android:layout_width="match_parent"
    android:layout_height="match_parent">

    <Button
        android:id="@+id/btn"
        android:layout_width="match_parent"
        android:layout_height="wrap_content"
        android:text="loadPic" />

    <ImageView
        android:id="@+id/image"
        android:layout_width="match_parent"
        android:layout_height="match_parent"
        android:layout_below="@+id/btn" />
</RelativeLayout>
```

上述代码添加一个 Button 监听单击事件发起网络请求，下载的 Image 显示在 ImageView 中。

MainActivity.java 代码如下：

```java
public class MainActivity extends AppCompatActivity {
    private Button mButton;
    private ImageView mImageView;
    private HttpURLConnection mHttpURLConnection;
    private static final String URLSTRING = "http://hiphotos.baidu.com/"
            + "doc/pic/item/d53f8794a4c27d1e02f0f5671cd5ad6edcc438bb.jpg";

    private Handler mHandler = new Handler() {
        @Override
        public void handleMessage(Message msg) {
            super.handleMessage(msg);
            Bitmap bitmap = (Bitmap) msg.obj;
            mImageView.setImageBitmap(bitmap);
        }
    };

    @Override
    protected void onCreate(Bundle savedInstanceState) {
        super.onCreate(savedInstanceState);
        setContentView(R.layout.activity_main);
        mButton = (Button) findViewById(R.id.btn);
        mButton.setOnClickListener(new View.OnClickListener() {
            @Override
            public void onClick(View v) {
                new Thread(new Runnable() {
                    @Override
                    public void run() {
                        Bitmap bitmap = getConnectionContent();
                        Message message = new Message();
                        message.obj = bitmap;
                        mHandler.sendMessage(message);
                    }
                }).start();
            }
        });
        mImageView = (ImageView) findViewById(R.id.image);
    }

    private Bitmap getConnectionContent() {
        InputStream inputStream = null;
        try {
            URL url = new URL(URLSTRING);
            mHttpURLConnection = (HttpURLConnection) url.openConnection();
            mHttpURLConnection.setConnectTimeout(5 * 1000);
            mHttpURLConnection.setReadTimeout(5 * 1000);
            mHttpURLConnection.setRequestMethod("GET");
            inputStream = mHttpURLConnection.getInputStream();
            Bitmap bitmap = BitmapFactory.decodeStream(inputStream);
            return bitmap;
        } catch (MalformedURLException e) {
```

```
                e.printStackTrace();
                return null;
            } catch (IOException e) {
                e.printStackTrace();
                return null;
            } finally {
                if (inputStream != null) {
                    try {
                        inputStream.close();
                    } catch (IOException e) {
                        e.printStackTrace();
                    }
                }
                if (mHttpURLConnection != null) {
                    mHttpURLConnection.disconnect();
                }
            }
        }
    }
```

上述代码中 getConnectionContent（自定义方法）返回一个 Bitmap 对象，这里是通过 BitmapFactory 的静态方法 decodeStream 将获得的 InputStream 对象转换成 Bitmap 对象。在单击事件中调用 Handler 的 sendMessage 方法将这个 Bitmap 对象发送出去，在 handleMessage 方法中调用了 setImageBitmap 方法将 Bitmap 在 ImageView 中显示出来。

最后不要忘记在 AndroidManifest.xml 中配置权限，代码如下：

```
<uses-permission android:name="android.permission.INTERNET" />
```

运行实例，如图 9.15 所示。

图 9.15　Android 网络之下载图片

可以看出，网络的图片被下载并显示到应用中。下面通过一个实例看一下如何保存这个下载下来的图片。

9.3.3　HttpURLConnection 保存图片

新建项目，主布局文件（activity_main.xml）代码如下：

```xml
<?xml version="1.0" encoding="utf-8"?>
<LinearLayout xmlns:android="http://schemas.android.com/apk/res/android"
    android:layout_width="match_parent"
    android:layout_height="match_parent"
    android:orientation="vertical">

    <Button
        android:id="@+id/btnLoad"
        android:layout_width="match_parent"
        android:layout_height="wrap_content"
        android:text=" 下载图片 " />

    <Button
        android:id="@+id/btnSave"
        android:layout_width="match_parent"
        android:layout_height="wrap_content"
        android:text=" 保存图片 " />

    <ImageView
        android:id="@+id/image"
        android:layout_width="wrap_content"
        android:layout_height="wrap_content"
        android:adjustViewBounds="true" />

</LinearLayout>
```

MainActivity.java 代码如下：

```java
public class MainActivity extends Activity {

    private final static String SAVE_PATH
            = Environment.getExternalStorageDirectory() + "/download_test/";
    private ImageView mImageView;
    private Button mBtnSave, mBtnLoad;
    private ProgressDialog mSaveDialog = null;
    private Bitmap mBitmap;
    private String mFileName;
    private String mMessage;
    String filePath = "http://hiphotos.baidu.com/doc/" +
            "pic/item/d53f8794a4c27d1e02f0f5671cd5ad6edcc438bb.jpg";

    private Handler mHandler = new Handler() {
        @Override
        public void handleMessage(Message msg) {
            switch (msg.what) {
                case 1:
                    if (mBitmap != null) {
                        mImageView.setImageBitmap(mBitmap);
                    }
```

```java
                    break;
                case 2:
                    mSaveDialog.dismiss();
                    Toast.makeText(MainActivity.this, mMessage,
                            Toast.LENGTH_SHORT).show();
                    break;
                default:
                    break;
            }
        }
    };

    @Override
    protected void onCreate(Bundle savedInstanceState) {
        super.onCreate(savedInstanceState);
        setContentView(R.layout.activity_main);
        mImageView = (ImageView) findViewById(R.id.image);
        mBtnSave = (Button) findViewById(R.id.btnSave);
        mBtnLoad = (Button) findViewById(R.id.btnLoad);
        if (shouldAskPermissions()) {
            askPermissions();
        }

        mBtnLoad.setOnClickListener(new View.OnClickListener() {
            @Override
            public void onClick(View v) {
                new Thread(new Runnable() {
                    @Override
                    public void run() {
                        try {
                            mFileName = "test.jpg";
                            mBitmap = BitmapFactory.decodeStream(
                                    getImageStream(filePath));
                            mHandler.sendEmptyMessage(1);
                        } catch (Exception e) {
                            mMessage = "下载失败";
                            mHandler.sendEmptyMessage(2);
                            e.printStackTrace();
                        }
                    }
                }).start();
            }
        });

        mBtnSave.setOnClickListener(new View.OnClickListener() {
            @Override
            public void onClick(View v) {
                mSaveDialog = ProgressDialog.show(MainActivity.this,
                        "保存图片", "图片正在保存中,请稍等...", true);
                new Thread(new Runnable() {
```

```java
                    @Override
                    public void run() {
                        try {
                            // 注意，子线程中不能使用 Toast
                            // 这里也是通过 Handler 接收到消息后显示 Toast
                            if (mBitmap == null) {
                                mMessage = "图片保存失败！";
                            } else {
                                saveFile(mBitmap, mFileName);
                                mMessage = "图片保存成功！";
                            }
                            mSaveDialog.dismiss();
                        } catch (IOException e) {
                            mMessage = "图片保存失败！";
                            e.printStackTrace();
                        }
                        mHandler.sendEmptyMessage(2);
                    }
                }).start();
            }
        });
    }

    protected boolean shouldAskPermissions() {
        return (Build.VERSION.SDK_INT > Build.VERSION_CODES.LOLLIPOP_MR1);
    }

    @TargetApi(23)
    protected void askPermissions() {
        String[] permissions = {
                "android.permission.READ_EXTERNAL_STORAGE",
                "android.permission.WRITE_EXTERNAL_STORAGE"
        };
        int requestCode = 200;
        requestPermissions(permissions, requestCode);
    }

    public InputStream getImageStream(String path) throws Exception {
        URL url = new URL(path);
        HttpURLConnection conn = (HttpURLConnection) url.openConnection();
        conn.setConnectTimeout(5 * 1000);
        conn.setRequestMethod("GET");
        if (conn.getResponseCode() == HttpURLConnection.HTTP_OK) {
            return conn.getInputStream();
        }
        return null;
    }

    public void saveFile(Bitmap bitmap, String fileName) throws IOException {
        File dirFile = new File(SAVE_PATH);
```

```
            if (!dirFile.exists()) {
                dirFile.mkdir();
            }
            File file = new File(SAVE_PATH + fileName);
            BufferedOutputStream bufferedOutputStream = new BufferedOutputStream(
                    new FileOutputStream(file));
            bitmap.compress(Bitmap.CompressFormat.JPEG, 80,
            bufferedOutputStream);
            bufferedOutputStream.flush();
            bufferedOutputStream.close();
        }
    }
```

保存图片到存储卡需要申请权限，这个权限属于危险权限，当版本大于 LOLLIPOP_MR1 时需要在代码中进行动态获取，这里使用 shouldAskPermissions 方法进行判断。

askPermissions 方法用于请求权限，主要是通过 Activity 的 requestPermissions 方法进行获取，这个方法需要传入两个参数：权限字符串数组和请求码，这里添加了读写权限。

这里添加了两个 Button——"下载图片"和"保存图片"，考虑到图片下载和图片保存都属于耗时操作，因此都开启了新的线程去执行这些操作并将结果发送到 Handler 中进行 UI 更新操作。

在"下载图片"的单击事件监听中调用了 getImageStream 方法返回一个输入流，然后调用 BitmapFactory 的 decodeStream 方法将 InputStream 转换成 Bitmap，然后调用 Handler 的 sendEmptyMessage 方法传入一个 what 值为 1，通知下载完成。异常情况时调用 sendEmptyMessage 方法传入一个 what 值为 2，通知显示 Toast 提示信息（注意 Toast 不能再子线程中执行，因此也是通过 Handler 的方式实现）。

在"保存图片"的单击事件监听中调用了自定义的 saveFile 方法，这个方法传入了两个参数：下载图片获得的 Bitmap 对象和要保存的图片名，SAVE_PATH 是保存图片的路径，先判断路径是否存在，若不存在则调用 mkdir 方法创建路径。保存图片的方法这里调用了 compress 和 flush 方法，compress 方法需要传入三个参数：图片格式、图片质量和输入流，最后调用 flush 清空缓存区，调用 close 方法关闭输出流。

同样需要在 AndroidManifest.xml 中配置一些权限，代码如下：

```
<uses-permission android:name="android.permission.INTERNET" />
<uses-permission android:name="android.permission.WRITE_EXTERNAL_
STORAGE"/>
<uses-permission android:name="android.permission.MOUNT_UNMOUNT_
FILESYSTEMS"/>
<uses-sdk android:minSdkVersion="11"/>
```

运行实例，如图 9.16 所示。单击 ALLOW 按钮允许读写权限，然后单击"下载图片"按钮，图片下载并显示出来，如图 9.17 所示。

单击"保存图片"按钮，Toast 显示图片保存成功，这时打开 Android Studio 的文件管理器，可以看到图片被下载到指定的目录中，如图 9.18 所示。

查看动态图，请扫描图 9.19 中的二维码。

图 9.16　Android 网络之下载并保存图片一

图 9.17　Android 网络之下载并保存图片二

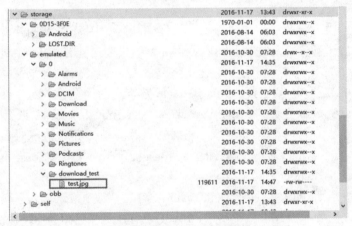

图 9.18　Android 网络之下载并保存图片查看

图 9.19　Android 网络之下载并保存图片二维码

9.4　Android Handler 消息处理机制

上一节已经介绍了 Handler 的一些简单用法，本节将系统地介绍这个类及 Android 的消息处理机制。Android 为什么要提供这样一套使用起来不是很方便的机制呢？熟悉 Android 系统的朋友都知道，耗时的操作是不能在主线程中运行的，这样可能会阻塞主线程而造成系统 ANR（类似 Windows 中的未响应），因此在开发中要避免这样的事情发生。Android 制定了一条规则：耗时的操作必须在子线程中运行。另一方面，出于性能优化考虑，Android UI 线程并不是线程安全的，因此多个线程操作 UI 线程可能会造成同步问题，考虑到这一问题，Android 又制定了一条规则：子线程不能操作主线程 UI。

综合考虑这两条规则，若在子线程中耗时的操作完成之后必须要更新到主线程，就要用到这里的消息处理机制。这一套机制中有几个关键概念：

消息类 Message：所有的消息都是封装在 Message 对象中进行传递的，一个 Message 对象就是一个消息实体。

消息队列 MessageQueue：多个消息实体构成了一个消息队列，按照 FIFO（先进先出）原则进行排序。

消息管理类 Looper：负责将在 MessageQueue 排队的 Message 取出来交由 Handler 进行处理。一个 MessageQueue 需要一个消息管理类 Looper。

消息处理类 Handler：负责发送（sendMessage）和接收处理（handleMessage）消息 Message。

9.4.1 消息类 Message

所有的消息都会用 Message 类进行封装然后才进行传递。Message 的常用方法和属性如表 9.2 所示。

表 9.2　Message 的常用方法和属性

方法和属性	说　　明
obj	包装要传递的数据对象
what	用来区分不同的 Message 对象
getTarget	获得发送此消息的 Handler 对象

可以通过 Message 的 obtain 方法可以获得 Message 对象，同样也可以通过 new 的方式获得一个 Message 对象。

9.4.2 消息处理类 Handler

Message 只是负责封装消息，而消息的发送和接收则需要用到 Handler 类，其常用方法如表 9.3 所示。

表 9.3　Handler 的常用方法

方　　法	说　　明
handleMessage（Message msg）	处理消息，创建 Handler 对象要覆写此方法
removeMessage（int what）	删除指定的消息
obtainMessage (int what)	获得一个 Message 对象
sendMessage (Message msg)	发送消息
sendEmptyMessage (int what)	发送一个空消息
sendEmptyMessageDelayed (int what, long delayMillis)	延迟发送一个空消息
postDelayed (Runnable r, long delayMillis)	延迟进行某项操作
removeCallbacks (Runnable r)	移出某项操作

下面通过实例看一下这些方法的使用。

主布局文件（activity_main.xml）代码如下：

```xml
<?xml version="1.0" encoding="utf-8"?>
<LinearLayout xmlns:android="http://schemas.android.com/apk/res/android"
    android:layout_width="match_parent"
    android:layout_height="match_parent"
    android:orientation="vertical">

    <TextView
        android:id="@+id/textView"
        android:layout_width="match_parent"
```

```xml
        android:layout_height="80dp"
        android:gravity="center"
        android:textSize="16sp" />

    <Button
        android:layout_width="match_parent"
        android:layout_height="wrap_content"
        android:onClick="sendEmpty"
        android:text="发送空消息" />

    <Button
        android:layout_width="match_parent"
        android:layout_height="wrap_content"
        android:onClick="sendWhat0"
        android:text="发送 WHAT 为 0 的 Message" />

    <Button
        android:layout_width="match_parent"
        android:layout_height="wrap_content"
        android:onClick="sendWhat1"
        android:text="发送 WHAT 为 1 的 Message" />

    <Button
        android:layout_width="match_parent"
        android:layout_height="wrap_content"
        android:onClick="sendDelay"
        android:text="延迟发送信息" />
</LinearLayout>
```

上述代码中添加 TextView 用于显示接收的信息，四个 Button 用来响应不同的操作请求。MainActivity.java 代码如下：

```java
public class MainActivity extends AppCompatActivity {
    private TextView mTextView;
    private Handler mHandler = new Handler() {
        @Override
        public void handleMessage(Message msg) {
            super.handleMessage(msg);
            switch (msg.what) {
                case -1:
                    mTextView.setText("接收到空消息");
                    break;
                case 0:
                    mTextView.setText("接收到 what 为 0 的消息，消息内容为："
                            + msg.getData().getString("key"));
                    break;
                case 1:
                    mTextView.setText("接收到 what 为 1 的消息，消息内容为："
                            + msg.getData().getString("key"));
                    break;
                case 2:
                    mTextView.setText("接收到延迟发送的消息，消息内容为："
```

```
                    + msg.getData().getString("key"));
                break;
        }
    }
};

@Override
protected void onCreate(Bundle savedInstanceState) {
    super.onCreate(savedInstanceState);
    setContentView(R.layout.activity_main);
    mTextView = (TextView) findViewById(R.id.textView);
}

public void sendWhat0(View view) {
    Message message = Message.obtain();
    message.what = 0;
    Bundle bundle = new Bundle();
    bundle.putString("key", "这是what为0的消息");
    message.setData(bundle);
    mHandler.sendMessage(message);
}

public void sendWhat1(View view) {
    Message message = Message.obtain();
    message.what = 1;
    Bundle bundle = new Bundle();
    bundle.putString("key", "这是what为1的消息");
    message.setData(bundle);
    mHandler.sendMessage(message);
}

public void sendDelay(View view) {
    Message message = new Message();
    message.what = 2;
    Bundle bundle = new Bundle();
    bundle.putString("key", "这是what为2的延迟消息");
    message.setData(bundle);
    mHandler.sendMessageDelayed(message, 2000);
}

public void sendEmpty(View view) {
    mHandler.sendEmptyMessage(-1);
}
}
```

上述代码在第一个Button"发送空消息"的单击事件方法sendEmpty中，直接调用了Handler的sendEmptyMessage方法发送一个空消息，这个方法需要传入一个what值，这里传入-1。

在第二个Button"发送WHAT为0的MESSAGE"的单击事件方法sendWhat0中，通过Message的静态方法obtain获得一个Message对象，设置了Message的what属性值为 0，对于消息的包装，这里调用 setData 方法包装了一个 Bundle 对象，最后调用 Handler 的 sendMessage 方法发送这个 Message 对象。

第三个 Button 的单击事件方法和第二个 Button 的基本一致，不同的是其属性 what 值为 1。第四个 Button 的单击事件方法 sendDelay 中调用了 Handler 的 sendMessageDelayed 延迟发送一个消息，这个方法需要传入两个参数：Message 对象和延迟的时间。

上述代码创建了一个 Handler 对象并覆写了其 handleMessage 方法，根据 Message 的 what 属性可以分辨不同的消息。运行实例并单击第一个 Button，TextView 中显示"接收到空消息"，如图 9.20 所示。单击第二个 Button 按钮，如图 9.21 所示。

这时 TextView 的信息刷新了，同时也接收到了包装在 Message 中的字符串信息。单击"延迟发送消息"按钮，这时延迟 2s 才会刷新 TextView 中的信息。查看动态图，请扫描图 9.22 中的二维码。

图 9.20　Android Handler 实例一

图 9.21　Android Handler 实例二

图 9.22　Android Handler 实例二维码

9.4.3　Handler 实现倒计时功能

上面讲解了 Handler 发送和接收信息的方法，这里讲解一个 Handler 有创意的用法——实现倒计时功能。

主布局文件（activity_main.xml）代码如下：

```xml
<?xml version="1.0" encoding="utf-8"?>
<RelativeLayout xmlns:android="http://schemas.android.com/apk/res/android"
    android:layout_width="match_parent"
    android:layout_height="match_parent">

    <TextView
        android:id="@+id/txttime"
        android:layout_width="wrap_content"
        android:layout_height="wrap_content"
        android:layout_centerInParent="true"
        android:text="倒计时开始"
```

```xml
            android:textColor="@android:color/holo_green_light"
            android:textSize="32sp" />
</RelativeLayout>
```

上述代码添加一个 TextView 控件用来显示当前倒计时。

MainActivity.java 代码如下:

```java
public class MainActivity extends Activity {
    private int secondLeft = 6;
    private TextView mTextView;

    public void onCreate(Bundle savedInstanceState) {
        super.onCreate(savedInstanceState);
        setContentView(R.layout.activity_main);
        mTextView = (TextView) findViewById(R.id.txttime);
        mTextView.setOnClickListener(new View.OnClickListener() {
            @Override
            public void onClick(View v) {
                if (secondLeft != 6) return;
                // Message
                Message message = handler.obtainMessage(1);
                handler.sendMessage(message);
            }
        });
    }

    final Handler handler = new Handler() {
        // handle message
        public void handleMessage(Message msg) {
            switch (msg.what) {
                case 1:
                    secondLeft--;
                    mTextView.setText("" + secondLeft);
                    if (secondLeft > 0) {
                        Message message = handler.obtainMessage(1);
                        // send message
                        handler.sendMessageDelayed(message, 1000);
                    } else {
                        mTextView.setText("倒计时结束");
                        secondLeft = 6;
                    }
            }
            super.handleMessage(msg);
        }
    };
}
```

在 TextView 的单击事件监听中，调用 Handler 的 obtainMessage 方法创建一个 what 属性为 1 的 Message 对象，然后调用 Handler 的 sendMessage 方法将这个对象发送出去。在 Handler 的 handleMessage 方法中对全局变量 secondLeft 自减 1 并将这时的变量通过 TextView 显示出来。若 secondLeft 大于 1 则调用 sendMessageDelayed 方法延迟 1s 发送这

个消息，否则 TextView 显示倒计时结束。

运行实例，如图 9.23 所示。单击这个 TextView 则倒计时开始，如图 2.24 所示。

倒计时结束 TextView 将显示"倒计时结束"的信息，如图 9.25 所示。

图 9.23　Android Handler 实现　　图 9.24　Android Handler 实现　　图 9.25　Android Handler 实现
倒计时一　　　　　　　　　　倒计时二　　　　　　　　　　倒计时三

倒计时一旦开始可不可以停止呢？可以借助 Handler 的 removeMessages 方法来实现。在上面主布局文件（activity_main.xml）中再添加一个按钮，代码如下：

```xml
<Button
    android:layout_width="match_parent"
    android:layout_height="wrap_content"
    android:background="@null"
    android:onClick="removeMessage"
    android:text=" 停止倒计时 "
    android:textSize="20sp" />
```

上述代码为这个 Button 添加了 onClick 属性，其值为 removeMessage。在 MainActivity 中添加对应的 removeMessage 方法，代码如下：

```java
public void removeMessage(View view) {
    handler.removeMessages(1);
}
```

上述代码中调用 Handler 的 removeMessages 方法传入 1 则移除 what 属性值为 1 的 Message 对象。运行实例并单击"停止倒计时"按钮，如图 9.26 所示，倒计时将停止。查看动态图，请扫描图 9.27 中的二维码。

图 9.26　Android Handler 实现倒计时四　　　　图 9.27　Android Handler 实现倒计时二维码

9.4.4　Handler 延迟操作

在某些特殊情况下有些操作需要延时执行，可以调用 Handler 的 postDelayed 方法来实现。

主布局文件（activity_main.xml）代码如下：

```xml
<?xml version="1.0" encoding="utf-8"?>
<LinearLayout xmlns:android="http://schemas.android.com/apk/res/android"
    android:layout_width="match_parent"
    android:layout_height="match_parent"
    android:orientation="vertical">

    <Button
        android:layout_width="match_parent"
        android:layout_height="wrap_content"
        android:background="@null"
        android:onClick="postDelayedClick"
        android:text="postDelayed 方法延迟操作" />
</LinearLayout>
```

上述代码添加一个 Button 用来启动 postDelayed 方法。

MainActivity.java 代码如下：

```java
public class MainActivity extends AppCompatActivity {
    private static final String TAG = "MainActivity";
    private Handler mHandler = new Handler();
    private long time = 0;

    @Override
    protected void onCreate(Bundle savedInstanceState) {
        super.onCreate(savedInstanceState);
        setContentView(R.layout.activity_main);
    }

    public void postDelayedClick(View view) {
        mHandler.postDelayed(new MyRunnable(), 3000);
        time = System.currentTimeMillis();
    }

    private class MyRunnable implements Runnable {

        @Override
        public void run() {
            Log.d(TAG, "run: time delay = " + (System.currentTimeMillis()
                - time));
        }
    }
}
```

调用 postDelayed 方法需要一个 Handler 对象，因此首先通过 new 的方式创建一个 Handler 对象。在单击事件监听方法 postDelayedClick 中调用 Handler 的 postDelayed 方法进行延时操作，这个方法需要传入两个参数：Runnable 对象和延时时间，对于 Runnable

对象，这里通过自定义类实现 Runable 接口，在覆写的 run 方法中打印 Log 信息并计算延迟的时间。

运行实例并单击 Button 按钮查看 Log 如下：

```
D/MainActivity: run: time delay = 3000
```

可以看出，在单击 Button 之后延迟 3000ms 之后 Log 才进行打印，这就达到了延时进行某项操作的目的。和 sendMessageDelayed 方法类似，用 postDelayed 方法启动的延时操作同样可以取消，这里是通过 removeCallbacks 方法取消延时操作。在上面的主布局文件（activity_main.xml）中再添加一个 Button 按钮用于取消延时操作，代码如下：

```xml
<Button
    android:layout_width="match_parent"
    android:layout_height="wrap_content"
    android:onClick="removeCallbacksClick"
    android:text="removeCallbacks 方法取消延迟操作" />
```

修改 MainActivity.java 代码如下：

```java
public class MainActivity extends AppCompatActivity {
    private static final String TAG = "MainActivity";
    private Handler mHandler = new Handler();
    private long time = 0;
    private MyRunnable mMyRunnable;

    @Override
    protected void onCreate(Bundle savedInstanceState) {
        super.onCreate(savedInstanceState);
        setContentView(R.layout.activity_main);
    }

    public void postDelayedClick(View view) {
        mMyRunnable = new MyRunnable();
        mHandler.postDelayed(mMyRunnable, 3000);
        time = System.currentTimeMillis();
    }

    public void removeCallbacksClick(View view) {
        mHandler.removeCallbacks(mMyRunnable);
        Log.d(TAG, "removeCallbacksClick: ");
    }

    private class MyRunnable implements Runnable {

        @Override
        public void run() {
            Log.d(TAG, "run: time delay = " + (System.
                currentTimeMillis()- time));
        }
    }
}
```

在"removeCallbacks 方法延迟操作"按钮的单击事件监听中调用了 Handler 的 removeCallbacks 方法移除 Runnable 对象,也就移除了这个延时操作。

运行实例单击"postDelayed 方法延迟操作"按钮,然后再单击"removeCallbacks 方法延迟操作"按钮,查看 Log 如下:

```
D/MainActivity: removeCallbacksClick:
```

可以看出,只是打印了移除的 Log,并没有打印延时完成时的 Log,说明延时操作被成功移除。

9.4.5 Handler postDelay 实现循环调用

上面的倒计时功能可以理解成 1000ms 调用某方法一次的循环,用 postDelay 可不可以实现循环调用的功能呢?分析下面的实例,很巧妙地实现循环调用的功能。

主布局文件(activity_main.xml)如下:

```xml
<?xml version="1.0" encoding="utf-8"?>
<LinearLayout xmlns:android="http://schemas.android.com/apk/res/android"
    android:layout_width="match_parent"
    android:layout_height="match_parent"
    android:orientation="vertical">

    <Button
        android:layout_width="match_parent"
        android:layout_height="wrap_content"
        android:onClick="begin"
        android:text=" 开始循环 " />

    <Button
        android:layout_width="match_parent"
        android:layout_height="wrap_content"
        android:onClick="stop"
        android:text=" 结束循环 " />
</LinearLayout>
```

上述代码添加了两个 Button,同时为每个 Button 设置了不同的 onClick 属性值。MainActivity.java 代码如下:

```java
public class MainActivity extends AppCompatActivity {
    private static final String TAG = "MainActivity";
    private Handler mHandler = new Handler();
    private MyRunnable mMyRunnable;
    private int count;

    @Override
    protected void onCreate(Bundle savedInstanceState) {
        super.onCreate(savedInstanceState);
        setContentView(R.layout.activity_main);
    }

    public void begin(View view) {
```

```
        count = 0;
        mMyRunnable = new MyRunnable();
        mHandler.postDelayed(mMyRunnable, 2000);
    }

    public void stop(View view) {
        mHandler.removeCallbacks(mMyRunnable);
    }

    private class MyRunnable implements Runnable {
        @Override
        public void run() {
            count++;
            Log.d(TAG, "run: " + count);
            mHandler.postDelayed(mMyRunnable, 2000);
        }
    }
}
```

实现循环的核心理念就是在覆写的 run 方法中再次调用 Handler 的 postDelayed 的方法（递归的理念）。若需要停止循环就调用 Handler 的 removeCallbacks 方法移除 Runnable 对象。

运行实例并单击"开始循环"按钮，查看 Log 信息如下：

```
02-05 09:19:47.299 29138-29138/ad.handlerpostdelay D/MainActivity: run: 1
02-05 09:19:49.302 29138-29138/ad.handlerpostdelay D/MainActivity: run: 2
02-05 09:19:51.304 29138-29138/ad.handlerpostdelay D/MainActivity: run: 3
02-05 09:19:53.308 29138-29138/ad.handlerpostdelay D/MainActivity: run: 4
02-05 09:19:55.310 29138-29138/ad.handlerpostdelay D/MainActivity: run: 5
```

可以看出，Log 信息每 2000ms 打印一次，这也就实现了循环调用的功能。单击"结束循环"按钮，Log 信息停止打印，循环调用也就结束了。

Handler 的 postDelayed 方法和 sendMessageDelayed 方法都可以进行延迟操作，这两个方法有什么不同呢？可以看出，前者的延迟操作在子线程中运行，后者的延迟操作在 handleMessage 中处理。

9.4.6 Looper 用法

上面讲解了 Handler 和 Message 的基本用法，读者不免疑惑，没有 Looper 的参与也正常实现了 Handler 接收和发送 Message，那么 Looper 又有什么用处呢？原来上面所有的 Handler 创建都是在主线程中，而在主线程中 Android 会默认生成一个 Looper 对象，因此就算开发者没有使用 Looper 也正常实现了 Message 的发送和接收。若在子线程中创建使用 Handler 对象，就不得不使用 Looper 了，否则会出现问题。下面通过一个实例看一下在子线程中如何使用 Handler 对象发送和接收消息。

新建主布局文件（activity_main.xml）代码如下：

```
<?xml version="1.0" encoding="utf-8"?>
<RelativeLayout xmlns:android="http://schemas.android.com/apk/res/android"
    android:layout_width="match_parent"
```

```xml
        android:layout_height="match_parent">

    <Button
        android:layout_width="match_parent"
        android:layout_height="wrap_content"
        android:onClick="testHandler"
        android:text="向子线程发送消息" />
</RelativeLayout>
```

MainActivity.java 代码如下:

```java
public class MainActivity extends AppCompatActivity {
    private Handler mHandler;

    @Override
    protected void onCreate(Bundle savedInstanceState) {
        super.onCreate(savedInstanceState);
        setContentView(R.layout.activity_main);
        new Thread(new Runnable() {
            @Override
            public void run() {
                mHandler = new Handler() {
                    @Override
                    public void handleMessage(Message msg) {
                        super.handleMessage(msg);
                        switch (msg.what) {
                            case 1:
                                Toast.makeText(MainActivity.this, "接收到了消息",Toast.LENGTH_SHORT).show();
                                break;
                        }
                    }
                };
            }
        }).start();
    }

    public void testHandler(View view) {
        mHandler.sendEmptyMessage(1);
    }
}
```

上述代码调用 onCreate 方法创建了一个子线程，在覆写的 run 方法中创建了一个 Handler 对象。Button 事件的单击监听事件方法中调用 Handler 的 sendEmptyMessage 方法发送消息。

这时运行实例会发生 crash，查看 Log 如下:

```
FATAL EXCEPTION: Thread-159
    Process: project.first.com.looperdemo, PID: 16841
    java.lang.RuntimeException: Can't create handler inside thread that has not called Looper.prepare()
```

```
        at android.os.Handler.<init>(Handler.java:200)
        at android.os.Handler.<init>(Handler.java:114)
        at project.first.com.looperdemo.MainActivity$1$1.<init>(MainActivi
ty.java:20)
        at project.first.com.looperdemo.MainActivity$1.run(MainActivity.
java:20)
        at java.lang.Thread.run(Thread.java:818)
```

通过 Log 信息可以看出：在子线程中创建 Handler 对象必须先调用 Looper 的 prepare 方法创建一个 Looper 对象。

设置修改 MainActivity.java 代码如下：

```
public class MainActivity extends AppCompatActivity {
    private Handler mHandler;

    @Override
    protected void onCreate(Bundle savedInstanceState) {
        super.onCreate(savedInstanceState);
        setContentView(R.layout.activity_main);
        new Thread(new Runnable() {
            @Override
            public void run() {
                Looper.prepare();
                mHandler = new Handler() {
                    @Override
                    public void handleMessage(Message msg) {
                        super.handleMessage(msg);
                        switch (msg.what) {
                            case 1:
                                Toast.makeText(MainActivity.this, "接收
                                    到了消息", Toast.LENGTH_SHORT).show();
                                break;
                        }
                    }
                };
                Looper.loop();
            }
        }).start();
    }

    public void testHandler(View view) {
        mHandler.sendEmptyMessage(1);
    }
}
```

上述代码在子线程中调用 Looper 的 prepare 方法创建一个 Looper，但是要想让消息循环起来还必须调用 Looper 的 loop 方法，从消息队列（MessageQueue）里取消息、处理消息。

再次运行实例，如图 9.28 所示。

可以看出，项目正常运行并在子线程中接收到了消息。

图 9.28　Android Looper 用法

9.5　Android 异步操作类 AsyncTask

上一节所讲解的 Handler 可以实现子线程和主线程之间的消息传递，即异步操作。除了这种异步操作的方式外，Android 还提供了一个封装好的异步操作类 AsyncTask，这个类较 Handler 更轻量级，使用也更简单。AsyncTask 类是一个抽象类，代码如下：

```
public abstract class AsyncTask<Params, Progress, Result>
```

可以看出，这个类要指定三个泛型参数：Params 为启动时传入的参数；Progress 为后台执行进度百分比；Result 为执行完毕后的返回结果。我们都知道抽象类不能直接实例化，需要创建一个子类来继承，继承抽象类可以覆写抽象类中的抽象方法，AsyncTask 中的抽象方法如下：

```
@WorkerThread
protected abstract Result doInBackground(Params... params);
```

由上面的注解（@WorkerThread）可以看出，这个方法在工作现场线程也就是子线程中运行，耗时的操作可以在这个方法中执行，由于子线程不能操作主线程 UI，在这个方法中可以调用 publishProgress 方法更新进度（类似 Handler 的 sendMessage 方法）：

```
@MainThread
protected void onProgressUpdate(Progress... values) {
}
```

上面的 publishProgress 方法用于更新进度，想要在 UI 中显示当前进度就要覆写 onProgressUpdate 方法，由注解（@MainThread）可以看出，这个方法将运行在主线程，因此，可以直接更新主线程 UI。

在异步操作之前可能需要一些准备工作，想要知道异步何时开始，就需要覆写 onPreExecute 方法，代码如下：

```
@MainThread
protected void onPreExecute() {
}
```

onPreExecute 方法在主线程中运行，在异步操作前回调。同时除了要知道异步何时开始，可能还需要知道异步何时结束，在结束时进行一下消息的通知工作，这时就需要覆写 **onPostExecute** 方法，代码如下：

```
@MainThread
protected void onPostExecute(Result result) {
}
```

onPostExecute 方法在主线程中运行，异步操作完成后回调此方法。操作过程中可能会被取消，取消时可能会进行一些资源释放的处理，我们可以覆写 onCancelled 方法来监听操作是否被取消。

```
@MainThread
protected void onCancelled() {
}
```

onCancelled 方法在主线程中运行，异步操作被取消时回调此方法。

9.5.1 AsyncTask 基本用法

下面通过实例来看一下这些方法的用法。

主布局文件（activity_main.xml）代码如下：

```xml
<?xml version="1.0" encoding="utf-8"?>
<LinearLayout xmlns:android="http://schemas.android.com/apk/res/android"
    android:layout_width="match_parent"
    android:layout_height="match_parent"
    android:orientation="vertical">

    <ProgressBar
        android:id="@+id/progress"
        style="@android:style/Widget.Holo.Light.ProgressBar.Horizontal"
        android:layout_width="match_parent"
        android:layout_height="wrap_content" />

    <Button
        android:layout_width="match_parent"
        android:layout_height="wrap_content"
        android:onClick="begin"
        android:text="开始下载" />
</LinearLayout>
```

上述代码添加一个 ProgressBar 显示异步操作的进度，单击 Button 按钮开始模拟异步操作。

MainActivity.java 代码如下：

```java
public class MainActivity extends AppCompatActivity {
    private ProgressBar mProgressBar;
    private int i = 0;

    @Override
    protected void onCreate(Bundle savedInstanceState) {
```

```java
        super.onCreate(savedInstanceState);
        setContentView(R.layout.activity_main);
        mProgressBar = (ProgressBar) findViewById(R.id.progress);
    }

    public void begin(View view) {
        new ProgressAsyncTask().execute(10);
    }

    private class ProgressAsyncTask extends AsyncTask<Integer, Integer, String> {
        @Override
        protected void onPostExecute(String s) {
            super.onPostExecute(s);
            Toast.makeText(MainActivity.this, s, Toast.LENGTH_SHORT).
             show();
        }

        @Override
        protected void onProgressUpdate(Integer... values) {
            super.onProgressUpdate(values);
            mProgressBar.setProgress(values[0]);
        }

        @Override
        protected String doInBackground(Integer... params) {
            while (i < 100) {
                i++;
                publishProgress(i);
                try {
                    Thread.sleep(params[0]);
                } catch (InterruptedException e) {
                    e.printStackTrace();
                }
            }
            return "操作完成！";
        }
    }
}
```

这里创建一个内部类 ProgressAsyncTask，继承自 AsyncTask<Integer,Integer,String>，有三个泛型参数，第一个参数为 Integer 型，为每次操作的间隔；第二个参数为 Integer 型，为当前进度的百分比；第三个参数为 String 型，这里返回的是结束后的提示信息。覆写了 onPostExecute 方法，这个方法在异步操作结束后回调，这个方法中的参数即为 doInBackground 的返回值，这里通过 Toast 显示出来；覆写了 onProgressUpdate 方法，这个方法会不断回调，此方法在主线程中运行，因此可以直接调用 ProgressBar 的 setProgress 方法更新进度；覆写了 doInBackground 方法，这个方法在子线程中进行，一般用于耗时操作，这里通过一个 while 循环模拟耗时操作并不断地调用 publishProgress 方法"向外"发送当前进度，这个方法的参数即为启动操作时传入的值，这里是一次 publishProgress 方法执行的间隔，doInBackground 方法的返回值即操作完成后要提示的信息。

在 Button 的单击事件监听方法 begin 中创建了自定义类 ProgressAsyncTask 对象并调用其 execute 方法开始异步操作,这个方法传入的值即每次子操作的间隔时间。

运行实例,如图 9.29 所示。单击"开始下载"按钮,进度条即不断更新,进度条走完之后将 Toast 提示"操作完成!"。查看动态图,请扫描图 9.30 中的二维码。

图 9.29　Android AsyncTask 模拟下载

图 9.30　Android AsyncTask 模拟下载二维码

9.5.2　AsyncTask 实用实例

上面是模拟了 AsyncTask 类主要接口和方法的用法,下面通过一个具体的下载实例来看一下它的用法。

新建项目,主布局文件(activity_main.xml)代码如下:

```xml
<?xml version="1.0" encoding="utf-8"?>
<RelativeLayout xmlns:android="http://schemas.android.com/apk/res/android"
    android:layout_width="match_parent"
    android:layout_height="match_parent">

    <ImageView
        android:id="@+id/image"
        android:layout_width="match_parent"
        android:layout_height="match_parent"
        android:layout_above="@+id/progressBar"
        android:scaleType="fitXY" />

    <ProgressBar
        android:id="@+id/progressBar"
        style="?android:attr/progressBarStyleHorizontal"
        android:layout_width="match_parent"
        android:layout_height="wrap_content"
        android:layout_above="@+id/btnDownload"
        android:maxHeight="10dip"
        android:minHeight="10dip" />
```

```xml
    <Button
        android:id="@+id/btnDownload"
        android:layout_width="match_parent"
        android:layout_height="wrap_content"
        android:layout_alignParentBottom="true"
        android:text="下载图片" />

</RelativeLayout>
```

上述代码添加一个 ImageView 控件用来显示下载的图片，添加 ProgressBar 控件用来显示下载进度，添加 Button 按钮用来响应下载请求。

MainActivity.java 代码如下：

```java
public class MainActivity extends Activity {

    private Button mButtonDownload;
    private ProgressBar mProgressBar;
    private ImageView mImageView;
    private LoadImage loadImage;
    private static final String IMAGE_URL = "http://sjbz.fd.zol-img.
    com.cn/t_s1080x1920c/g5/M00/00/04/ChMkJlfJWF6IEFcNABcFwKa6JQUAAU-
    JQMX2oEAFwXY671.jpg";

    @Override
    protected void onCreate(Bundle savedInstanceState) {
        super.onCreate(savedInstanceState);
        setContentView(R.layout.activity_main);
        initViews();
    }

    private void initViews() {
        mButtonDownload = (Button) findViewById(R.id.btnDownload);
        mProgressBar = (ProgressBar) findViewById(R.id.progressBar);
        mButtonDownload.setOnClickListener(new View.OnClickListener() {
            @Override
            public void onClick(View v) {
                loadImage = new LoadImage();
                loadImage.execute(IMAGE_URL);
            }
        });
        mProgressBar.setVisibility(View.INVISIBLE);
        mImageView = (ImageView) findViewById(R.id.image);
    }
```

上面的代码主要对控件进行初始化并为 Button 添加了单击事件监听，在覆写的 onClick 方法中调用了自定义类 LoadImage 的 execute 方法开始下载，这个方法的参数 IMAGE_URL 即为下载图片的地址。

自定义类 LoadImage 继承自类 AsyncTask：

```java
private class LoadImage extends AsyncTask<String, Integer, Bitmap> {
    @Override
    protected void onPreExecute() {
```

```java
            mProgressBar.setVisibility(View.VISIBLE);
            mProgressBar.setProgress(0);
            super.onPreExecute();
        }

        @Override
        protected Bitmap doInBackground(String... params) {
            String imageUrl = params[0];
            try {
                URL url;
                HttpURLConnection httpURLConnection = null;
                InputStream inputStream = null;
                OutputStream outputStream = null;
                String filename = "load_image";
                try {
                    url = new URL(imageUrl);
                    httpURLConnection = (HttpURLConnection) url.
                    openConnection();
                    httpURLConnection.setConnectTimeout(5 * 1000);
                    inputStream = httpURLConnection.getInputStream();
                    outputStream = openFileOutput(filename,
                            Context.MODE_PRIVATE);
                    long total = httpURLConnection.getContentLength();
                    byte[] data = new byte[1024];
                    int length;
                    long current = 0;
                    while ((length = inputStream.read(data)) != -1) {
                        outputStream.write(data, 0, length);
                        current += length;
                        int progress = (int) ((float) current / total *
                        100);
                        publishProgress(progress);
                    }
                } finally {
                    if (httpURLConnection != null) {
                        httpURLConnection.disconnect();
                    }
                    if (inputStream != null) {
                        inputStream.close();
                    }
                    if (outputStream != null) {
                        outputStream.close();
                    }
                }
                return BitmapFactory.decodeFile(getFileStreamPath(filena
                me).getAbsolutePath());
            } catch (MalformedURLException e) {
                e.printStackTrace();
```

```java
        } catch (IOException e) {
            e.printStackTrace();
        }
        return null;
    }

    @Override
    protected void onProgressUpdate(Integer... values) {
        mProgressBar.setProgress(values[0]);
        super.onProgressUpdate(values);
    }

    @Override
    protected void onPostExecute(Bitmap bitmap) {
        super.onPostExecute(bitmap);
        if (bitmap != null) {
            mImageView.setImageBitmap(bitmap);
        }
        mProgressBar.setVisibility(View.INVISIBLE);
    }
}
```

AsyncTask 类的三个泛型参数为：String（下载图片的 URL 地址）、Integer（下载进度百分比）、Bitmap（下载完成后的 Bitmap 对象）。覆写了 AsyncTask 的四个方法：

- onPreExecute 方法：这个方法最先被调用，在主线程中运行，调用 ProgressBar 的 setVisibility 方法显示 ProgressBar。
- doInBackground 方法：这个方法在子线程中运行，在这个方法参数数组的第一个元素即要下载图片的 Url 地址，然后调用 HttpURLConnection 类下载图片。对于更新下载进度，这里首先调用 HttpURLConnection 类的 getConnectLength 方法获得要下载文件的总长度，然后在 while 方法中不断地计算当前进度（current+=length），最后调用 publishProgress 方法将进度更新到 UI 线程。
- onProgressUpdate 方法：这个方法在主线程中运行，可以直接调用 ProgressBar 的 setProgress 方法更新进度，这个方法的参数值 values[0] 即为 publishProgress 方法更新的进度值。
- onPostExecute 方法：这个方法在下载结束后被回调，这个方法的参数即为 doInBackground 方法的返回值（下载的图片）。在主线程中运行，调用 ImageView 的 setImageBitmap 方法将这个下载的 Bitmap 方法显示出来，同时调用 ProgressBar 的 setVisibility 方法隐藏这个 ProgressBar。

最后记得添加网络权限，代码如下：

```xml
<uses-permission android:name="android.permission.INTERNET" />
```

运行实例，如图 9.31 所示。单击"下载图片"按钮，如图 9.32 所示。

可以看出，图片下载成功并在 ImageView 中显示出来，没有看到进度条，是因为下载完成后进度条被隐藏了。查看进度条和动态图，请扫描图 9.33 中的二维码。

图 9.31　Android AsyncTask 下载图片一　　　　图 9.32　Android AsyncTask 下载图片二

图 9.33　Android AsyncTask 下载图片二维码

第 10 章　Android 手机基本功能及多媒体操作实战

智能机相对传统的功能机而言，仿佛拥有无穷无尽的功能和亮点待用户发觉。一部智能机必须具备哪些基本功能呢？除了基本的拨打电话、发送短信等功能之外，还应该具备播放音乐、播放视频、录制音频拍照等常用功能，Android 系统提供了丰富的接口供开发者调用来实现这些基本功能。

10.1　Android 拨打电话功能实例

Android 系统提供了基本的电话拨打功能，通过如图 10.1 中的面板就可以拨打电话了。

图 10.1　Android 手机拨号盘

但是在开发中时常需要自定义电话拨打功能，Android 中提供两种电话拨打方式：一种是跳转到如图 10.1 所示的面板中，单击面板中的拨打按钮来拨打电话；另一种则是直接拨打电话。下面通过实例看一下这两种方式的异同。

主布局文件（activity_main.xml）代码如下：

```xml
<?xml version="1.0" encoding="utf-8"?>
<LinearLayout xmlns:android="http://schemas.android.com/apk/res/android"
    android:layout_width="match_parent"
    android:layout_height="match_parent"
    android:orientation="vertical">

    <TextView
        android:id="@+id/edit"
        android:layout_width="match_parent"
        android:layout_height="wrap_content"
        android:layout_margin="5dp"
```

```xml
        android:gravity="center"
        android:textSize="34sp" />

    <LinearLayout
        android:layout_width="match_parent"
        android:layout_height="wrap_content"
        android:orientation="horizontal">

        <Button
            android:id="@+id/btn1"
            android:layout_width="0dp"
            android:layout_height="wrap_content"
            android:layout_weight="1"
            android:background="#00000000"
            android:onClick="btnNum"
            android:text="1"
            android:textSize="28sp" />

        <Button
            android:id="@+id/btn2"
            android:layout_width="0dp"
            android:layout_height="wrap_content"
            android:layout_weight="1"
            android:background="#00000000"
            android:onClick="btnNum"
            android:text="2"
            android:textSize="28sp" />

        <Button
            android:id="@+id/btn3"
            android:layout_width="0dp"
            android:layout_height="wrap_content"
            android:layout_weight="1"
            android:background="#00000000"
            android:onClick="btnNum"
            android:text="3"
            android:textSize="28sp" />

    </LinearLayout>
    // 省略部分相似代码
    <LinearLayout
        android:layout_width="match_parent"
        android:layout_height="wrap_content"
        android:orientation="horizontal">

        <Button
            android:id="@+id/btnStar"
            android:layout_width="0dp"
            android:layout_height="wrap_content"
            android:layout_weight="1"
            android:background="#00000000"
            android:onClick="btnNum"
```

```xml
            android:text="*"
            android:textSize="28sp" />

        <Button
            android:id="@+id/btn0"
            android:layout_width="0dp"
            android:layout_height="wrap_content"
            android:layout_weight="1"
            android:background="#00000000"
            android:onClick="btnNum"
            android:text="0"
            android:textSize="28sp" />

        <Button
            android:id="@+id/btnWell"
            android:layout_width="0dp"
            android:layout_height="wrap_content"
            android:layout_weight="1"
            android:background="#00000000"
            android:onClick="btnNum"
            android:text="#"
            android:textSize="28sp" />

    </LinearLayout>

    <Button
        android:id="@+id/btnDial"
        android:layout_width="match_parent"
        android:layout_height="wrap_content"
        android:background="#00000000"
        android:onClick="btnNum"
        android:text="DIAL"
        android:textSize="28sp" />

    <Button
        android:id="@+id/btnCall"
        android:layout_width="match_parent"
        android:layout_height="wrap_content"
        android:background="#00000000"
        android:onClick="btnNum"
        android:text="CALL"
        android:textSize="28sp" />

</LinearLayout>
```

上述代码中父布局采用了线性布局并设置其 orientation 属性为 vertical（垂直布局），添加了四个 LinearLayout，每个 LinearLayout 中添加了三个 Button 作为数字按钮，设置了每个 Button 按钮的 layout_width 为 0，并为每个 Button 添加了 layout_weight 属性，值为 1，这样这三个 Button 就可以平分整个屏幕的宽（由于篇幅限制，中间省略了部分相似代码）。最下边放置了两个按钮，添加了 onClick 属性，分别用来响应跳转到拨号盘和直接拨打电话。

MainActivity.java 代码如下：

```java
public class MainActivity extends AppCompatActivity {
    private TextView mTextView;
    private StringBuilder stringBuilder = new StringBuilder();

    @Override
    protected void onCreate(Bundle savedInstanceState) {
        super.onCreate(savedInstanceState);
        setContentView(R.layout.activity_main);
        mTextView = (TextView) findViewById(R.id.edit);
    }

    public void btnNum(View view) {
        switch (view.getId()) {
            case R.id.btn0:
                stringBuilder.append(0);
                mTextView.setText(stringBuilder);
                break;
            case R.id.btn1:
                stringBuilder.append(1);
                mTextView.setText(stringBuilder);
                break;
                // 省略部分相似代码
                break;
            case R.id.btn9:
                stringBuilder.append(9);
                mTextView.setText(stringBuilder);
                break;
            case R.id.btnDial:
                Intent intentDial = new Intent(Intent.ACTION_DIAL);
                Uri uri = Uri.parse("tel:" + stringBuilder.toString());
                intentDial.setData(uri);
                stringBuilder.delete(0, stringBuilder.length());
                mTextView.setText(stringBuilder);
                startActivity(intentDial);
                break;
            case R.id.btnCall:
                Intent intentCall = new Intent(Intent.ACTION_CALL);
                Uri uriCall = Uri.parse("tel:" + stringBuilder.
                toString());
                intentCall.setData(uriCall);
                stringBuilder.delete(0, stringBuilder.length());
                mTextView.setText(stringBuilder);
                if (Build.VERSION.SDK_INT >= 23) {
                    int checkCallPhonePermission = ContextCompat
                        .checkSelfPermission(this, Manifest.
                            permission.CALL_PHONE);
                    if (checkCallPhonePermission
                            != PackageManager.PERMISSION_GRANTED) {
                        ActivityCompat.requestPermissions(this,
```

```
                        new String[]{Manifest.permission.CALL_
                            PHONE}, 1);
                    return;
                } else {
                    startActivity(intentCall);
                }
            } else {
                startActivity(intentCall);
            }
            break;
    }
}
```

这里使用了 onClick 属性来监听单击事件，通过 id 来区分哪一个 Button 被单击了，单击数字 Button 将追加对应按钮的数字。对于跳转到拨号盘的方式，创建一个 Intent，其 Action 为 Intent.ACTION_DIAL，通过 Uri 的静态方法 parse 将 String 包装成一个 Uri 字符串，然后调用 Intent 的 setData 方法将上面包装的 Uri 字符串作为参数传进去，最后调用 startActivity 启动 Intent。直接拨打电话的方式和跳转拨号盘的方式比较类似，所不同的只是创建 Intent 时传入的参数 Action 值（Intent.ACTION_CALL）不同。

需要注意的是，对于直接拨打电话的方式，Android 6.0 以上版本需要动态获取权限，首先判断是否拥有 CALL_PHONE 的权限，若没有则调用 ActivityCompat 的静态方法 requestPermissions 申请权限。但也不要忘记在 AndroidManifest.xml 中配置电话权限，代码如下：

```
<uses-permission android:name="android.permission.CALL_PHONE"/>
```

运行实例，如图 10.2 所示。单击 DIAL 按钮将跳转到拨号界面，如图 10.3 所示。单击 CALL 按钮将弹出权限确认对话框，如图 10.4 所示。

图 10.2　Android 拨打电话　　图 10.3　Android 拨打电话　　图 10.4　Android 拨打电话
　　　　　功能实例　　　　　　　　　　　跳转到拨号盘　　　　　　　　　直接拨打

单击 ALLOW 按钮就会直接拨打这个电话号码。

10.2　Android 发送短信功能实例

和拨打电话一样，发送短信也是手机最基本的功能之一，自从短信诞生以来都备受用户的喜爱，虽然随着即时通信工具的流行，其使用率越来越低，但其作为核心功能的地位还没有改变。Android 提供了丰富的接口供开发者调用来实现自定义的短信功能，下面通过实例来学习这些接口的使用。

和拨打电话相似，发送短信也有两种方式：一种是直接调用 SMS 接口发送短信，即直接发送短信；另一种是跳转到短信发送界面。

10.2.1 直接发送短信

主布局文件代码如下：

```xml
<?xml version="1.0" encoding="utf-8"?>
<LinearLayout xmlns:android="http://schemas.android.com/apk/res/android"
    android:layout_width="match_parent"
    android:layout_height="match_parent"
    android:orientation="vertical">

    <EditText
        android:id="@+id/edit_phone"
        android:layout_width="match_parent"
        android:layout_height="wrap_content"
        android:paddingLeft="20dp"
        android:paddingRight="20dp"
        android:phoneNumber="true" />

    <EditText
        android:id="@+id/edit_content"
        android:layout_width="match_parent"
        android:layout_height="90dp"
        android:paddingLeft="20dp"
        android:paddingRight="20dp" />

    <Button
        android:id="@+id/btn_send"
        android:layout_width="match_parent"
        android:layout_height="wrap_content"
        android:text="直接发送" />
</LinearLayout>
```

上述代码在一个线性布局中添加了两个 EditText 控件，第一个 EditText 用于输入电话号码，第二个 EditText 用于输入短信内容，同时添加了一个 Button 控件，单击这个 Button 控件发送短信。

MainActivity.java 代码如下：

```java
public class MainActivity extends AppCompatActivity {
    private Button mButtonSend;
    private EditText mEditPhone, mEditContent;

    @Override
    protected void onCreate(Bundle savedInstanceState) {
        super.onCreate(savedInstanceState);
        setContentView(R.layout.activity_main);
```

```java
            final SmsManager smsManager = SmsManager.getDefault();
            mButtonSend = (Button) findViewById(R.id.btn_send);
            mEditContent = (EditText) findViewById(R.id.edit_content);
            mEditPhone = (EditText) findViewById(R.id.edit_phone);
            mButtonSend.setOnClickListener(new View.OnClickListener() {
                @Override
                public void onClick(View v) {
                    String phone = mEditPhone.getText().toString();
                    String content = mEditContent.getText().toString();

                    if (Build.VERSION.SDK_INT >= 23) {
                        int checkCallPhonePermission = ContextCompat.
                                checkSelfPermission(MainActivity.this,
                                    Manifest.permission.SEND_SMS);
                        if (checkCallPhonePermission !=
                                PackageManager.PERMISSION_GRANTED) {
                            ActivityCompat.requestPermissions(MainActivity.
                            this,
                                new String[]{Manifest.permission.SEND_
                                SMS}, 1);
                            return;
                        } else {
                            smsManager.sendTextMessage(phone, null,
                            content, null, null);
                        }
                    } else {
                        smsManager.sendTextMessage(phone, null, content,
                        null, null);
                    }
                }
            });
        }
    }
```

对于发送短信功能，这里通过调用 SmsManager 的 sendTextMessage 方法，SmsManager 对象是通过 SmsManager 的静态方法 getDefault 方法来获得。

sendTextMessage 方法如下：

```
public void sendTextMessage(
        String destinationAddress, String scAddress, String text,
        PendingIntent sentIntent, PendingIntent deliveryIntent)
```

可以看出，调用这个方法需要传入五个参数：
- destinationAddress：要接收短信的手机号码；
- scAddress：短信中心号码，null 为默认中心号码；
- text：短信内容；
- sentIntent：发送是否成功的回调，用于监听短信是否成功发送；
- deliveryIntent：接收是否成功的回调，用于监听对方是否成功接收短信。

作为功能演示，它仅传入了接收短信的手机号码和短信内容。需要注意的是，发送短信也属于运行时权限，因此它在代码里进行了权限的申请。

同时还需要在 AndroidManifest.xml 中配置发送短信的权限，代码如下：

```xml
<uses-permission android:name="android.permission.SEND_SMS" />
```

注意，发送短信需要在插 SIM 卡的真机中进行测试。运行实例，如图 10.5 所示。为了验证短信发送是否成功，这里发送 "00" 到 "10010"，单击 "发送" 按钮将弹出权限申请提示框，这里单击 "允许" 按钮，稍等片刻收到 "10010" 回发的查询结果，如图 10.6 所示。

图 10.5　Android 发送短信实例一　　　　　图 10.6　Android 发送短信实例二

通过回发的联通短信可以证明此时短信已发送成功。

10.2.2　跳转到短信发送界面

在上面实例的主布局文件（activity_main.xml）中再次添加一个 Button 按钮，代码如下：

```xml
<Button
    android:layout_width="match_parent"
    android:layout_height="wrap_content"
    android:onClick="jumpToSms"
    android:text="跳转到短信发送界面" />
```

在 MainActivity.java 中添加单击事件的响应方法，代码如下：

```java
public void jumpToSms(View view) {
    Uri smsToUri = Uri.parse("smsto:" + mEditPhone.getText().toString());
    Intent intent = new Intent(Intent.ACTION_SENDTO, smsToUri);
    intent.putExtra("sms_body", mEditContent.getText().toString());
    startActivity(intent);
}
```

上述代码中，将短信的发送地址包裹到 Uri 中，前缀为 "smsto"，创建 Intent 时传入两个参数：第一个参数是 Action，其值为 Intent.ACTION_SENDTO，第二个参数为上面包裹的 Uri 地址；将短信内容通过 Extra 的形式添加到 Intent 中，其 key 为 sms_body，最后调用 startActivity 方法启动 Intent。

再次运行实例，如图 10.7 所示。单击 "跳转到短信发送界面" 按钮，如图 10.8 所示。

可以看出，成功跳转到了短信发送界面并将上个界面中输入的手机号码和短信内容带了过来。

图 10.7　Android 发送短信跳转短信发送界面一　　图 10.8　Android 发送短信跳转短信发送界面二

10.3　Android 播放音乐功能实例

智能手机多媒体功能丰富，播放音乐现已是 Android 手机的必备功能之一。Android 中提供了 MediaPlayer 类来操作音频文件，其常用方法如表 10.1 所示。

表 10.1　MediaPlayer 的常用方法

方　　法	说　　明
create（Context context, int resid）	创建一个 MediaPlayer 类
prepare	准备播放，在 start 方法前调用
start	开始播放
pause	暂停播放
stop	停止播放
release	释放资源
isPlaying	判断是否正在播放
getDuration	获得媒体长度
getCurrentPosition	获得当前播放进度
setOnCompletionListener（MediaPlayer.OnCompletionListener listener）	媒体完成播放监听
setOnSeekCompleteListener（MediaPlayer.OnSeekCompleteListener listener）	设置完进度时触发

下面结合表中的方法来实现一个简单的音乐播放器。
主布局文件如下：

```xml
<?xml version="1.0" encoding="utf-8"?>
<RelativeLayout xmlns:android="http://schemas.android.com/apk/res/android"
    android:layout_width="match_parent"
    android:layout_height="match_parent">

    <Button
        android:id="@+id/btn_play"
        android:layout_width="80dp"
```

```
            android:layout_height="80dp"
            android:layout_centerInParent="true"
            android:background="@drawable/play86" />

    <Button
        android:id="@+id/btn_stop"
        android:layout_width="80dp"
        android:layout_height="80dp"
        android:layout_below="@+id/btn_play"
        android:layout_centerInParent="true"
        android:background="@drawable/stop22" />

</RelativeLayout>
```

上述代码中添加了两个 Button，上面的按钮用来控制开始播放和暂停播放，下面的按钮用来控制停止播放。播放的文件是事先在项目中存入的一个 mp3 文件，首先在 res 目录下新建一个 raw 文件夹。在 res 文件夹下右击，在弹出的快捷菜单中选择 New → Android Resource Directory，在 Resource type 下拉列表框中选择 raw，如图 10.9 所示。

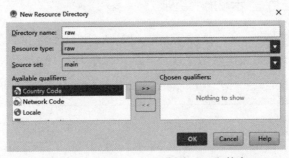

图 10.9　Android Studio 创建 raw 文件夹

单击 OK 按钮即可创建一个 raw 文件夹，在文件夹中保存要播放的 mp3 文件。
MainActivity.java 代码如下：

```
public class MainActivity extends Activity {
    private Button mPlayButton;
    private Button mStopButton;
    private MediaPlayer mMediaPlayer;
    private boolean mIsPlaying = false;

    @Override
    public void onCreate(Bundle savedInstanceState) {
        super.onCreate(savedInstanceState);
        setContentView(R.layout.activity_main);
        mPlayButton = (Button) findViewById(R.id.btn_play);
        mPlayButton.setBackgroundResource(R.drawable.play86);
        mStopButton = (Button) findViewById(R.id.btn_stop);
        mPlayButton.setOnClickListener(new View.OnClickListener() {
            @Override
            public void onClick(View v) {
                if (mIsPlaying) {
```

```java
                    mPlayButton.setBackgroundResource(R.drawable.play86);
                    mMediaPlayer.pause();
                    mIsPlaying = false;
                } else {
                    mPlayButton.setBackgroundResource(R.drawable.pause17);
                    if (mMediaPlayer == null) {
                        mMediaPlayer = MediaPlayer.create(
                                MainActivity.this, R.raw.test);
                        mMediaPlayer.setOnCompletionListener(
                                new MediaPlayer.OnCompletionListener() {
                                    @Override
                                    public void onCompletion(MediaPlayer mp) {
                                        mp.release();
                                        mMediaPlayer = null;
                                        mIsPlaying = false;
                                        mPlayButton.setBackgroundResource(
                                                R.drawable.play86);
                                        Toast.makeText(MainActivity.this,
                                                "播放结束", Toast.LENGTH_SHORT)
                                                .show();
                                    }
                                });
                        try {
                            mMediaPlayer.prepare();
                        } catch (IOException e) {
                            e.printStackTrace();
                        } catch (IllegalStateException e) {
                            e.printStackTrace();
                        }
                    }
                    mMediaPlayer.start();
                    mIsPlaying = true;
                }
            }
        });

        mStopButton.setOnClickListener(new View.OnClickListener() {
            @Override
            public void onClick(View v) {
                if (mMediaPlayer != null) {
                    mPlayButton.setBackgroundResource(R.drawable.play86);
                    mMediaPlayer.stop();
                    mMediaPlayer.release();
                    mMediaPlayer = null;
                    Toast.makeText(MainActivity.this,
                            "取消播放", Toast.LENGTH_SHORT).show();
                    mIsPlaying = false;
                }
            }
        });
    }
}
```

从代码可以看出，播放音乐的流程比较固定，首先判断 mMediaPlayer 是否为 null，若为 null，则调用 MediaPlayer 的 create 方法创建一个 MediaPlayer 对象，create 方法需要传入两个参数：上下文对象和要播放文件的 id。然后调用 MediaPlayer 的 prepare 方法进入准备状态，最后调用 MediaPlayer 的 start 方法即可播放音乐。若要停止播放，则首先调用 MediaPlayer 的 stop 方法停止播放，然后调用其 release 方法释放资源。

为了控制第一个 Button 是播放还是暂停，这里添加了一个布尔型的变量 mIsPlaying 记录状态。此外，本实例中还添加了一个监听 OnCompletionListener，覆写了其 onCompletion 方法，这个方法将在音乐文件播放完毕后调用，在这个方法中调用了 release 方法释放资源并进行了其他逻辑操作。

运行实例，如图 10.10 所示。单击上方的"播放"按钮，即可播放出悠扬的音乐，同时按钮的背景变成了播放的图片，如图 10.11 所示。

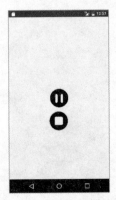

图 10.10　Android 音频播放实例一　　　图 10.11　Android 音频播放实例二

上述是在 MainActivity 中播放音乐，这样存在一个问题：即若此 Activity 退出音乐也就停止了，这样的体验很不好。因此，一般的音乐软件都会启动一个 Service，在 Service 中播放音乐，这样即使退出了音乐播放界面也可以继续播放音乐。下面实践如何在 Service 中播放音乐。

首先新建一个 Service 类 – MusicService，代码如下：

```java
public class MusicService extends Service {
    private static final String TAG = "MusicService";
    private MediaPlayer mp;

    @Override
    public void onCreate() {
        super.onCreate();
        Log.d(TAG, "onCreate: ");
        if (mp == null) {
            mp = MediaPlayer.create(this, R.raw.test);
        }
    }

    @Override
    public void onDestroy() {
        super.onDestroy();
        mp.release();
        mp = null;
```

```
            stopSelf();
        }

        @Override
        public int onStartCommand(Intent intent, int flags, int startId) {
            if (mp == null) {
                mp = MediaPlayer.create(this, R.raw.test);
            }
            boolean playing = intent.getBooleanExtra("playing", false);
            boolean stop = intent.getBooleanExtra("stop", false);
            if (stop && mp != null) {
                mp.stop();
                mp.release();
                mp = null;
                Toast.makeText(MusicService.this,
                        "停止播放", Toast.LENGTH_SHORT).show();
            }

            if (mp != null) {
                if (playing) {
                    try {
                        mp.prepare();
                    } catch (Exception e) {
                        e.printStackTrace();
                    }
                    mp.start();
                    Log.d(TAG, "onStartCommand:start ");
                } else {
                    mp.pause();
                    Log.d(TAG, "onStartCommand:pause ");
                }
            }
            return super.onStartCommand(intent, flags, startId);
        }

        @Override
        public IBinder onBind(Intent intent) {
            return null;
        }
    }
```

上述代码中 onCreate 方法仅在第一次启动 Service 时调用，在这个方法中创建一个 MediaPlayer 对象；onStartCommand 方法每次启动 Service 时都会调用，这里根据 Intent 传递过来的值来实现对应的开始播放、暂停播放和停止播放的功能；onDestroy 方法在 Service 销毁时会调用，这里调用了 MediaPlayer 的 release 方法释放资源。

MainActivity.java 代码如下：

```
public class MainActivity extends Activity {
    private static final String TAG = "MainActivity";
    private Button mPlayButton;
    private Button mStopButton;
```

```java
        private boolean mIsPlaying = false;

        @Override
        public void onCreate(Bundle savedInstanceState) {
            super.onCreate(savedInstanceState);
            setContentView(R.layout.activity_main);
            mPlayButton = (Button) findViewById(R.id.btn_play);
            mPlayButton.setBackgroundResource(R.drawable.play86);
            mStopButton = (Button) findViewById(R.id.btn_stop);
            mPlayButton.setOnClickListener(new View.OnClickListener() {
                @Override
                public void onClick(View v) {
                    if (mIsPlaying) {
                        mPlayButton.setBackgroundResource(R.drawable.play86);
                        Intent intent = new Intent(MainActivity.this,
                            MusicService.class);
                        intent.putExtra("playing", false);
                        startService(intent);
                        mIsPlaying = false;
                    } else {
                        mPlayButton.setBackgroundResource(R.drawable.pause17);
                        Intent intent = new Intent(MainActivity.this,
                            MusicService.class);
                        intent.putExtra("playing", true);
                        startService(intent);
                        mIsPlaying = true;
                    }
                }
            });
            mStopButton.setOnClickListener(new View.OnClickListener() {
                @Override
                public void onClick(View v) {
                    Intent intent = new Intent(MainActivity.this,
                        MusicService.class);
                    intent.putExtra("stop", true);
                    startService(intent);
                    mPlayButton.setBackgroundResource(R.drawable.play86);
                    mIsPlaying = false;

                }
            });
        }

        @Override
        protected void onDestroy() {
            super.onDestroy();
            Log.d(TAG, "onDestroy: ");
        }
    }
```

可以看出，MainActivity 中已经没有 MediaPlayer 类，仅通过 Intent 启动 Service 并传递布尔型的参数控制音乐的播放、暂停和停止等。

最后记得在 AndroidManifest.xml 的 application 标签中配置自定义的 Service，代码如下：

```
<service android:name=".MusicService"/>
```

运行实例并单击"播放"按钮，这时单击返回键销毁 Activity，然而音乐还将继续在后台播放。

10.4 Android 播放视频功能实例

同播放音乐一样，播放视频也是 Android 手机最常用、最基本的功能之一，Android 手机也提供了内置的视频播放器，同样也提供了视频播放器开发的接口，开发者可以灵活地调用这些接口实现功能强大的播放器。要实现一个播放器的基本功能，同样也可以借助上一节介绍的 MediaPlayer 类。与播放音频不同的是，播放视频需要一个容器来显示视频内容，这里采用了 SurfaceView 组件，这个组件可以对视频进行解码，从而得到一帧帧的图片，这些图片按照一定的顺序和速率播放出来就变成了视频。播放视频也需要一些方法的支持，常用的方法如表 10.2 所示。

表 10.2 MediaPlayer 的常用方法

方　　法	说　　明
setDataSource（String path）	为 MediaPlayer 添加数据源
setDisplay（SurfaceHolder sh）	设置播放容器
setAudioStreamType（int streamtype）	设置音频类型
prepareAsync	进入准备状态
getDuration	获得文件播放时长
seekTo（int msec）	跳转到指定位置
setOnPreparedListener（MediaPlayer.OnPreparedListener listener）	设置准备状态的监听，准备完毕后回调

下面通过实例介绍如何创建一个简单的视频播放器。

主布局文件如下：

```xml
<?xml version="1.0" encoding="utf-8"?>
<RelativeLayout xmlns:android="http://schemas.android.com/apk/res/android"
    android:layout_width="match_parent"
    android:layout_height="match_parent">

    <TextView
        android:id="@+id/text"
        android:layout_width="match_parent"
        android:layout_height="48dp"
        android:layout_toLeftOf="@+id/btn_select"
        android:gravity="center"
        android:text=" 指示文本 "
        android:textSize="18sp" />

    <Button
        android:id="@+id/btn_select"
        android:layout_width="98dp"
        android:layout_height="48dp"
```

```xml
        android:layout_alignParentRight="true"
        android:text="选择文件" />

    <SurfaceView
        android:id="@+id/surfaceView"
        android:layout_width="match_parent"
        android:layout_height="match_parent"
        android:layout_above="@+id/ll_btns"
        android:layout_below="@+id/btn_select" />

    <LinearLayout
        android:id="@+id/ll_btns"
        android:layout_width="match_parent"
        android:layout_height="wrap_content"
        android:layout_alignParentBottom="true">

        <Button
            android:id="@+id/btn_play"
            android:layout_width="0dp"
            android:layout_height="wrap_content"
            android:layout_weight="1"
            android:text="播放" />

        <Button
            android:id="@+id/btn_pause"
            android:layout_width="0dp"
            android:layout_height="wrap_content"
            android:layout_weight="1"
            android:text="暂停" />

        <Button
            android:id="@+id/btn_stop"
            android:layout_width="0dp"
            android:layout_height="wrap_content"
            android:layout_weight="1"
            android:text="停止" />

        <Button
            android:id="@+id/btn_replay"
            android:layout_width="0dp"
            android:layout_height="wrap_content"
            android:layout_weight="1"
            android:text="重播" />
    </LinearLayout>
</RelativeLayout>
```

上述代码在最上方添加了一个 TextView 作为文本指示器，用于显示当前的播放状态，在 TextView 的右边添加了一个 Button 控件用于选择要播放的文件；添加了一个 SurfaceView 控件用于视频的播放，在最下方添加了四个 Button 分别用于控制视频的播放、暂停/继续、停止和重播功能。

MainActivity.java 代码如下（由于代码较长，这里拆分进行讲解）：

```java
public class MainActivity extends Activity {
    private static final String TAG = "MainActivity";
    private SurfaceView sv;
    private Button mButtonPlay, mButtonPause, mButtonSelect,
     mButtonStop, mButtonReplay;
    private MediaPlayer mMediaPlayer;
    private Uri mUri;
    private TextView mTextView;

    @Override
    protected void onCreate(Bundle savedInstanceState) {
        super.onCreate(savedInstanceState);
        setContentView(R.layout.activity_main);
        sv = (SurfaceView) findViewById(R.id.surfaceView);
        mButtonPlay = (Button) findViewById(R.id.btn_play);
        mButtonPause = (Button) findViewById(R.id.btn_pause);
        mButtonSelect = (Button) findViewById(R.id.btn_select);
        mButtonStop = (Button) findViewById(R.id.btn_stop);
        mButtonReplay = (Button) findViewById(R.id.btn_replay);
        mButtonPlay.setOnClickListener(onClickListener);
        mButtonPause.setOnClickListener(onClickListener);
        mButtonSelect.setOnClickListener(onClickListener);
        mButtonReplay.setOnClickListener(onClickListener);
        mButtonStop.setOnClickListener(onClickListener);
        mTextView = (TextView) findViewById(R.id.text);
    }
```

上面的代码对布局中的控件进行了初始化并调用了 setOnClickListener 为 Button 添加单击事件监听。

实现监听的代码如下：

```java
private View.OnClickListener onClickListener = new View.OnClickListener() {

    @Override
    public void onClick(View v) {

        switch (v.getId()) {
            case R.id.btn_play:
                play(0);
                break;
            case R.id.btn_pause:
                pause();
                break;
            case R.id.btn_select:
                showFileChooser();
                break;
            case R.id.btn_stop:
                stop();
                break;
            case R.id.btn_replay:
                replay();
```

```
                break;
            default:
                break;
        }
    }
};

private void showFileChooser() {
    Intent intent = new Intent(Intent.ACTION_GET_CONTENT);
    intent.setType("*/*");
    intent.addCategory(Intent.CATEGORY_OPENABLE);
    try {
        startActivityForResult(Intent.createChooser(intent, "选择视频文件播放"), 1);
    } catch (ActivityNotFoundException ex) {
        mTextView.setText("打开失败");
    }
}

@Override
protected void onActivityResult(int requestCode, int resultCode, Intent data) {
    switch (requestCode) {
        case 1:
            if (resultCode == RESULT_OK) {
                mUri = data.getData();
                mTextView.setText("文件选择成功,请单击播放按钮");
            }
            break;
    }
    super.onActivityResult(requestCode, resultCode, data);
}
```

这里创建了 setOnClickListener 方法中传入的变量 onClickListener, 覆写了 View.OnClickListener 接口的 onClick 方法, 由 View 的 getId 方法获得被单击按钮的 id, 不同 id 则调用不同方法的响应。

showFileChooser 方法会通过 Intent 的方式打开文件选择器, 创建 Intent 时传入了 Intent.ACTION_GET_CONTENT 常量, 为这个 Intent 添加了 Catagory 为 Intent.CATEGORY_OPENABLE, 最后调用 startActivityForResult 方法启动这个 Intent, 这个方法传入两个参数: Intent 对象和请求码。

在 Activity 中覆写了 onActivityResult 方法, 这个方法将会在启动 Intent 后被回调。由请求码可以判断是哪一次 Intent 启动的回调, 同时判断 resultCode 为 RESULT_OK 则认为 Intent 启动成功, 这时通过 onActivityResult 方法的参数 data 并调用其 getData 方法即可获得选中文件的 Uri 地址。

下面的代码是四个按钮(播放、暂停/继续、停止和重播)单击事件的具体实现:

```
private void stop() {
    if (mMediaPlayer != null) {
        mTextView.setText("停止播放");
```

```java
            mMediaPlayer.stop();
            mMediaPlayer.release();
            mMediaPlayer = null;
            mButtonPlay.setEnabled(true);
        }
    }

    private void play(final int msec) {
        if (mUri == null) {
            mTextView.setText("请先选择要播放的文件");
        } else {
            try {
                if (mMediaPlayer == null) {
                    mMediaPlayer = new MediaPlayer();
                    mMediaPlayer.setAudioStreamType(AudioManager.
                    STREAM_MUSIC);
                    mMediaPlayer.setDataSource(MainActivity.this, mUri);
                    mMediaPlayer.setDisplay(sv.getHolder());
                    mMediaPlayer.prepareAsync();
                }
                mMediaPlayer.setOnPreparedListener(new MediaPlayer.
                    OnPreparedListener() {

                    @Override
                    public void onPrepared(MediaPlayer mp) {
                        mTextView.setText("开始播放");
                        mMediaPlayer.start();
                        mMediaPlayer.seekTo(msec);
                        mButtonPlay.setEnabled(false);
                    }
                });
                mMediaPlayer.setOnCompletionListener(new MediaPlayer.
                    OnCompletionListener() {

                    @Override
                    public void onCompletion(MediaPlayer mp) {
                        mButtonPlay.setEnabled(true);
                        mTextView.setText("已播放完");
                        mMediaPlayer.release();
                        mMediaPlayer = null;
                    }
                });

                mMediaPlayer.setOnErrorListener(new MediaPlayer.
                OnErrorListener() {

                    @Override
                    public boolean onError(MediaPlayer mp, int what,
                    int extra) {
                        play(0);
                        return false;
                    }
```

```java
            });
        } catch (Exception e) {
            e.printStackTrace();
        }
    }
}

private void replay() {
    if (mMediaPlayer != null && mMediaPlayer.isPlaying()) {
        mMediaPlayer.seekTo(0);
        mTextView.setText("重新播放");
        mButtonPause.setText("暂停");
        return;
    }
    play(0);
}

private void pause() {
    if (mButtonPause.getText().toString().trim().
            equals("继续") && mMediaPlayer != null) {
        mButtonPause.setText("暂停");
        mMediaPlayer.start();
        mTextView.setText("继续播放");
        return;
    }
    if (mMediaPlayer != null && mMediaPlayer.isPlaying()) {
        mMediaPlayer.pause();
        mButtonPause.setText("继续");
        mTextView.setText("暂停播放");
    }
}
```

这里四个方法对应最下方四个 Button 的逻辑。

stop 方法即停止播放，首先调用了 MediaPlayer 方法的 stop 方法停止播放，然后调用其 release 方法释放资源，最后将 MediaPlayer 置空。

play 方法用于播放视频文件，首先判断 mUri 是否为空，若为空则认为没有进行播放资源的选取，否则创建一个 MediaPlayer 对象并为这个对象设置一些基本的方法。setAudioStreamType 方法设置音频类型；setDataSource 方法设置数据源，这个方法需要传入两个参数：上下文对象和要播放资源的 Uri 地址，这个 Uri 地址即上面选取文件的 Uri 地址；setDisplay 方法添加播放容器，这个方法传入的 ViewHolder 对象可以通过 SurfaceView 的 getHolder 方法获得；prepareAsync 方法视频将进入准备状态。为了避免没有准备完成就调用 start 方法而导致的异常，这里调用 setOnPreparedListener 方法为 MeidaPlayer 添加准备状态的监听，在回调方法 onPrepared（准备完毕后回调）中调用 start 方法开始播放。调用 setOnCompletionListener 方法为 MediaPlayer 添加是否播放完毕的监听，播放完毕回调 onCompletion 方法，在这个方法中调用 release 方法释放资源。调用 setOnErrorListener 方法添加监听播放错误，若播放错误则回调 onError 方法，在 onError 方法中调用 play 方法重启播放。

replay 方法执行时将重播视频，这里调用了 MediaPlayer 的 seekTo 方法传入"0"，将进度移至最前方，然后调用 play 方法再次播放。

pause 方法将视频暂停或继续播放，调用 isPlaying 方法判断当前视频播放的状态，若此方法返回 true 则认为正在播放，这时调用 pause 方法暂停播放；反之，则认为视频播放暂停，则调用 start 方法继续播放。

运行实例，如图 10.12 所示。单击"选择文件"按钮选择要播放的文件，如图 10.13 所示。

图 10.12　Android 视频播放实例一

图 10.13　Android 视频播放实例二

选中"test.mp4"并单击"播放"按钮，开始播放，如图 10.14 所示。查看其他操作效果，可以扫描图 10.15 中的二维码查看动态图。

图 10.14　Android 视频播放实例三

图 10.15　Android 视频播放实例二维码

10.5　Android 录制音频功能实例

前面几节讲解了音频的播放、视频的播放，本节将讲解音频的录制功能。音频录制在实际生活中应用场景也比较多，录音机 APP 也是 Android 手机中必备的应用之一。播放音频和播放视频需要借助 MediaRecord 类，录制音频则需要借助 MediaRecorder 类。

MeidaRecorder 类的使用方式也比较固定，示例代码如下：

```
MediaRecorder recorder = new MediaRecorder();
recorder.setAudioSource(MediaRecorder.AudioSource.MIC);
recorder.setOutputFormat(MediaRecorder.OutputFormat.THREE_GPP);
recorder.setAudioEncoder(MediaRecorder.AudioEncoder.AMR_NB);
recorder.setOutputFile(PATH_NAME);
recorder.prepare();
recorder.start();   // Recording is now started
...
recorder.stop();
recorder.release(); // Now the object cannot be reused
```

上述代码中可总结其步骤如下：

- 通过 new 的方式创建一个 MediaPlayer 对象。
- 调用 setAudioSource 方法为 MediaPlayer 设置声音源。
- 调用 setOutputFormat 方法设置音频输出格式。
- 调用 setAudioEncoder 方法设置用于录制的音频编码器。
- 调用 setOutputFile 方法设置录制音频的保存位置。
- 调用 prepare 方法进入准备状态。
- 调用 start 方法开始录制。
- 调用 stop 方法停止录制。
- 调用 release 方法释放资源。

下面通过实例实践如何使用 MediaRecorder 类录制音频。

主布局文件（activity_main.xml）代码如下：

```xml
<?xml version="1.0" encoding="utf-8"?>
<RelativeLayout xmlns:android="http://schemas.android.com/apk/res/android"
    android:layout_width="match_parent"
    android:layout_height="match_parent">

    <LinearLayout
        android:id="@+id/ll"
        android:layout_width="match_parent"
        android:layout_height="wrap_content"
        android:orientation="horizontal">

        <Button
            android:id="@+id/btn_start"
            android:layout_width="0dp"
            android:layout_height="wrap_content"
            android:layout_weight="1"
            android:text=" 开始录音 " />

        <Button
            android:id="@+id/btn_stop"
            android:layout_width="0dp"
            android:layout_height="wrap_content"
```

```xml
            android:layout_weight="1"
            android:text="停止录音" />
    </LinearLayout>

    <ListView
        android:id="@+id/list"
        android:layout_width="match_parent"
        android:layout_height="wrap_content"
        android:layout_below="@id/ll" />

</RelativeLayout>
```

上述代码在最上方添加了两个 Button 控制音频的录制和停止，下方添加了一个 ListView 用于显示录制音频文件的列表。

ListView 子项的布局文件（item.xml）代码如下：

```xml
<?xml version="1.0" encoding="utf-8"?>
<LinearLayout xmlns:android="http://schemas.android.com/apk/res/android"
    android:layout_width="match_parent"
    android:layout_height="match_parent"
    android:orientation="horizontal">

    <TextView
        android:id="@+id/recorder_file_name"
        android:layout_width="match_parent"
        android:layout_height="wrap_content"
        android:gravity="center"
        android:text="2323"
        android:textSize="20sp" />

</LinearLayout>
```

上述代码中只放置了一个 TextView 用来显示录音文件名。

MainActivity.java 代码如下（代码较长，分段进行讲解）：

```java
public class MainActivity extends Activity {

    private Button mButtonStart;
    private Button mButtonStop;
    private ListView mListView;
    private MediaRecorder mMediaRecorder;
    private String pathRoot;
    private String paths = pathRoot;
    private File saveFilePath;
    private String[] listFile = null;
    private RecorderListAdpter recorderListAdpter;

    @Override
    protected void onCreate(Bundle savedInstanceState) {
        super.onCreate(savedInstanceState);
        setContentView(R.layout.activity_main);
        mButtonStart = (Button) findViewById(R.id.btn_start);
        mButtonStop = (Button) findViewById(R.id.btn_stop);
```

```java
        mListView = (ListView) findViewById(R.id.list);
        recorderListAdpter = new RecorderListAdpter();
        if (Environment.getExternalStorageState().equals(
                Environment.MEDIA_MOUNTED)) {
            try {
                pathRoot = Environment.getExternalStorageDirectory()
                        .getCanonicalPath().toString()
                        + "/MediaRecorder";
                File files = new File(pathRoot);
                if (!files.exists()) {
                    files.mkdir();
                }
                listFile = files.list();
            } catch (IOException e) {
                e.printStackTrace();
            }
        }
```

上面代码主要是对布局中的控件和数据集进行初始化，在存储卡中创建了一个名为 MediaRecorder 的文件夹来保存所有的录音文件（调用 File 类的 exists 方法判断文件夹是否存在，若不存在则调用 mkdir 方法新建文件夹），调用 File 的 list 方法将返回文件夹中所有文件名的字符串数组，这个数组将作为 ListView 的数据源。

下面的代码是对两个 Button（"开始录音"和"停止录音"）单击事件的实现：

```java
    mButtonStart.setOnClickListener(new View.OnClickListener() {
        @Override
        public void onClick(View v) {
            try {
                initRecorder();
                paths = pathRoot
                        + "/"
                        + new SimpleDateFormat(
                        "yyyyMMddHHmmss").format(System
                        .currentTimeMillis())
                        + ".amr";
                saveFilePath = new File(paths);
                mMediaRecorder.setOutputFile(saveFilePath
                        .getAbsolutePath());
                saveFilePath.createNewFile();
                mMediaRecorder.prepare();
                mMediaRecorder.start();
                mButtonStart.setEnabled(false);
                mButtonStart.setText("录音中");
            } catch (Exception e) {
                e.printStackTrace();
            }
        }
    });

    mButtonStop.setOnClickListener(new View.OnClickListener() {
        @Override
```

```java
            public void onClick(View v) {
                if (saveFilePath.exists() && saveFilePath != null) {
                    mMediaRecorder.stop();
                    mMediaRecorder.release();
                    mMediaRecorder = null;
                    new AlertDialog.Builder(MainActivity.this)
                            .setTitle("是否保存该录音")
                            .setPositiveButton("OK", new DialogInterface.
                                    OnClickListener() {
                                @Override
                                public void onClick(DialogInterface dialog,
                                        int which) {
                                    File files = new File(pathRoot);
                                    listFile = files.list();
                                    recorderListAdpter.notifyDataSetChanged();
                                }
                            })
                            .setNegativeButton("Cancel",
                                    new DialogInterface.OnClickListener() {
                                        @Override
                                        public void onClick(DialogInterface
                                        dialog,int which) {
                                            saveFilePath.delete();
                                        }
                                    }).show();

                }
                mButtonStart.setText("开始录音");
                mButtonStart.setEnabled(true);
            }
        });
        if (listFile != null) {
            mListView.setAdapter(recorderListAdpter);
        }
    }
```

在"开始录音"按钮的监听事件中，首先调用 initRecorder 方法创建 MediaRecorder 对象，初始化了录音文件的保存路径，这里使用录音的当前时间作为录音文件的文件名。然后调用 setOutputFile 方法设置录音文件的保存路径，调用 createNewFile 方法创建这个录音文件。最后一次调用 MediaRecorder 的 prepare、start 方法开始音频的录制。

在"停止录音"按钮的监听事件中，首先调用 MediaRecorder 类的 stop 方法停止录制，调用 release 方法释放资源。然后弹出一个对话框由用户确认是否保存这个录音文件，在"PositiveButton（确定按钮）"的单击事件中调用 File 的 list 方法获得最新的文件列表，由 notifyDataSetChanged 方法刷新文件列表。在"NegativeButton（取消按钮）"的单击事件中调用 delete 删除这个文件。

MainActivity 最后一段代码如下：

```java
private void initRecorder() {
    if (mMediaRecorder == null) {
```

```java
            mMediaRecorder = new MediaRecorder();
            mMediaRecorder.setAudioSource(MediaRecorder.AudioSource.
            DEFAULT);
            mMediaRecorder.setOutputFormat(MediaRecorder.OutputFormat.
            DEFAULT);
            mMediaRecorder.setAudioEncoder(MediaRecorder.AudioEncoder.
            DEFAULT);
        }
    }

    class RecorderListAdpter extends BaseAdapter {

        @Override
        public int getCount() {
            return listFile.length;
        }

        @Override
        public Object getItem(int id) {
            return listFile[id];
        }

        @Override
        public long getItemId(int id) {
            return id;

        }

        @Override
        public View getView(final int postion, View arg1, ViewGroup
        arg2) {
            View view = LayoutInflater.from(MainActivity.this).inflate(
                    R.layout.item, null);
            TextView filename = (TextView) view.findViewById(R.
             id.recorder_file_name);
            filename.setText(listFile[postion]);
            return view;
        }
    }

    @Override
    protected void onDestroy() {
        if (mMediaRecorder == null) return;
        mMediaRecorder.release();
        mMediaRecorder = null;
        super.onDestroy();
    }
}
```

上述代码中 initRecorder 方法创建一个 MediaRecorder 对象，调用它的基本方法为其设置录音源、输出格式、录音编码等参数。内部类 RecorderListAdapter 继承自

BaseAdapter，它是录音文件显示控件 ListView 的适配器。最后覆写了 onDestroy 方法，在这个方法中释放 MediaRecorder 对象。

录制音频和保存文件都需要申请权限，在 AndroidManifest.xml 添加如下代码：

```
<uses-permission android:name="android.permission.MOUNT_FORMAT_FILESYSTEMS"/>
<uses-permission android:name="android.permission.WRITE_EXTERNAL_STORAGE"/>
<uses-permission android:name="android.permission.RECORD_AUDIO"/>
```

运行实例（在真机上调试运行），如图 10.16 所示。可以看出，MediaRecorder 目录下已经有几个音频文件，单击"开始录音"按钮会弹出权限提示框，由于用于演示手机的 Android 版本较低，这里没有在代码中动态申请权限，若在 6.0 以上版本中调试，需要动态申请权限，这部分知识前面已经介绍过，这里不再进行讲解。单击"允许"按钮即开始录音，单击"停止录音"按钮会弹出对话框，如图 10.17 所示。单击 CANCEL 按钮取消录音文件的保存，单击 OK 按钮则保存音频文件，这里单击 OK 按钮，如图 10.18 所示。

图 10.16　Android 录音功能实例一

图 10.17　Android 录音功能实例二

图 10.18　Android 录音功能实例三

可以看出，这里增加了名为"20161207214558"的音频文件，即新录制的音频文件。到此录音功能介绍完毕，读者可以拓展一下这个 APP，结合前面讲解的音频文件的播放，在列表中添加按钮，单击按钮即播放此条录音文件。

10.6　Android 拍照功能实例

现如今智能手机的拍照功能越来越强大，各大厂商也十分喜欢在手机拍照效果上做文章，因此，详细学习基础的 Camera 接口对开发者来讲十分重要，考虑到兼容性，这里主要结合 API1 进行讲解。开发拍照功能也有两种途径：一种是通过 Intent 方式直接跳转到系统 Camera；另一种是借助 Camera 类开发自己的 Camera 应用。下面分别讲解。

10.6.1 Intent 方式

主布局文件代码如下：

```xml
<?xml version="1.0" encoding="utf-8"?>
<RelativeLayout xmlns:android="http://schemas.android.com/apk/res/android"
    android:layout_width="match_parent"
    android:layout_height="match_parent">

    <Button
        android:id="@+id/btn"
        android:layout_width="match_parent"
        android:layout_height="wrap_content"
        android:onClick="launchCamera"
        android:text="launchCamera" />

    <ImageView
        android:id="@+id/image"
        android:layout_width="match_parent"
        android:layout_height="match_parent"
        android:layout_below="@+id/btn" />
</RelativeLayout>
```

上述代码添加 Button 控件并为其添加了 onClick 属性，其值为 launchCamera，用于监听按钮的单击事件，添加了 ImageView 控件显示拍照的图片。

MainActivity.java 代码如下：

```java
public class MainActivity extends Activity {
    private static final int REQUEST_CODE = 1;
    private ImageView mImageView;

    @Override
    protected void onCreate(Bundle savedInstanceState) {
        super.onCreate(savedInstanceState);
        setContentView(R.layout.activity_main);
        mImageView = (ImageView) findViewById(R.id.image);
    }

    public void launchCamera(View view) {
        Intent intent = new Intent(MediaStore.ACTION_IMAGE_CAPTURE);
        startActivityForResult(intent, REQUEST_CODE);
    }

    @Override
    protected void onActivityResult(int requestCode, int resultCode,
    Intent data) {
        super.onActivityResult(requestCode, resultCode, data);
        if (resultCode == Activity.RESULT_OK) {
            if (!Environment.getExternalStorageState().
                    equals(Environment.MEDIA_MOUNTED)) {
                Toast.makeText(MainActivity.this,
                        "未能打开SD卡", Toast.LENGTH_SHORT).show();
                return;
```

```
            }
            String name = new DateFormat().format("yyyyMMdd_hhmmss",
                Calendar.getInstance(Locale.CHINA)) + ".jpg";
            Bundle bundle = data.getExtras();
            Bitmap bitmap = (Bitmap) bundle.get("data");

            FileOutputStream fileOutputStream = null;
            File file = new File("/sdcard/Camera/");
            if (!file.exists()) {
                file.mkdirs();
            }
            String path = "/sdcard/Camera/" + name;

            try {
                fileOutputStream = new FileOutputStream(path);
                bitmap.compress(Bitmap.CompressFormat.JPEG, 100,
                    fileOutputStream);
            } catch (FileNotFoundException e) {
                e.printStackTrace();
            } finally {
                try {
                    fileOutputStream.flush();
                    fileOutputStream.close();
                } catch (IOException e) {
                    e.printStackTrace();
                }
            }
            mImageView.setImageBitmap(bitmap);
        }
    }
}
```

上述代码在按钮的单击事件监听方法 launchCamera 中创建了一个 Intent 对象，在初始化这个 Intent 对象时传入 MediaStore.ACTION_IMAGE_CAPTURE 作为 Action（第一个参数），调用 startActivityForResult 方法启动 Intent，第二个参数 REQUEST_CODE 作为请求码，用于在回调中分辨是哪个 Intent 的返回。

调用 startActivityForResult 方法启动 Intent，自然需要覆写 onActivityResult 方法处理 Intent 的返回，因为要保存拍照的图片，所以这里首先判断 SD 卡是否可用，然后创建保存图片的文件名，和上一节保存音频文件类似，这里也是采用当前时间来命名拍照文件。如何获得拍照文件呢？这里通过 onActivityResult 的参数 Intent 来获得，拍照文件（Bitmap）是包装在这个 Intent 中传递过来的，根据其 key（data）即可获得。接下来创建一个输出流（FileOutputStream 对象）用于文件的保存，在 compress 方法中，第二个参数为 100 表示不压缩此图片，最后调用 FileOutputStream 的 flush 方法清空缓存区保存图片。

拍摄照片和保存文件都需要权限，在 AndroidManifest.xml 进行配置，代码如下：

```
<uses-permission android:name="android.permission.MOUNT_FORMAT_
FILESYSTEMS"/>
```

```
<uses-permission  android:name="android.permission.WRITE_EXTERNAL_
STORAGE"/>
<uses-permission android:name="android.permission.CAMERA"/>
```

同样若在 Android 6.0 以上的版本运行代码需要在代码中动态申请权限，这部分内容前面的章节有讲解，这里不再讲解。

运行实例（在真机上运行），如图 10.19 所示。单击 LAUNCHCAMERA 按钮将弹出权限提示框，单击"允许"按钮将跳转到相机界面，如图 10.20 所示。

单击"拍照"按钮并选择图片，回到 MainActivity 界面并将拍摄的照片显示出来，如图 10.21 所示。

图 10.19　Android 跳转到相机　　图 10.20　Android 跳转到相机　　图 10.21　Android 跳转到相机
　　　　　　应用拍照一　　　　　　　　　　　　应用拍照二　　　　　　　　　　　　应用拍照三

10.6.2　借助 Camera 类

相机有两项最基本的功能，即拍照和录像，Android 提供了 Camera API 供开发者使用来实现丰富的相机功能。下面介绍 Camera 类（考虑到兼容性，这里介绍 API1）。

Camera 类提供了接口操作 Camera 底层来拍摄照片，并可以控制预览的开启和停止。拍摄一张图片的流程如下：

- 创建一个 SurfaceView 对象，设置 Surface 的基础属性并为其添加回调监听。
- 在 SurfaceHolder.Callback 的 surfaceCreate 方法中调用 Camera 的 open 方法打开摄像头，然后调用 Camera 的 startPreview 方法启动预览，注意这些方法会抛出异常，需要添加 try-catch 语句捕获异常。
- 拍照则调用 Camera 的 takePicture(Camera.ShutterCallback, Camera.PictureCallback, Camera.PictureCallback, Camera.PictureCallback) 方法，这个方法需要传入三个参数，第三个参数的回调中包含拍照的图片。
- Camera 会占用较多资源，不使用时需要调用其 release 方法释放资源。

Camera 的常用方法如表 10.3 所示。

表 10.3 Camera 的常用方法

方　法	说　明
open	打开 Camera，返回一个 Camera 对象
getParameters	获得 Camera.Parameter 对象
setParameters（Camera.Parameters params）	为 Camera 设置属性
setPreviewDisplay（SurfaceHolder holder）	设置预览容器
startPreview	启动预览
stopPreview	停止预览
takePicture（Camera.ShutterCallback shutter, Camera.PictureCallback raw, Camera.PictureCallback jpeg）	拍摄一张照片
enableShutterSound（boolean enabled）	开启拍照声音
autoFocus（Camera.AutoFocusCallback cb）	注册自动对焦
cancelAutoFocus	取消 AutoFocus

结合上面的方法来实现一个简单的相机功能，包括拍照和图片保存功能。

主布局文件如下：

```xml
<RelativeLayout xmlns:android="http://schemas.android.com/apk/res/android"
    android:id="@+id/buttonLayout"
    android:layout_width="match_parent"
    android:layout_height="match_parent">

    <SurfaceView
        android:id="@+id/surfaceView"
        android:layout_width="match_parent"
        android:layout_height="match_parent" />

    <ImageView
        android:layout_width="80dp"
        android:layout_height="80dp"
        android:layout_alignParentBottom="true"
        android:layout_centerHorizontal="true"
        android:layout_marginBottom="20dp"
        android:background="@null"
        android:onClick="capture"
        android:src="@drawable/camera" />

</RelativeLayout>
```

布局文件很是简单，一个占据整个屏幕的 SurfaceView 控件，用于充当预览界面。ImageView 显示 "拍照" 按钮，单击这个按钮即拍下一张照片。

MainActivity.java 代码如下（分段讲解）：

```java
public class MainActivity extends Activity {
    private Camera mCamera;
    private Camera.Parameters mParameters;
    private boolean mIsCallBackReturn = true;

    @Override
    public void onCreate(Bundle savedInstanceState) {
```

```java
        super.onCreate(savedInstanceState);
        requestWindowFeature(Window.FEATURE_NO_TITLE);
        getWindow().setFlags(WindowManager.LayoutParams.FLAG_FULLSCREEN,
                WindowManager.LayoutParams.FLAG_FULLSCREEN);
        setContentView(R.layout.activity_main);

        initSurfaceView();
    }

    private void initSurfaceView() {
        SurfaceView surfaceView = (SurfaceView) this.findViewById(R.
        id.surfaceView);
        surfaceView.getHolder()
                .setType(SurfaceHolder.SURFACE_TYPE_PUSH_BUFFERS);
        // 屏幕常亮
        surfaceView.getHolder().setKeepScreenOn(true);
        // 添加回调监听
        surfaceView.getHolder().addCallback(new SurfaceCallback());
    }
```

为了获得更好的用户体验，这里调用 requestWindowFeature 方法传入 Window.FEATURE_NO_TITLE 消除标题栏，调用 Window 类的 setFlags 方法隐藏顶部状态栏。调用 initSurfaceView 方法创建了一个 SurfaceView 对象，调用其 getHolder 方法可以获得一个 SurfaceHolder 对象，并为这个对象设置了一些基本属性和方法。

```java
class SurfaceCallback implements SurfaceHolder.Callback {

    @Override
    public void surfaceChanged(SurfaceHolder holder, int format, int
    width,int height) {
    }

    @Override
    public void surfaceCreated(SurfaceHolder holder) {
        try {
            // 打开摄像头
            mCamera = Camera.open();
            // 设置用于显示拍照影像的 SurfaceHolder 对象
            mCamera.setPreviewDisplay(holder);
            // 获得 parameter 对象
            mParameters = mCamera.getParameters();
            // 设置照片方向
            mParameters.setRotation(90);
            // 设置参数
            mCamera.setParameters(mParameters);
            // 设置预览方向
            mCamera.setDisplayOrientation(90);
            // 开始预览
            mCamera.startPreview();
        } catch (Exception e) {
            e.printStackTrace();
        }
```

```
    }

    @Override
    public void surfaceDestroyed(SurfaceHolder holder) {
        if (mCamera != null) {
            // 释放资源
            mCamera.release();
            mCamera = null;
        }
    }
}

public void capture(View view) {
    if (!mIsCallBackReturn) return;
    mIsCallBackReturn = false;
    mCamera.takePicture(new MyShutterCallback(), null, new
    TakePictureCallback());
}
```

上述代码中内部类 SurfaceCallback 实现了 SurfaceHolder.Callback 接口并覆写了它的三个抽象方法，在 surfaceCreated 方法中调用 Camera 的 open 方法打开摄像头；调用 setPreviewDsiplay 方法设置预览的容器；调用 getParameters 方法获得 Camera.Parameters 对象，这个对象可以为 Camera 设置一些基本属性，添加属性后不要忘记调用 setParameters 方法对底层进行设置，最后调用 startPreview 开启预览。

MainActivity 最后一段代码如下：

```
public void capture(View view) {
    if (!mIsCallBackReturn) return;
    mIsCallBackReturn = false;
    mCamera.takePicture(new MyShutterCallback(),
                        null, new TakePictureCallback());
}

class MyShutterCallback implements Camera.ShutterCallback {

    @Override
    public void onShutter() {

    }
}

class TakePictureCallback implements Camera.PictureCallback {

    @Override
    public void onPictureTaken(byte[] data, Camera camera) {
        try {
            // 保存图片到 SD 卡中
            saveToSDCard(data);
            // 拍完照后，重新启动预览
            camera.startPreview();
            mIsCallBackReturn = true;
```

```
        } catch (Exception e) {
            e.printStackTrace();
        }
    }
}

    private void saveToSDCard(byte[] data) throws IOException {
        Date date = new Date();
        // 格式化时间
        SimpleDateFormat format = new SimpleDateFormat("yyyyMMddHHmmss");
        String filename = "DCIM" + format.format(date) + ".jpg";
        File fileFolder = new File(Environment.
        getExternalStorageDirectory()
                + "/CameraTest/");
        if (!fileFolder.exists()) {
            fileFolder.mkdir();
        }
        File jpgFile = new File(fileFolder, filename);
        // 文件输出流
        FileOutputStream outputStream = new FileOutputStream(jpgFile);
        // 写入 SD 卡中
        outputStream.write(data);
        // 关闭输出流
        outputStream.close();
    }
}
```

这里创建了两个内部类 MyShutterCallback 和 TakePictureCallback：MyShutterCallback 实现了 Camera.ShutterCallback 接口，覆写了 onShutter 方法；TakePictureCallback 类实现了 Camera.PictureCallback 接口并覆写了 onPictureTaken 方法，这个方法会在拍照成功后回调，其参数 data 即为拍摄的照片，因此这个方法中调用了 saveToSDCard 方法来保存图片，同时调用 startPreview 方法再次启动预览流。为了防止多次单击"拍照"按钮，多次发送拍照请求而导致的 crash，这里添加了标志位 mIsCallBackReturn，保证在 onPictureTaken 回调之前只会调用一次 takePicture 方法。这两个内部类的对象是作为 takePicture 方法的第一个和第三个参数。

saveToSDCard 方法用来保存拍照照片，这里同样使用格式化的时间作为文件名，在 SD 卡中添加一个名为"CameraTest 的文件夹"来保存照片，创建 FileOutputStream 文件输入流调用其 write 方法将字节流写入到 SD 卡中，最后记得调用其 close 方法关闭输出流。

打开相机、保存文件都需要相关权限，代码如下：

```xml
<?xml version="1.0" encoding="utf-8"?>
<manifest xmlns:android="http://schemas.android.com/apk/res/android"
    package="project.first.com.camerademo">

    <uses-permission android:name="android.permission.CAMERA" />
    <uses-permission android:name="android.permission.MOUNT_UNMOUNT_
    FILESYSTEMS" />
    <uses-permission android:name="android.permission.WRITE_EXTERNAL_
    STORAGE" />
```

```
    <application
        // 省略部分代码
        <activity android:name=".ShowPicActivity"/>
    </application>
</manifest>
```

Android 6.0 以上系统还需要在代码中添加权限。

运行实例，如图 10.22 所示。首先弹出权限请求对话框，单击"允许"按钮，打开 Camera，如图 10.23 所示。

图 10.22 Android 相机实例一

图 10.23 Android 相机实例二

Camera 打开，可以看到实时的预览图像，并且隐藏了顶部的状态栏，这时单击"拍照"按钮即可拍下一张照片，并听到快门音（takePicture 方法第一个参数传入 null 则没有快门音）。不过此时的相机功能是不完善的，它并没有持续对焦和单击拍照对焦的功能，下面再丰富一下这个 Camera 应用。

在 surfaceCreated 中为相机设置参数的地方，调用 setFocusMode 方法为其设置对焦模式，代码如下：

```
@Override
public void surfaceCreated(SurfaceHolder holder){
    try{
        // 打开摄像头
        mCamera=Camera.open();
        // 设置用于显示拍照影像的 SurfaceHolder 对象
        mCamera.setPreviewDisplay(holder);
        // 获得 parameter 对象
        mParameters=mCamera.getParameters();
        // 设置照片方向
        mParameters.setRotation(90);
        mParameters.setFocusMode("continuous-picture");
        // 设置参数
        mCamera.setParameters(mParameters);
```

```
        // 设置预览方向
        mCamera.setDisplayOrientation(90);
        // 开始预览
        mCamera.startPreview();
        }catch(Exception e){
        e.printStackTrace();
        }
    }
```

setFocusMode 传入参数 continuous-picture 表示一种持续对焦的方式（底层算法会根据当前画面质量来动态调整焦距，保证画面一直是清晰的状态），该方式也是手机自带相机应用的常规对焦方式，打开相机应用就会启动这一方式。

修改 capture 方法，代码如下：

```
public void capture(View view) {
    if (!mIsCallBackReturn) return;
    mIsCallBackReturn = false;
    mCamera.autoFocus(new Camera.AutoFocusCallback() {
        @Override
        public void onAutoFocus(boolean success, Camera camera) {
            Log.d(TAG, "onAutoFocus: " + success);
            mCamera.takePicture(new MyShutterCallback(),
                    null, new TakePictureCallback());
            mCamera.cancelAutoFocus();
        }
    });
}
```

这里调用了 Camera 的 autoFocus 方法，这个方法将会触发一次对焦，一般相机在单击"拍照"按钮时都会触发一次对焦，这样可以提高拍摄照片的清晰度，这个方法需要传入一个 Camera.AutoFocusCallback 对象作为参数，实现这个接口需要覆写 onAutoFocus 方法，onAutoFocus 方法在对焦成功后会回调，这时再调用 takePicture 方法拍摄一张照片，可以提高照片质量。最后不要忘记调用 Camera 的 cancelAutoFocus 取消一次对焦模式，若不调用这个方法就不会再进入持续对焦的模式，有兴趣的读者可以自行测试。

再次运行项目，前后移动手机，可以看到这时相机会持续地对焦，在模糊的情况下单击"拍照"按钮并不会马上拍照，而是等到一次对焦完成后才拍照。

那么拍下来的照片到哪儿去了呢？继续丰富这个程序，添加图库图标，单击这个图标将显示拍摄的照片。在布局文件中添加一个 ImageView 显示这个图标，代码如下：

```
<ImageView
    android:id="@+id/imageViewGallery"
    android:layout_width="40dp"
    android:layout_height="40dp"
    android:layout_alignParentLeft="true"
    android:layout_alignParentBottom="true"
    android:layout_centerHorizontal="true"
    android:layout_marginBottom="40dp"
    android:background="@null"
    android:layout_marginLeft="40dp"
```

```
android:scaleType="centerCrop"
android:onClick="goToGallery"
android:src="@drawable/gallery1" />
```

scaleType 可以控制图片显示的样式，这里选择 centerCrop 表示当 ImageView 的大小很难显示整张图片时将居中截取图片显示，这种显示方式常用于显示空间有限的场景。

在 MainActivity 中添加上面图片的单击监听事件，代码如下：

```java
public void goToGallery(View view) {
    String path = Environment.getExternalStorageDirectory()
            + "/CameraTest/" + mFileName;
    File file = new File(path);
    if (null == file || !file.exists()) {
        return;
    }
    Intent intent = new Intent(Intent.ACTION_VIEW);
    intent.addCategory(Intent.CATEGORY_DEFAULT);
    intent.addFlags(Intent.FLAG_ACTIVITY_NEW_TASK);
    intent.setDataAndType(Uri.fromFile(file), "image/*");
    try {
        startActivity(intent);
    } catch (ActivityNotFoundException e) {
        e.printStackTrace();
    }
}
```

上述代码根据当前拍摄图片的路径 path 获得 File 对象，创建一个 Intent 对象传入 Intent.ACTION_VIEW 作为参数，调用 Intent 的 setDataAndType 方法添加数据和数据类型，第一个参数 Uri 对象由前面的 File 对象获得，最后调用 startActivity 启动 Intent。

添加方法 showImage，在保存图片的同时将图片显示出来，代码如下：

```java
private void saveToSDCard(byte[] data) throws IOException {
    Date date = new Date();
    // 格式化时间
    SimpleDateFormat format = new SimpleDateFormat("yyyyMMddHHmmss");
    mFileName = "DCIM" + format.format(date) + ".jpg";
    File fileFolder = new File(Environment.getExternalStorageDirectory()
            + "/CameraTest/");
    if (!fileFolder.exists()) {
        fileFolder.mkdir();
    }
    File jpgFile = new File(fileFolder, mFileName);
    // 文件输出流
    FileOutputStream outputStream = new FileOutputStream(jpgFile);
    // 写入SD卡中
    outputStream.write(data);
    // 关闭输出流
    outputStream.close();
    showImage();
}
```

```
private void showImage() {
    String path = Environment.getExternalStorageDirectory()
            + "/CameraTest/" + mFileName;
    mImageView.setImageBitmap(BitmapFactory.decodeFile(path));
    ObjectAnimator.ofFloat(mImageView,"alpha",0,1).setDuration(500).
    start();
    ObjectAnimator.ofFloat(mImageView,"rotation",0,360).
    setDuration(500).start();
}
```

上述代码中，showImage 方法用于显示当前拍摄的图片，由路径 path 调用 BitmapFactory 类的静态方法 decodeFile 将得到一个 Bitmap 对象，调用 ImageView 的 setImageBitmap 方法即可将这个 Bitmap 显示出来。为了提高用户体验，这里为 ImageView 添加了两个属性动画（透明度动画和旋转动画）。

运行实例，如图 10.24 所示。默认在左下角显示图库图标，单击拍摄一张照片，左下角将显示刚才拍摄的照片，如图 10.25 所示。

单击左下角的照片图标将显示大图，如图 10.26 所示。

图 10.24　Android 相机实例三　　图 10.25　Android 相机实例四　　图 10.26　Android 相机实例五

还可以选择分享、编辑等功能处理照片。